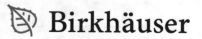
Birkhäuser

Operator Theory: Advances and Applications
Volume 257

Founded in 1979 by Israel Gohberg

More information about this series at http://www.springer.com/series/4850

András Bátkai • Marjeta Kramar Fijavž
Abdelaziz Rhandi

Positive Operator Semigroups

From Finite to Infinite Dimensions

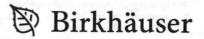 Birkhäuser

András Bátkai
Pädagogische Hochschule Vorarlberg
Feldkirch, Austria

Abdelaziz Rhandi
Università di Salerno
Fisciano, Salerno, Italy

Marjeta Kramar Fijavž
Faculty of Civil and Geodetic Engineering
University of Ljubljana
Ljubljana, Slovenia

and

Institute for Mathematics, Physics and Mechanics
Ljubljana, Slovenia

ISSN 0255-0156   ISSN 2296-4878   (electronic)
Operator Theory: Advances and Applications
ISBN 978-3-319-82670-7   ISBN 978-3-319-42813-0   (eBook)
DOI 10.1007/978-3-319-42813-0

Mathematics Subject Classification (2010): 47D06, 47B65, 15B48

This book is published under the trade name Birkhäuser, www.birkhauser-science.com
The registered company is Springer International Publishing AG
The registered company address is: Gewerbestrasse 11, 6330 Cham, Switzerland

To Asma, Brigitta, and Gašper

Foreword

Oscar Perron's theorem from 1907 on strictly positive matrices allows the following interpretation in terms of matrix exponentials: If a real $n \times n$ matrix A with spectral bound 0 generates a *semigroup* $\left(e^{tA}\right)_{t \geq 0}$ *consisting entirely of positive matrices*, and if for some $t > 0$ the matrix e^{tA} is even strictly positive, then e^{tA} converges, for $t \to \infty$, to a positive projection P of rank one onto the kernel of A.

One does not need much phantasy to imagine that in infinite dimensions similar convergence results hold for positive semigroups acting on, say, Banach lattices. Actually, Einar Hille, in the Foreword to the first edition of his treatise *Semigroups and Functional Analysis*, explicitly expressed the need to develop an adequate theory of transformation semigroups acting on partially ordered spaces, announcing however to leave this task to "more competent hands". The subtle irony in this remark – who would dare, in 1948, to call him/herself more competent than E. Hille in matters concerning semigroups – may have discouraged potential candidates. Be this as it may, for a while the rapidly developing theory of operators on Banach lattices steered clear of the semigroup approach to, e.g., linear partial differential equations.

The first major attempt to systematically connect the theory of positive operators to the theory of strongly continuous semigroups resulted in the publication, in 1986, of *Springer Lecture Notes in Mathematics* 118, by nine brave people who combined their competence to meet Hille's constraint. The plan to extend this publication into a manuscript covering both theory and applications was never realized. Instead, to make a new start, we (R.N. and U.S.) organized, in 2000/01, an *Internet Seminar on Evolution Equations (ISem)* with a reading course containing the finite-dimensional situation. Again, the plan to extend this to a comprehensive book manuscript was given up.

Finally, the authors of the present book organized, in 2013/14, another *ISem* using parts of the one of 2000/01 and including serious applications to partial differentential equations. Subsequently this was extended to yield the present manuscript. We are very happy that, in the end, our intentions have been realized.

Tübingen, May 2016   *Rainer Nagel, Ulf Schlotterbeck*

Contents

Part II   Infinite Dimensions

Part III Advanced Topics and Applications

Contents

Preface

This book is devoted to positive linear dynamical systems appearing frequently in mathematical models of various real-life problems. Here we present an operator theoretic approach and treat these systems using the theory of one-parameter semi-groups. We study quantitative and qualitative properties of positive semigroups both in finite and infinite dimensions.

Our starting point is the initial value problem (or Cauchy problem)

$$u'(t) = Au(t), \quad t \geq 0,$$
$$u(0) = u_0 \in D(A),$$

where A generates a C_0-semigroup $(T(t))_{t \geq 0}$ on a Banach lattice E, u_0 is a positive initial value, and the solution to this problem is given by $u(t) = T(t)u_0$.

An important special case is when $E = \mathbb{R}^n$ and $A = (a_{ij})$ is a real constant $n \times n$ matrix. Then the solution to the initial value problem above is given by $u(t) = e^{tA}u_0$. It is well known that the matrix exponential e^{tA} is positive for all $t \geq 0$ if and only if $A - \operatorname{diag}(A) \geq 0$, that is, $a_{ij} \geq 0$ for $i \neq j$. O. Perron (1907) and F.G. Frobenius (1909) discovered remarkable spectral properties of positive matrices with striking consequences on the asymptotic behavior of the solutions to the Cauchy problem. This theory has found many and sophisticated applications.

In infinite dimensions, W. Feller (1952) and R.S. Phillips (1962) first obtained results characterizing the generators of positive semigroups. Based on the theory of positive operators on ordered Banach spaces developed in the 60s and 70s, many applications of positive semigroups to concrete evolution equations from transport theory, mathematical biology, and physics have been made available. Most of what was known around 1985 about this subject can be found in the monograph R. Nagel (ed.) [101], written by the functional analysis group from Tübingen. This led to further progress during the last decades.

What is in this book?

Our aim is to present these developments in a streamlined and unified way. Though our ultimate goal is the infinite-dimensional situation, we feel that it is helpful to focus first on the finite-dimensional case without the need of functional analytic technicalities. This is why we spend quite some time in the realm of linear algebra. We develop the Perron–Frobenius theory in great detail and supply several recent applications, dealing, for example, with the Google matrix and the Leslie matrix.

The main body of this book is Part II where we give a detailed introduction to positive operator semigroups. Since we do not want to assume too many prerequisites, we start with a crash-course on operator semigroups and on positive operators on Banach lattices. After that, generation theorems, spectral theory and perturbation theory for positive semigroups are treated in detail.

In Part III some more advanced topics reflecting our research interests are presented. We treat advanced topics from spectral theory, investigate positive delay equations, Koopman and Perron–Frobenius semigroups, linear Boltzmann equations, flows on networks, and age structured population equations. The last chapter also gives an outlook on related research topics.

Each chapter has a set of references at the end, and a large stack of exercises, which should help the reader to obtain a fuller understanding of the topics treated.

What is not in this book?

Our presentation concentrates on those parts of the general theory of operator semigroups which are relevant for positivity. Other aspects, like for example the Lumer–Phillips theorem, are not covered. The monographs by Engel and Nagel [43, 44] are our main references.

Because of length constraints, we could not cover analytic semigroups, approximation problems or functional calculi. The monograph by Haase [62] is an excellent reference. The Beurling–Deny theory of Dirichlet forms and the corresponding rich theory is also missing. Here we refer to Ouhabaz [108].

Positivity plays an important role in the treatment of semilinear parabolic partial differential equations as well, and the interested reader is referred to Cazenave and Haraux [22].

Scope

In the first, finite-dimensional part we use the material of an introductory linear algebra course and make scattered references to topics usually presented in vector calculus and in an introductory ODE course. The typical reader we have in mind should have completed an introductory course on functional analysis, which is needed for the second part. In the examples and applications occasional references to basic facts from probability theory, graph theory, and measure theory are made. In the last part we use some more advanced results from functional analysis to be able to present striking applications.

The majority of the material covered in the first 8 chapters can be used after an introductory linear algebra course. We feel that this is a good opportunity to introduce the viewpoint of functional analysis and order theory to undergraduates.

The second part of our book addresses more advanced students assuming an introductory functional analysis course. Topics from the third part may be used in a follow-up seminar or in an advanced special course.

The material was used as lecture notes for the 17th Internet Seminar on Evolution Equations during the academic year 2013–2014, and for several years also in our own classes.

Budapest and Feldkirch,   *András Bátkai*
Ljubljana,   *Marjeta Kramar Fijavž*
Salerno,   *Abdelaziz Rhandi*
May 2016

Acknowledgements

We thank all the participants of the 17th Internet Seminar on Evolution Equations for remarks and constructive criticism. We are, especially grateful to Jürgen Voigt (Dresden), Johannes Eilinghoff (Karlsruhe), Reinhard Stahn (Dresden), Hendrik Vogt (Dresden), Raffael Hagger (Hamburg), and Marten Wortel (Kent) for pointing out mistakes in the lecture notes and suggesting improvements.

We gratefully acknowledge financial support for the Internet Seminar workshop by the GAMM "Applied Operator Theory" Special Interest Group.

Many colleagues read parts of the manuscript and suggested improvements. We thank Anna Canale (Salerno), Bálint Farkas (Wuppertal), Gašper Fijavž (Ljubljana), Federica Gregorio (Salerno), Marko Kandić (Ljubljana), Abdallah Maichine (Salerno), Mustapha Mokhtar-Kharroubi (Besançon), Aljoša Peperko (Ljubljana), Agnes Radl (Leipzig), Eszter Sikolya (Budapest), Felix Schwenninger (Wuppertal), Cristian Tacelli (Salerno), Sabrina Vatovec (Ljubljana), Jens Wintermayr (Wuppertal), and Aljaž Zalar (Ljubljana).

A.B. was funded during the writing of the manuscript by a DAAD guest professorship at the University of Wuppertal. He thanks Bálint Farkas for being a great host.

In March 2015 the authors spent two fruitful weeks and worked on the book manuscript at the MFO funded by the Oberwolfach Research in Pairs programme.

Last but not least we have to mention our dear friends and colleagues Rainer Nagel (Tübingen) and Ulf Schlotterbeck (Tübingen). The majority of the first part of our monograph is heavily inspired by their unfinished script on positive matrices. For many years they have served as our mentors and reliable guides on our mathematical journey. We owe them many fruitful discussions and valuable advices. Without their support this book would not exist.

Notations

For the convenience of our readers, we collect here some of our notations.

$\mathbb{N}$ The set of natural numbers, $\mathbb{N} = \{1, 2, 3, \ldots\}$

$\mathbb{N}_0$ The set $\mathbb{N} \cup \{0\}$.

(x_k) The sequence with elements $x_1, x_2, x_3, \ldots$

$\mathcal{L}(X)$ The set of bounded (continuous) linear operators on the normed space X.

$\mathbb{1}$ The vector with all coordinates equal to one, $\mathbb{1} = (1, \ldots, 1)$, or constant one function.

χ_S The characteristic function of the set S.

u_k The k-th canonical basis vector in $\mathbb{R}^n$, $u_k = (0, 0, \ldots, 1, 0, \ldots, 0)$.

$(x|y)$ The inner product of two vectors in Hilbert space.

$\langle \cdot, \cdot \rangle$ The duality pairing, $\langle f, g^* \rangle = g^*(f)$.

Let us make a point in our terminology. The matrices called "positive" in this book were first called "matrices with non-negative entries", which is a rather long name. As it turned out that they were important, the name got shortened to "non-negative matrices". Since they induce positive operators on the ordered vector space $\mathbb{R}^n$, we will not use this terminology but call them *positive matrices* to make the presentation consistent with the infinite-dimensional theory.

Part I

Finite Dimensions

Chapter 1

An Invitation to Positive Matrices

In this chapter we set the stage for our story. We fix our notation and summarize the linear algebraic background that will be needed for the first part of the book. We present some motivating examples of positive matrices at the very beginning and shall return to these examples later on.

1.1 Motivating Examples

Let us start with three different situations where positive matrices make an appearance. We will see that relevant properties can be described by the behavior of such matrices.

The Fibonacci Sequence

Consider the following, well-known sequence

$$f_0 = 0, \ f_1 = 1, \quad f_{k+1} = f_k + f_{k-1} \quad \text{for } k \geq 1.$$

Introducing the new sequence $g_k := f_{k-1}$, we obtain the system

$$f_{k+1} = f_k + g_k, \quad g_{k+1} = f_k, \quad \text{with } f_1 = 1 \text{ and } g_1 = 0.$$

Hence, using vectorial notation and the well-established connection between systems of linear equations and matrices, we can rewrite the system above as

$$\begin{pmatrix} f_{k+1} \\ g_{k+1} \end{pmatrix} = \begin{pmatrix} 1 & 1 \\ 1 & 0 \end{pmatrix} \begin{pmatrix} f_k \\ g_k \end{pmatrix}$$

for $k \in \mathbb{N}$. Repeating the above argument, we see that

$$\begin{pmatrix} f_{k+1} \\ g_{k+1} \end{pmatrix} = \begin{pmatrix} 1 & 1 \\ 1 & 0 \end{pmatrix} \begin{pmatrix} f_k \\ g_k \end{pmatrix} = \begin{pmatrix} 1 & 1 \\ 1 & 0 \end{pmatrix}^2 \begin{pmatrix} f_{k-1} \\ g_{k-1} \end{pmatrix} = \cdots = \begin{pmatrix} 1 & 1 \\ 1 & 0 \end{pmatrix}^k \begin{pmatrix} f_1 \\ g_1 \end{pmatrix}. \quad (1.1)$$

Hence, the properties of the numerical sequence[1] (f_k) are related to the properties of a matrix sequence (A^k) with $A = \begin{pmatrix} 1 & 1 \\ 1 & 0 \end{pmatrix}$.

---

[1] We consider sequences as functions and write (a_k) instead of $(a_k)_{k \in \mathbb{N}}$

Graphs and Markov chains

A *graph* $G = (V, E)$ models connections between objects. The set of objects is represented by the set of *vertices* V, while every *edge* $e \in E$ corresponds to a pairwise connection between two vertices $u, v \in V$: $e = (u, v)$. If the set of connections is symmetric we call the graph *undirected* and write $e = \{u, v\}$ instead. Otherwise the graph is called *directed*. For an edge $(u, v) \in E$ we call u its *head* and v its *tail*.

The simplest way to represent a graph is by drawing a picture, see Figures 1.1, 1.2, and 1.3.

To study properties of a graph or even a dynamical process on it, it is useful to encode it in terms of matrices. A positive $n \times n$ matrix $A = (a_{ij})$ encodes a graph $G = (V, E)$ if its set of edges satisfies

$$E = \{(v_i, v_j) \ : \ a_{ij} > 0\},$$

where $V = \{v_1, \ldots, v_n\}$ is the set of vertices of G. We call A an *adjacency matrix* of the graph G. If the nonzero elements of A all equal 1, we say that G is an *unweighted* graph, otherwise G is *weighted* with *weights* a_{ij} corresponding to the edges (v_i, v_j). Note that the adjacency matrix of an undirected graph is always symmetric. As examples see adjacency matrices of three graphs depicted in Figures 1.1, 1.2, and 1.3.

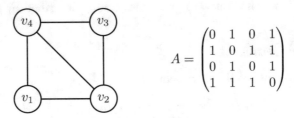

$$A = \begin{pmatrix} 0 & 1 & 0 & 1 \\ 1 & 0 & 1 & 1 \\ 0 & 1 & 0 & 1 \\ 1 & 1 & 1 & 0 \end{pmatrix}$$

Figure 1.1: An undirected graph and its adjacency matrix.

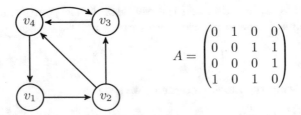

$$A = \begin{pmatrix} 0 & 1 & 0 & 0 \\ 0 & 0 & 1 & 1 \\ 0 & 0 & 0 & 1 \\ 1 & 0 & 1 & 0 \end{pmatrix}$$

Figure 1.2: A directed graph and its adjacency matrix.

Let $u, v \in V$. A *walk* from u to v in G is an alternating sequence of vertices and edges,

$$W := v_{i_0} e_{i_0} v_{i_1} e_{i_1} v_{i_2} \cdots v_{k-1} e_{i_{k-1}} v_{i_k},$$

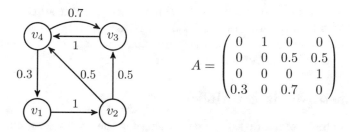

Figure 1.3: A weighted directed graph and its adjacency matrix.

where $u = v_{i_0}$, $v = v_{i_k}$, and for every $\ell = 0, \ldots, k-1$ the edge $e_{i_\ell} = (v_{i_\ell}, v_{i_{\ell+1}}) \in E$. The *length of a walk* is the number of edges in the sequence. The *weight of a walk* is the product of all weights $a_{i_\ell i_{\ell+1}}$ corresponding to the edges in the walk. A walk is *closed* if $v_{i_0} = v_{i_k}$ and a closed walk is a *cycle* if the vertices v_{i_ℓ}, $\ell = 0, \ldots, k-1$, are all different.

An interesting property of the adjacency matrix of an unweighted graph is that its powers count the number of different walks between vertices.

Proposition 1.1. *Let A be an $n \times n$ adjacency matrix of an unweighted graph. Then $(A^k)_{ij}$ is the number of walks of length $k \in \mathbb{N}$ from vertex v_i to vertex v_j, $i, j \in \{1, \ldots, n\}$.*

Proof. We prove the statement by induction. The base case $k = 1$ follows immediately since a_{ij} counts the edges (= walks of length 1) from v_i to v_j.

Now assume that the statement holds for some $k > 1$. For every walk from v_i to v_j of length $k+1$, there is a vertex v_ℓ reached one step before v_j. This means that $a_{\ell j} = 1$, and since the number of walks from v_i to v_ℓ of length k equals $(A^k)_{i\ell}$ by the induction hypothesis, we can count the number of walks from v_i to v_j of length $k+1$ as

$$\sum_{\ell=1}^{n} (A^k)_{i\ell} a_{\ell j} = (A^{k+1})_{ij}$$

by the matrix multiplication rule. □

Remark 1.2. Observe that in the case of a weighted graph $(A^k)_{ij}$ is the sum of the weights of all the walks of length $k \in \mathbb{N}$ from vertex v_i to vertex v_j.

Consider now a discrete, finite, homogeneous *Markov chain* with state space V and *transition matrix* $P = (p_{ij})$. Its entries $p_{ij} \in [0, 1]$ represent *transition probabilities* to move from state v_i to v_j in one step. The matrix P is positive and *row stochastic*, that is, $\sum_{j=1}^{n} p_{ij} = 1$ for all $i = 1, \ldots, n$.

Actually, P is an adjacency matrix of a directed weighted graph $G = (V, E)$. Note that G might have loops, i.e., edges where head and tail coincide (corresponding to the probability p_{ii} of remaining in the state v_i). According to Remark

1.2, $(P^k)_{ij}$ yields the probability of reaching state v_j from state v_i in k steps. Moreover, various long-term properties of a Markov process, such as periodicity, ergodicity, the existence of a stationary distribution, etc., can be seen from the entries of P^k as $k \to \infty$. We shall discuss this in Chapter 6.

The Competitive Market Model

Suppose n similar commodities are competing for the consumer's money. Excess demand (that is, demand minus supply) for commodity i will be denoted by f_i, and it is assumed to be approximately a linear function of prices p_j less equilibrium prices p_j^0 for all $j = 1, \ldots, n$. In formula, this means that

$$f_i \approx \sum_{j=1}^{n} a_{ij}(p_j - p_j^0).$$

Since higher prices for one commodity will increase excess demand for the others, $a_{ij} \geq 0$ for $i \neq j$ and $a_{ii} < 0$.

Once equilibrium is disturbed, the rate of price readjustment must be proportional to excess demand, leading to the differential equation

$$\dot{p}(t) = KA(p(t) - p^0), \tag{1.2}$$

where $K = \mathrm{diag}(k_1, \ldots, k_n)$ is a diagonal matrix of positive *adjustment speeds*. Here we introduced the shorthand notations $A = (a_{ij})$ and $p = (p_j)$. Thus future prices are given by

$$p(t) = p^0 + e^{tKA}c, \tag{1.3}$$

where $c = p(0) - p^0$.

If we start with a (non-equilibrium) price $p(0)$, will prices eventually return to the equilibrium p^0 or will (some) prices stay away from p^0, oscillate, or become unbounded? We will see that the answers to these questions depend on the spectral properties of the matrix KA.

One of our main topics will be the behavior of matrix sequences of the form (A^k) as $k \to +\infty$, or the behavior of matrix exponentials of the form "e^{tA}" as $t \to +\infty$. Special attention will be paid to *positive matrices*. We will apply the results to various biological, economical, and mathematical problems.

1.2   Convergence

Since the behavior of A^k as $k \to \infty$ and of e^{tA} as $t \to \infty$ is one of our central topics, we need to understand the concept of convergence in $X = \mathbb{C}^n$ and in $\mathcal{L}(X) = M_n(\mathbb{C})$.

The most natural convergence in $\mathbb{C}^n$ is coordinatewise convergence, or convergence in every coordinate, a notion which explains itself. However, the question immediately arises if a change of coordinates affects this sort of convergence. Moreover, sequences of $n \times n$ matrices, a formally different kind of convergence seems natural, namely, pointwise convergence on $\mathbb{C}^n$. This means for a sequence (A_k) in $\mathcal{L}(X)$ the convergence of $(A_k x)$ for every $x \in \mathbb{C}^n$.

The following proposition tells that these concepts all amount to the same thing, and are related to convergence with respect to a norm. For basic definitions and examples of norms, see Appendix A.1. Before stating and proving the proposition, we need an elementary lemma.

Lemma 1.3. *For any $n \times n$ matrix $A = (a_{ij}) \in \mathcal{L}(X)$, the inequalities*

$$\|A\|_{\max} \leq \|A\|_1 \leq n \cdot \|A\|_{\max}$$

hold.

Proof. First observe that

$$\|A\|_{\max} = |a_{i_0 j_0}| \leq \sum_{i=1}^{n} |a_{ij_0}| \leq \max_{1 \leq j \leq n} \sum_{i=1}^{n} |a_{ij}| = \|A\|_1.$$

On the other hand, for every $j \in \{1, \ldots, n\}$ we have

$$\sum_{i=1}^{n} |a_{ij}| \leq n \cdot \max_{1 \leq i \leq n} |a_{ij}| \leq n \cdot \|A\|_{\max},$$

hence $\|A\|_1 \leq n \cdot \|A\|_{\max}$. $\square$

Here is the previously announced proposition.

Proposition 1.4. *Let $A_k := \left(a_{ij}^{(k)} \right)$, $k \in \mathbb{N}$, be a matrix sequence in $\mathcal{L}(X)$ and $A_0 = \left(a_{ij}^{(0)} \right) \in \mathcal{L}(X)$ a given matrix. Then the following statements are equivalent.*

(i) $A_k x \to A_0 x$ *coordinatewise, as $k \to \infty$, for every $x \in X = \mathbb{C}^n$.*
(ii) $a_{ij}^{(k)} \to a_{ij}^{(0)}$ *as $k \to \infty$, for all $1 \leq i, j \leq n$.*
(iii) $\|A_k - A_0\|_{\max} \to 0$ *as $k \to \infty$.*
(iv) $\|A_k - A_0\|_1 \to 0$ *as $k \to \infty$.*

Proof. The implications (ii) $\implies$ (i) and (ii) $\implies$ (iii) are straightforward, while the equivalence (iii) $\iff$ (iv) holds by Lemma 1.3.

(i) $\implies$ (ii): Choose arbitrary $1 \leq i, j \leq n$. Now taking $x = u_j$, the canonical unit basis vector of X, in (i) yields (ii).

(iii) $\implies$ (i): Take any $x \in X$ and use the triangle inequality for the absolute value and the maximum norm $\|x\|_\infty := \max_{1 \le i \le n} |x_i|$ to obtain

$$
\begin{aligned}
|(A_k x)_i - (A_0 x)_i| &= \left| \sum_{j=1}^n \left(a_{ij}^{(k)} - a_{ij}^{(0)} \right) x_j \right| \\
&\le \|x\|_\infty \sum_{j=1}^n \left| a_{ij}^{(k)} - a_{ij}^{(0)} \right| \\
&\le \|x\|_\infty \cdot n \cdot \|A_k - A_0\|_{\max}.
\end{aligned}
$$

Therefore $A_k x \to A_0 x$ coordinatewise as $k \to \infty$. $\square$

There are many possibilities to define norms, hence, there exists a large variety of potentially different notions of convergence. Proposition 1.4, however, indicates that the latter might always coincide. This is actually true (at least in finite-dimensional vector spaces) by the following result, due to A. Tikhonov, which will allow us to define convergence without specific reference to coordinates (or entries).

Theorem 1.5 (Tikhonov). *Let X be a finite-dimensional (complex) vector space, and let $\| \cdot \|$ and $\|\cdot\|$ be norms on X. Then there exist constants $m, M > 0$ such that*

$$
m \|x\| \le \|x\| \le M \|x\| \tag{1.4}
$$

for all $x \in X$.

We call two norms $\| \cdot \|$ and $\|\cdot\|$ *equivalent* if (1.4) holds for suitable constants $m, M > 0$.

Proof. Clearly, it is enough to prove the statement of the theorem in the case where $X = \mathbb{C}^n$ and $\|x\| = \|x\|_\infty = \max_{1 \le i \le n} |x_i|$. We will make use of the fact that X is complete in the $\| \cdot \|_\infty$-norm and that the Weierstrass extreme value theorem holds, meaning that continuous functions defined on compact sets are bounded.

Let us introduce the constant

$$
C := \max_{1 \le i \le n} \|u_i\|,
$$

where u_i are the canonical unit basis vectors of X. We obtain one side of the required inequality immediately from the estimate

$$
\|x\| = \|x_1 u_1 + x_2 u_2 + \cdots + x_n u_n\| \le C \left(|x_1| + |x_2| + \cdots + |x_n| \right) \le C n \|x\|_\infty.
$$

Consider the closed and bounded set

$$
K := \{ x \in X \ : \ \|x\|_\infty = 1 \},
$$

and the function

$$f : K \longrightarrow \mathbb{R}, \quad f(x) = \|x\|.$$

Then f is Lipschitz continuous since, using the estimate above, we obtain that

$$|f(x) - f(y)| = |\,\|x\| - \|y\|\,| \leq \|x - y\| \leq Cn\|x - y\|_\infty.$$

Hence there are vectors $z, w \in K$ such that for all $x \in K$,

$$\|z\| = f(z) \leq f(x) = \|x\| \leq f(w) = \|w\|.$$

Finally, to conclude our proof, take $x \in X$ such that $x \neq 0$. Then $x = \|x\|_\infty x'$ with $x' = \frac{x}{\|x\|_\infty} \in K$. By the previous inequality, and using the notation $m := \|z\|$, $M := \|w\|$, we obtain

$$m \leq \|x'\| \leq M.$$

Multiplying this inequality by $\|x\|_\infty$, we obtain the desired statement. $\qquad\square$

Note, however, that Tikhonov's theorem does not hold in spaces of infinite dimension.

Equivalent norms possess the same convergent sequences and the same Cauchy sequences. Therefore, in finite dimensions, Theorem 1.5 allows us to define convergence as follows.

Definition 1.6. Let X be a vector space of finite dimension and (x_k) a sequence in X.

a) We say that (x_k) is a *Cauchy sequence* if (x_k) is a Cauchy sequence for one, and hence for every norm on X.

b) We say that (x_k) *converges*, as $k \to \infty$, to an element $x_0 \in X$, if $\|x_k - x_0\| \to 0$ as $k \to \infty$ for one, and hence for every norm on X.

We will now give a more specific description of convergence, which will be the clue to our subsequent discussions. To this end, let (x_k) be a sequence in X. If $\{z_1, \dots, z_n\}$ is a basis for X, there exist uniquely determined scalars $\xi_i^{(k)}$, $1 \leq i \leq n$, such that

$$x_k = \sum_{i=1}^{n} \xi_i^{(k)} z_i.$$

We call the sequences $\left(\xi_1^{(k)}\right), \dots, \left(\xi_n^{(k)}\right)$, the *coordinate sequences* of (x_k) with respect to the basis $\{z_1, \dots, z_n\}$.

Theorem 1.7. *For a sequence (x_k) in X, the following are equivalent.*

(i) *(x_k) is a Cauchy sequence.*

(ii) *(x_k) converges, as $k \to \infty$, to an element $x_0 \in X$.*

(iii) *For every basis $\{z_1, \dots, z_n\}$ of X, the coordinate sequences of (x_k) with respect to $\{z_1, \dots, z_n\}$ converge, as $k \to \infty$.*

In this case, the element x_0 in item (ii) *is uniquely determined and the coordinates of x_0 with respect to $\{z_1, \dots, z_n\}$ are the limits of the respective coordinate sequences.*

Proof. The statement is an immediate consequence of the fact that for each basis $\{z_1, \ldots, z_n\}$ of X,

$$X \ni x \longmapsto \max_{1 \le i \le n} \left\{ |\xi_i| \,:\, x = \sum_{i=1}^{n} \xi_i z_i \right\}$$

is a norm. $\square$

We introduce yet another notion related to norms. Again, X is a vector space of finite dimension.

Definition 1.8. A subset $\Omega \subset X$ is called *bounded* if for one, hence for every norm $\| \cdot \|$ on X,

$$\sup_{x \in \Omega} \|x\| < \infty.$$

It is clear that boundedness of a sequence (x_k) in X is equivalent to boundedness of all coordinate sequences of (x_k) with respect to one, hence all, bases of X.

Although there appears to be no need to discriminate between different norms on a finite-dimensional vector space, the equivalence of all norms on such a space gives us the possibility to discuss convergence of a specific sequence by using a particular favorable norm. This was the key point in the proof of the previous theorem. Apart from that, we will occasionally have to distinguish, even on the finite-dimensional vector space $\mathcal{L}(X)$, between norms in general and norms that come from the underlying vector space X.

Definition 1.9. Let $A \in \mathcal{L}(X)$ and let $\| \cdot \|$ be a norm on X. The corresponding *operator norm* on $\mathcal{L}(X)$, again denoted by $\| \cdot \|$, is defined as

$$\|A\| := \sup\{\|Ax\| \,:\, x \in X, \|x\| \le 1\} = \max\{\|Ax\| \,:\, x \in X, \|x\| \le 1\}.$$

Alternative descriptions of the operator norm coming from a norm $\| \cdot \|$ on X are the following:

$$\|A\| = \sup\{\|Ax\| : \|x\| = 1\}$$
$$= \sup\left\{ \frac{\|Ax\|}{\|x\|} : x \ne 0 \right\}$$
$$= \min\{M : \|Ax\| \le M\|x\| \text{ for all } x \in X\}.$$

The relation

$$\|Ax\| \le \|A\|\|x\|, \quad x \in X,$$

is a reformulation of the last two expressions above. In particular, for $A, B \in \mathcal{L}(X)$,

$$\|ABx\| = \|A(Bx)\| \le \|A\| \, \|Bx\| \le \|A\| \, \|B\| \, \|x\|$$

for all $x \in X$. Hence, $\|AB\| \leq \|A\| \, \|B\|$ whenever $\| \cdot \|$ is an operator norm. We note in particular that for any operator norm $\| \cdot \|$ on $\mathcal{L}(X)$, the identity must have norm 1. Thus, if $\| \cdot \|$ is an operator norm, the norm

$$A \longmapsto \mu \|A\|, \quad \mu > 0,$$

cannot be an operator norm unless $\mu = 1$.

Another interesting feature of operator norms is that $|\lambda| \leq \|A\|$ holds for every eigenvalue λ. This is obvious since for any eigenvector x belonging to λ we have $\|Ax\| = |\lambda| \, \|x\|$. For later reference, we record this as a corollary.

Corollary 1.10. *Let* $\| \cdot \|$ *be an operator norm on* $\mathcal{L}(X)$, $A \in \mathcal{L}(X)$, *and* λ *an eigenvalue of* A. *Then* $\|A\| \geq |\lambda|$.

We give two important examples where the operator norm can be calculated explicitly. Recall that for $1 \leq p < \infty$ and $x \in X = \mathbb{C}^n$,

$$\|x\|_p := \left(\sum_{i=1}^n |x_i|^p \right)^{1/p}$$

is called the *p-norm* on X, while the ∞-*norm* is defined as

$$\|x\|_\infty := \max_{1 \leq i \leq n} |x_i|.$$

Example 1.11.

a) The operator norm of a matrix $A = (a_{ij}) \in \mathcal{L}(X)$ corresponding to the ∞-norm on $\mathbb{C}^n$ is the *row norm*

$$\|A\|_\infty := \max\{\|Ax\|_\infty : x \in \mathbb{C}^n, \|x\|_\infty \leq 1\} = \max_{1 \leq i \leq} \sum_{j=1}^n |a_{ij}|.$$

b) The operator norm on $\mathcal{L}(X)$ corresponding to the 1-norm on $\mathbb{C}^n$ is the *column norm*

$$\|A\|_1 := \max\{\|Ax\|_1 : x \in \mathbb{C}^n, \|x\|_1 \leq 1\} = \max_{1 \leq j \leq n} \sum_{i=1}^n |a_{ij}|.$$

Proposition 1.12. *For a matrix sequence* $(A_k) \subset \mathcal{L}(X)$ *the following assertions are equivalent.*

(i) $(A_k x)$ *converges in* X *for all* $x \in X$.

(ii) (A_k) *converges in* $\mathcal{L}(X)$.

(iii) *For any basis of* X *and the corresponding matrix representation* $A_k = \left(a_{ij}^{(k)} \right)$ *the entries* $a_{ij}^{(k)}$ *converge in* $\mathbb{C}$ *for every pair* (i, j) *as* $k \to \infty$.

Proof. By Theorem 1.5, it is enough to prove the statement for the case where $\|x\| = \|x\|_\infty$. Since (iii) $\Longrightarrow$ (ii) follows by Proposition 1.4 and (ii) $\Longrightarrow$ (i) by the estimate $\|A_k x - A_m x\| \leq \|A_k - A_m\|\, \|x\|$, it remains to show the implication (i) $\Longrightarrow$ (iii).

Let $\{u_1, \dots, u_n\}$ be any basis of X and denote by $a_{ij}^{(k)}$ the entries of the matrix of A_k in this basis. Then

$$\left| a_{ij}^{(k)} - a_{ij}^{(m)} \right| \leq \max_{1 \leq p \leq n} \sum_{q=1}^{n} \left| \left(a_{pq}^{(k)} - a_{pq}^{(m)} \right) (u_j)_q \right| = \| (A_k - A_m) u_j \|_\infty ,$$

hence the convergence follows by (i). $\square$

1.3   Notes and Remarks

We recall some necessary results and notations from linear algebra in the Appendix. Tikhonov's theorem (Theorem 1.5) was proved in [142]. For more notions and results from graph theory we refer to Bondy and Murty [19], Godsil and Royle [52], or West [154], and for Markov chains to Norris [107] or Seneta [129]. The competitive market model is taken from MacCluer [90, p.492].

1.4   Exercises

1. Determine the eigenvalues and eigenvectors of the matrix $A = \left(\begin{smallmatrix} 1 & 1 \\ 1 & 0 \end{smallmatrix}\right)$, corresponding to the Fibonacci sequence (f_k) (see Section 1.1). Using this, find a formula for A^k and f_k.

2. An undirected graph G is depicted in Figure 1.4.

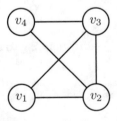

Figure 1.4: Graph G for Exercise 2.

   a) Write the adjacency matrix A of G and compute A^2 and A^3.

   b) Verify that $(A^k)_{ij}$ gives the number of walks of length k from vertex v_i to vertex v_j.

   c) Choose directions of the edges of G to obtain a directed graph $\widetilde{G}$. Repeat a) and b) for $\widetilde{G}$ instead of G.

3. Let A be the adjacency matrix of a simple undirected graph G (by simple we mean unweighted, without loops or multiple edges). Show the following spectral properties of A.

    a) The sum of all eigenvalues of A equals 0.

    b) The sum of all squares of eigenvalues of A equals $2m$, where m is the number of edges in G.

    c) The sum of all cubes of eigenvalues of A equals $6t$, where t is the number of triangles in G.

4.   a) Verify the following relations among the norms $\|\cdot\|_1$, $\|\cdot\|_2$, and $\|\cdot\|_\infty$:

$$\|x\|_2 \leq \|x\|_1 \leq \sqrt{n}\|x\|_2,$$
$$\|x\|_\infty \leq \|x\|_2 \leq \sqrt{n}\|x\|_\infty,$$
$$\|x\|_\infty \leq \|x\|_1 \leq n\|x\|_\infty.$$

    b) Show that $\lim_{p\to\infty} \|x\|_p = \|x\|_\infty$.

5. Verify the equivalence of the alternative descriptions of the operator norm in Definition 1.9 and prove the statements of Example 1.11.

6. Consider the following matrices:

$$P_1 := \begin{pmatrix} 1 & 0 \\ 0 & 0 \end{pmatrix} \quad \text{and} \quad P_2 := \frac{1}{3}\begin{pmatrix} -1 & 2 \\ -2 & 4 \end{pmatrix}.$$

Show that $P_i^2 = P_i$ ($i = 1, 2$), that is, they are projections. Can you find $\operatorname{im} P_i$ and $\ker P_i$ ($i = 1, 2$)? Can you give a geometric interpretation of the action of the matrices P_i?

Chapter 2

Functional Calculus

Our second chapter develops an abstract functional calculus for matrices. It seems of great advantage to have this abstract tool at hand even if we will not need the full power of this calculus later. This will take us to known statements of linear algebra using a coordinate-free approach, allowing to introduce some ideas which will be rather beneficial when we treat infinite-dimensional problems.

2.1 Polynomials

In this section we propose a method of constructing functions of a linear operator $A \in \mathcal{L}(X)$, where X is a finite-dimensional vector space. Of special interest are, in view of reducing subspaces (see Appendix A.2), projections commuting with A. We will look for such projections among the *polynomials* in A, i.e., the linear combinations of powers of A. Indeed, if such a projection is a polynomial, it automatically commutes with A and hence its range is a reducing subspace.

We denote the set of all *polynomials in A* by $\mathcal{P}_A$, i.e.,

$$\mathcal{P}_A := \left\{ \sum_{i=0}^{m} \alpha_i A^i : \alpha_i \in \mathbb{C}, m \in \mathbb{N}_0 \right\} \subset \mathcal{L}(X), \tag{2.1}$$

and if $p(x) = \sum_{i=0}^{m} \alpha_i x^i$ is a polynomial, we write

$$p(A) := \sum_{i=0}^{m} \alpha_i A^i$$

and say that the operator $p(A)$ is obtained by evaluating p at A, or just by *plugging A into p*. From this definition it is clear that the mapping

$$\Phi_A : \mathbb{C}[x] \longrightarrow \mathcal{P}_A, \quad p \longmapsto p(A),$$

is a homomorphism from the algebra $\mathbb{C}[x]$ of all polynomials onto $\mathcal{P}_A$. For those readers who are acquainted with the notion of the factor map, we remark that

$$\hat{\Phi}_A : {}^{\mathbb{C}[x]}\!/\ker \Phi_A \to \mathcal{P}_A \tag{2.2}$$

becomes an algebra isomorphism.

Example 2.1. For the matrix $A := \begin{pmatrix} 0 & 1 \\ 0 & 0 \end{pmatrix}$, we have $\dim \mathcal{P}_A = 2$.

In order to find polynomials such that $p(A) = (p(A))^2 = p^2(A)$, i.e., $p(A)$ is a projection, we have a closer look at the algebraic structure of $\ker \Phi_A$.

Proposition 2.2. *There exists a unique polynomial $m_A \in \mathbb{C}[x]$ of degree ≥ 1 and with leading coefficient 1 such that the following holds:*

A polynomial $p \in \mathbb{C}[x]$ belongs to $\ker \Phi_A$, i.e., $p(A) = 0$, if and only if $p = m_A \cdot q$ for some $q \in \mathbb{C}[x]$.

Proof. Take a nonzero polynomial $m \in \ker \Phi_A$ with minimal degree and leading coefficient 1. If $p = m \cdot q$ for some $q \in \mathbb{C}[x]$, then $p(A) = m(A) \cdot q(A) = 0$, hence $p \in \ker \Phi_A$.

On the other hand, if $p \in \mathbb{C}[x]$ is any polynomial with $p(A) = 0$, then the division of p by m gives

$$p = m \cdot q + r$$

for polynomials q and r satisfying $\deg r < \deg m$.

Thus

$$0 = p(A) = m(A) \cdot q(A) + r(A) = r(A)$$

and $r \in \ker \Phi_A$. Since m has minimal degree in $\ker \Phi_A$, we conclude that $r = 0$. This shows that $m_A := m$ is the desired polynomial.

Uniqueness follows since the leading coefficient was supposed to be 1. $\square$

We call this polynomial m_A generating the kernel of Φ_A the *minimal polynomial* of A. For polynomials $p, q, r \in \mathbb{C}[X]$ we use the notation

$$p \equiv q \mod r \iff p - q = s \cdot r \text{ for some } s \in \mathbb{C}[x].$$

The following is an immediate consequence of Proposition 2.2.

Corollary 2.3. *For $p, q \in \mathbb{C}[x]$, we have $p(A) = q(A)$ if and only if*

$$p \equiv q \mod m_A.$$

In particular, $p(A)$ is a projection if and only if $p^2 \equiv p \mod m_A$.

Denote by

$$\lambda_1, \ldots, \lambda_m$$

the zeros of the minimal polynomial m_A, and by

$$\nu_1, \ldots, \nu_m$$

their respective multiplicities (as zeros of m_A). Then m_A decomposes as

$$m_A(z) = (z - \lambda_1)^{\nu_1}(z - \lambda_2)^{\nu_2} \cdots (z - \lambda_m)^{\nu_m}. \qquad (2.3)$$

Lemma 2.4. *For $p, q \in \mathbb{C}[x]$, we have*

$$p \equiv q \mod m_A \quad \Longleftrightarrow \quad p^{(\nu)}(\lambda_i) = q^{(\nu)}(\lambda_i) \qquad (2.4)$$

for $i = 1, \ldots, m$ and $\nu = 0, \ldots, \nu_i - 1$.

Proof. This follows directly from Lemma A.12. $\qquad\qquad\qquad\qquad\square$

We use this characterization of the equality $p(A) = q(A)$ to extend the domain of Φ_A, and hence to be able to define more general functions of A. Before proceeding, let us make the following observation.

Remark 2.5. For the exponential function

$$f(x) = e^x = \sum_{k=0}^{\infty} \frac{x^k}{k!}$$

it seems to be natural to define the exponential function of a matrix A by the formula

$$e^A := \sum_{k=0}^{\infty} \frac{A^k}{k!}.$$

This is completely legitimate and we will justify this formula later on (see Corollary 2.14). However, since $\mathcal{L}(X)$ is an n^2-dimensional vector space (where $n = \dim X$), in the above infinite series there are at most n^2 linearly independent terms. Hence, e^A is actually a polynomial in A, of course with rather complicated coefficients. This observation justifies our procedure to look for the operator $f(A) = e^A$ among the polynomials of A.

Before going on with the abstract argument, let us illustrate our last point on three examples of simple 2×2 matrices where we are able to calculate the power series.

Examples 2.6.

a) Let $A := \begin{pmatrix} 0 & 1 \\ 0 & 0 \end{pmatrix}$. Then, using the above formula and the fact that $A^2 = 0$, we see that

$$e^{tA} = I + tA = \begin{pmatrix} 1 & t \\ 0 & 1 \end{pmatrix}.$$

This is clearly a polynomial in A.

b) Let $A := \begin{pmatrix} 1 & 0 \\ 0 & 2 \end{pmatrix}$. Then, using the above formula and the fact that

$$(tA)^k = \begin{pmatrix} t^k & 0 \\ 0 & (2t)^k \end{pmatrix},$$

we see that

$$e^{tA} = \begin{pmatrix} e^t & 0 \\ 0 & e^{2t} \end{pmatrix} = e^t \begin{pmatrix} 1 & 0 \\ 0 & 0 \end{pmatrix} + e^{2t} \begin{pmatrix} 0 & 0 \\ 0 & 1 \end{pmatrix} = e^t(2I - A) + e^{2t}(A - I).$$

This is again a polynomial in A.

c) Let $A = \begin{pmatrix} 0 & 1 \\ -1 & 0 \end{pmatrix}$. Then we see from

$$A^0 = I, \quad A^1 = A, \quad A^2 = -I, \quad A^3 = -A, \quad A^4 = I, \ldots,$$

that the powers of A alternate periodically. Hence,

$$e^{tA} = \sum_{i=0}^{\infty} \frac{(tA)^i}{i!} = \sum_{k=0}^{\infty} \frac{(tA)^{2k}}{(2k)!} + \sum_{k=0}^{\infty} \frac{(tA)^{2k+1}}{(2k+1)!}$$

$$= I \sum_{k=0}^{\infty} (-1)^k \frac{t^{2k}}{(2k)!} + A \sum_{k=0}^{\infty} (-1)^k \frac{t^{2k+1}}{(2k+1)!} = \cos t \cdot I + \sin t \cdot A,$$

which is a polynomial in A.

2.2 Smooth Functions

Let $A \in \mathcal{L}(X)$ be an operator on the finite-dimensional vector space X with minimal polynomial m_A. As before, $\lambda_1, \ldots, \lambda_m$ are the roots of the minimal polynomial with corresponding multiplicities $\nu_1, \ldots, \nu_m$. We denote the set of functions that are defined and are infinitely differentiable on an open neighborhood of $\{\lambda_1, \ldots, \lambda_m\}$ by

$$C_A^{\infty} := \{f : D(f) \to \mathbb{C} : \exists U \subset D(f) \text{ open}, \{\lambda_1, \ldots, \lambda_m\} \subset U, f|_U \in C^{\infty}\}.$$

Here $D(f) \subset \mathbb{C}$ denotes the domain of f. We consider C_A^{∞} as the set of functions for which we would like to define the functional calculus.

Definition 2.7 (Functional Calculus). Let $f \in C_A^\infty$. We then set

$$f(A) := \Phi_A(p_f) = p_f(A),$$

where p_f is an *interpolation polynomial*[2] for f in the sense that the derivatives of f satisfy

$$f^{(\nu)}(\lambda_i) = p_f^{(\nu)}(\lambda_i)$$

for $i = 1, \ldots, m$ and $\nu = 0, \ldots, \nu_i - 1$.

We say that $f(A)$ is the result of *plugging A into the function f*. In this way we extended our function Φ_A, defined on $\mathbb{C}[x]$, to the function $\widetilde{\Phi}_A$ on C_A^∞ by the formula

$$\widetilde{\Phi}_A(f) := \Phi_A(p_f) = p_f(A).$$

Now we collect useful properties of $\widetilde{\Phi}_A$ that follow directly from Definition 2.7, Corollary 2.3, and Lemma 2.4.

Lemma 2.8. *With the above notation the following holds.*

a) *The definition of $\widetilde{\Phi}_A(f)$ does not depend on the particular choice of the interpolation polynomial p_f.*

b) *The map $\widetilde{\Phi}_A$ is an extension of Φ_A.*

c) *The map $\widetilde{\Phi}_A$ is an algebra homomorphism in the sense that*

$$\widetilde{\Phi}_A(\lambda f + \mu g) = \lambda \widetilde{\Phi}_A(f) + \mu \widetilde{\Phi}_A(g),$$
$$\widetilde{\Phi}_A(f \cdot g) = \widetilde{\Phi}_A(f) \cdot \widetilde{\Phi}_A(g)$$

for $\lambda, \mu \in \mathbb{C}$ and functions $f, g \in C_A^\infty$.

At first glance, not much seems to be gained since the range of $\widetilde{\Phi}_A$ is still $\mathcal{P}_A$. However, the domain of $\widetilde{\Phi}_A$ is now much larger and contains many more functions. For example, any characteristic function χ_U of an open set $U \subset \mathbb{C}$ such that none of the points $\lambda_1, \ldots, \lambda_m$ lies on the boundary of U is an idempotent and belongs to the domain of $\widetilde{\Phi}_A$. As a consequence of Corollary 2.3, $\chi_U(A)$ must be a projection contained in $\mathcal{P}_A$, hence commuting with A.

Now we pick a particular set of such projections. Let $U_1, \ldots, U_m$ be open subsets of $\mathbb{C}$ satisfying

a) $\lambda_i \in U_i$ for $i = 1, \ldots, m$, and

b) $U_i \cap U_j = \emptyset$ for $i \neq j$.

Writing χ_i for the characteristic function of U_i, we obtain projections

$$P_i := \chi_i(A) \in \mathcal{P}_A \text{ for } i = 1, \ldots, m, \tag{2.5}$$

---

[2]For more information on interpolation polynomials see Appendix A, Section A.3.

whose ranges will be denoted by

$$X_i := \operatorname{im} P_i. \tag{2.6}$$

We remark that P_i is independent of the specific choice of U_i, and $P_i \neq 0$ for all i.

The following is now the fundamental structure theorem for linear operators on finite-dimensional vector spaces.

Theorem 2.9. *Let X be a finite-dimensional vector space and consider $A \in \mathcal{L}(X)$ with minimal polynomial m_A having zeros $\lambda_1, \ldots, \lambda_m$ with respective multiplicities $\nu_1, \ldots, \nu_m$. If we take P_i and X_i as in (2.5) and (2.6), respectively, then*

$$X = X_1 \oplus \cdots \oplus X_m$$

is a direct sum decomposition into A-invariant subspaces such that the restriction of $\lambda_i - A$ to X_i is nilpotent of order ν_i for $i = 1, \ldots, m$.

Proof. Since $\chi_i(\sum_{j \neq i} \chi_j) = 0$ for any i, $P_i(\sum_{j \neq i} P_j) = 0$ and hence

$$X_i \cap \left[\bigoplus_{j \neq i} X_j \right] = \{0\}.$$

Moreover, since $\{\lambda_1, \ldots, \lambda_m\} \subset \bigcup_{i=1}^m U_i =: U$, we conclude that

$$\sum_{i=1}^m P_i = \widetilde{\Phi}_A \left(\sum_{i=1}^m \chi_i \right) = \widetilde{\Phi}_A(\chi_U) = I,$$

hence $X_1 \oplus \cdots \oplus X_m = X$.

We recall that a matrix B is called *nilpotent of order k* if $B^k = 0$, but $B^{k-1} \neq 0$. Observe now that, for any fixed i, $g_i(\lambda) = (\lambda_i - \lambda)^{\nu_i} \chi_i(\lambda)$ is a function in the domain of $\widetilde{\Phi}_A$ that coincides in each of the points $\lambda_1, \ldots, \lambda_m$ with the zero function $\mathbb{O}$, including all relevant derivatives. By the properties of $\widetilde{\Phi}_A$ stated in Lemma 2.8 we must have

$$(\lambda_i - A)^{\nu_i} P_i = \left((\lambda_i - \lambda)^{\nu_i} \chi_i \right)(A) = \mathbb{O}(A) = 0$$

for $i = 1, \ldots, m$ and $\lambda \in \mathbb{C}$. On the other hand, the function

$$f_i(\lambda) := (\lambda_i - \lambda)^{\nu_i - 1} \chi_i(\lambda), \quad \lambda \in \mathbb{C},$$

does *not* satisfy

$$f_i^{(\nu_i - 1)}(\lambda_i) = 0,$$

hence

$$(\lambda_i - A)^{\nu_i - 1} P_i = f_i(A) \neq 0,$$

proving that $\lambda_i - A$ is nilpotent of order ν_i on X_i. $\qquad \square$

The zeros $\lambda_1, \ldots, \lambda_m$ of the minimal polynomial m_A can now be identified with the *eigenvalues* of A, i.e., with those $\lambda \in \mathbb{C}$ for which $(\lambda - A)x = 0$ for some $0 \neq x \in X$.

Corollary 2.10. *Under the above assumptions the points $\lambda_1, \ldots, \lambda_m$ are exactly those $\lambda \in \mathbb{C}$ for which $\lambda - A$ is not invertible.*

Proof. Every λ_i is an eigenvalue of A with an eigenvector contained in X_i. Otherwise, $\lambda_i - A$ would be bijective (see Proposition A.11), hence could not be nilpotent on X_i (which is $\neq \{0\}$ since $P_i \neq 0$).

There are no other eigenvalues of A since for any $\mu \in \mathbb{C}$ distinct from all $\lambda_1, \ldots, \lambda_m$, the function $f(\lambda) := \frac{1}{\mu - \lambda}$ is the inverse of $f^{-1}(\lambda) := \mu - \lambda$, thus the operator $f(A)$ is the inverse of $f^{-1}(A) = \mu - A$, and hence $\mu - A$ is invertible. $\qquad\square$

After these preparations we are now able to express the operator $f(A)$ by the action of $A - \lambda_i$ on the subspaces X_i and by the values of (the derivatives of) f in λ_i.

Theorem 2.11. *Let us consider $A \in \mathcal{L}(X)$ with eigenvalues $\lambda_1, \ldots, \lambda_m$ and respective multiplicities $\nu_1, \ldots, \nu_m$, and define the projections P_i as in (2.5). For every function $f \in C_A^\infty$ one has*

$$f(A) = \sum_{i=1}^{m} \sum_{\nu=0}^{\nu_i - 1} \frac{f^{(\nu)}(\lambda_i)}{\nu!} (A - \lambda_i)^\nu P_i. \tag{2.7}$$

In particular, the following representations hold:

$$R(\mu, A) := (\mu - A)^{-1} = \sum_{i=1}^{m} \sum_{\nu=0}^{\nu_i - 1} \frac{(A - \lambda_i)^\nu}{(\mu - \lambda_i)^{\nu+1}} P_i \quad \text{for } \mu \notin \{\lambda_1, \ldots, \lambda_m\}. \tag{2.8}$$

$$e^{tA} = \sum_{i=1}^{m} \sum_{\nu=0}^{\nu_i - 1} \frac{e^{t\lambda_i} t^\nu}{\nu!} (A - \lambda_i)^\nu P_i \quad \text{for } t \in \mathbb{R}. \tag{2.9}$$

$$A^k = \sum_{i=1}^{m} \sum_{\nu=0}^{\min\{\nu_i - 1, k\}} \binom{k}{\nu} \lambda_i^{k-\nu} (A - \lambda_i)^\nu P_i \quad \text{for } k \in \mathbb{N}. \tag{2.10}$$

Proof. The function

$$g(\lambda) := \sum_{i=1}^{m} \sum_{\nu=0}^{\nu_i - 1} \frac{f^{(\nu)}(\lambda_i)}{\nu!} (\lambda - \lambda_i)^\nu \chi_i(\lambda), \quad \lambda \in \mathbb{C},$$

coincides with f, including all relevant derivatives, on all of the points $\lambda_1, \ldots, \lambda_m$. By Lemma 2.8 we therefore obtain $f(A) = g(A)$.

The special cases in formulas (2.8), (2.9), and (2.10) follow by taking $f(\lambda)$ as $(\mu - \lambda)^{-1}$, $e^{t\lambda}$, and λ^k, respectively. $\qquad\square$

Recall that the *characteristic polynomial* of a matrix $A \in \mathcal{L}(X)$ is defined as

$$p_A(\lambda) := \det(\lambda - A).$$

By Corollary 2.10, the zeros of p_A are exactly the eigenvalues $\lambda_1, \ldots, \lambda_m$ and, by definition, the minimal polynomial m_A divides p_A. By using formula (2.7) for p_A we immediately obtain the following well-known property.

Corollary 2.12 (Cayley–Hamilton). *If p_A is the characteristic polynomial of A, then $p_A(A) = 0$.*

The following result tells us that our functional calculus is in accordance with questions of convergence. It will lead us to alternatives to formulas (2.9) and (2.10) in Theorem 2.11 involving infinite series.

Proposition 2.13. *Let $A \in \mathcal{L}(X)$ and let (f_k) be a sequence of functions in C_A^∞ that converges pointwise, including the relevant derivatives, at each of the eigenvalues $\lambda_1, \ldots, \lambda_m$, to a function $f \in C_A^\infty$. Then the operators $f_k(A)$ converge to $f(A)$ in $\mathcal{L}(X)$.*

Proof. By hypothesis, we have

$$f_k^{(\nu)}(\lambda_i) \longrightarrow f^{(\nu)}(\lambda_i) \quad \text{as } k \to \infty$$

for all $1 \leq i \leq m$ and $0 \leq \nu \leq \nu_i - 1$.

Let $\|\cdot\|$ be a norm on $\mathcal{L}(X)$. Then, by equation (2.7),

$$\|f_k(A) - f(A)\| = \left\| \sum_{i=1}^{m} \sum_{\nu=0}^{\nu_i-1} \left(f_k^{(\nu)}(\lambda_i) - f^{(\nu)}(\lambda_i) \right) \frac{(A - \lambda_i)^\nu}{\nu!} P_i \right\|$$

$$\leq \left(\sup_{\substack{0 \leq \nu \leq \nu_i - 1 \\ 1 \leq i \leq m}} \left| f_k^{(\nu)}(\lambda_i) - f^{(\nu)}(\lambda_i) \right| \right) \sum_{i=1}^{m} \sum_{\nu=0}^{\nu_i-1} \left\| \frac{(A - \lambda_i)^\nu}{\nu!} P_i \right\|.$$

Since (f_k) converges, including the relevant derivatives, at each of the points λ_i to a function f, the right-hand side tends to 0 as $k \to \infty$.   □

Using this proposition, we can see that our construction of the functional calculus yields the same result as the power series for the exponential function.

Corollary 2.14. *Let $A \in \mathcal{L}(X)$ and $t \in \mathbb{R}$. Then*

$$e^{tA} = \sum_{i=0}^{\infty} \frac{t^i A^i}{i!} = \lim_{k \to \infty} \sum_{i=0}^{k} \frac{t^i A^i}{i!}.$$

Proof. Put $f_k(\lambda) := \sum_{i=0}^{k} \frac{t^i \lambda^i}{i!}, \lambda \in \mathbb{C}$. Since f_k converges, as $k \to \infty$, together with all its derivatives, to $e^{\lambda t} =: f(\lambda)$, the assertion follows from Proposition 2.13.   □

For later reference, let us introduce the number $r(A) := \max_{1 \le i \le m} |\lambda_i|$ and call it the *spectral radius* of the operator A.

Corollary 2.15. *Let $A \in \mathcal{L}(X)$ and $|\lambda| > \max_{1 \le i \le m} |\lambda_i| = r(A)$. Then*

$$R(\lambda, A) = \sum_{k=0}^{\infty} \frac{A^k}{\lambda^{k+1}}. \tag{2.11}$$

Proof. Under the hypothesis on λ, we have

$$\frac{1}{\lambda - z} = \frac{1}{\lambda} \sum_{i=0}^{\infty} \left(\frac{z}{\lambda}\right)^i = \sum_{i=0}^{\infty} \frac{z^i}{\lambda^{i+1}}$$

for $|z| < |\lambda|$. Let $f_k(z) := \sum_{i=0}^{k} \frac{z^i}{\lambda^{i+1}}$ and note that $f_k(z) \to 1/(\lambda - z)$ pointwise as $k \to \infty$ together with all its derivatives, hence the result follows again by Proposition 2.13. $\qquad \square$

The above expression (2.11) is called the *Neumann series* for $R(\lambda, A)$.

Later on, we will frequently use the following observations helping to read the formulas of Theorem 2.11.

Lemma 2.16.

a) *Take $i \in \{1, \dots, m\}$ and $0 \ne z \in X_i$. Then the set*

$$\{(A - \lambda_i)^\nu z : \nu = 0, \dots, \nu_i - 1\} \setminus \{0\}$$

is linearly independent in X_i.

b) *The set*

$$B_A := \{(A - \lambda_i)^\nu P_i : i = 1, \dots, m; \ \nu = 0, \dots, \nu_i - 1\}$$

is linearly independent in $\mathcal{L}(X)$.

Proof. a) Since $A - \lambda_i$ is nilpotent of order ν_i in X_i, there exists, for every $0 \ne z \in X_i$, an exponent $\nu_0 \le \nu_i$ with $(A - \lambda_i)^{\nu_0 - 1} z \ne 0$ and $(A - \lambda_i)^\nu z = 0$ for $\nu \ge \nu_0$. An equation of the form

$$0 = \sum_{\nu=0}^{\nu_0 - 1} \alpha_\nu (A - \lambda_i)^\nu z$$

leads, after multiplication by $(A - \lambda_i)^{\nu_0 - 1}$, to $\alpha_0 = 0$. Subsequent multiplication by decreasing powers of $A - \lambda_i$ shows that the other coefficients α_ν must also be zero.

b) As in the proof of a), one shows that the set

$$B_A^{(i)} := \{(A - \lambda_i)^\nu P_i : \nu = 0, \dots, \nu_i - 1\}$$

is linearly independent in $\mathcal{L}(X)$ for each $1 \leq i \leq m$. If

$$\sum_{i=1}^{m} \sum_{\nu=0}^{\nu_i-1} \alpha_\nu^{(i)}(A - \lambda_i)^\nu P_i = 0,$$

then for fixed $1 \leq i \leq m$ multiplication by P_i shows that the partial sum

$$\sum_{\nu=0}^{\nu_i-1} \alpha_\nu^{(i)}(A - \lambda_i)^\nu P_i = 0$$

must vanish, thus $\alpha_\nu^{(i)} = 0$ for $\nu = 0, \ldots, \nu_i - 1$ by the linear independence of $B_A^{(i)}$. $\qquad\square$

2.3 Spectral Theory

Now we introduce the usual terminology in connection with the symbols appearing in Theorem 2.11. Let λ_i be an eigenvalue of the matrix A. We call $\ker(\lambda_i - A)$ an *eigenspace*, each $0 \neq x \in \ker(\lambda_i - A)$ an *eigenvector*, the projection P_i defined in (2.5) the *spectral projection*, and $X_i := P_i X$ the *spectral subspace* corresponding to the eigenvalue λ_i.

The set $\{\lambda_1, \ldots, \lambda_m\}$ of all eigenvalues of A is denoted by $\sigma(A)$ and called the *spectrum* of A. Its complement is the *resolvent set* $\rho(A) := \mathbb{C}\backslash\sigma(A)$, and the *resolvent* of A is the map

$$\rho(A) \ni \lambda \longmapsto R(\lambda, A),$$

where we use the notation $R(\lambda, A) = (\lambda - A)^{-1}$ introduced in equation (2.8).

Remark 2.17. For those readers who are familiar with elementary complex function theory, we mention that formula (2.8) shows that the resolvent is a meromorphic function with poles of order ν_i in λ_i for $i = 1, \ldots, m$. Since the principal part of the Laurent expansion of $R(\lambda, A)$ around λ_i is

$$\sum_{\nu=0}^{\nu_i-1} \frac{(A - \lambda_i)^\nu}{(\lambda - \lambda_i)^{\nu+1}} P_i,$$

we infer that P_i is just the residue of the resolvent $R(\cdot, A)$ at λ_i. This interpretation of relation (2.8) is not essential in our discussion now. However, it will become important in the infinite-dimensional situation. Therefore we will adopt the corresponding terminology right away and henceforth speak of ν_i as the *order of the pole* λ_i of $R(\cdot, A)$.

The following is a useful characterization of X_i by means of A.

Lemma 2.18. *The* spectral subspace *of* A *corresponding to the eigenvalue* λ_i *is*

$$X_i = \ker(\lambda_i - A)^{\nu_i},$$

where ν_i *is the order of the pole* λ_i *of* $R(\cdot, A)$.

Proof. The inclusion $X_i \subset \ker(\lambda_i - A)^{\nu_i}$ follows since $\lambda_i - A$ is nilpotent of order ν_i on X_i by Theorem 2.9.

Suppose now that the inclusion is strict, that is, there exists a nonzero $x \in \ker(\lambda_i - A)^{\nu_i} \setminus X_i$. Note that for some $j \neq i$ there is a nonzero $y := P_j x \in X_j \cap \ker(\lambda_i - A)^{\nu_i}$. Take the largest $p \in \mathbb{N}$ such that

$$z := (\lambda_i - A)^p y \neq 0.$$

Then $z \in X_j$ and $Az = \lambda_i z$, thus

$$(A - \lambda_j)^{\nu_j} z = (\lambda_i - \lambda_j)^{\nu_j} z \neq 0,$$

which is a contradiction. $\qquad\square$

We formulate two special cases explicitly.

Corollary 2.19.

a) *If* $\nu_i = 1$, *then* $A|_{X_i} = \lambda_i I_{X_i}$.
b) *If all* $\nu_i = 1$, *then* A *is diagonalisable.*

Now we are going to find out more about A and $f(A)$ with the help of Theorems 2.9 and 2.11. To that purpose, we will frequently use the following result.

Theorem 2.20 (Spectral Mapping Theorem). *Let* $A \in \mathcal{L}(X)$ *and* $f \in C_A^\infty$. *Then*

$$\sigma\big(f(A)\big) = f\big(\sigma(A)\big) := \{f(\lambda) : \lambda \in \sigma(A)\}.$$

Proof. If $\mu \notin f(\sigma(A))$, then $\frac{1}{\mu - f(\lambda)} =: u(\lambda) \in C_A^\infty$ with $u \cdot (\mu - f) = 1$ on a neighborhood of $\sigma(A)$. Hence, plugging A into $u \cdot (\mu - f)$ gives

$$u(A)\big(\mu - f(A)\big) = \big(\mu - f(A)\big)u(A) = I,$$

meaning that $\mu - f(A)$ is invertible and $\mu \notin \sigma(f(A))$.

On the other hand, if x_i is an eigenvector of A belonging to λ_i for some $1 \leq i \leq m$, then $P_j x_i = 0$ for all $j \neq i$ by Theorem 2.9. Now Theorem 2.11 gives

$$f(A)x_i = f(\lambda_i)x_i,$$

hence $f(\lambda_i) \in \sigma(f(A))$. $\qquad\square$

The following is a consequence of Lemma 2.18 and Theorem 2.20.

Corollary 2.21. *The matrix $A \in \mathcal{L}(X)$ is nilpotent (i.e., $A^k = 0$ for a suitable $k \in \mathbb{N}$) if and only if $\sigma(A) = \{0\}$.*

Proof. If $A^k = 0$, then $\{0\} = \sigma(A^k) = [\sigma(A)]^k$, hence $\sigma(A) = \{0\}$. If, on the other hand, $\sigma(A) = \{0\}$, then the spectral projection corresponding to the only eigenvalue 0 must be equal to the identity and $A = A - 0$ is nilpotent on the corresponding spectral subspace, which is equal to X. $\square$

Corollary 2.22.

a) *The map $\lambda_i - A$ is bijective on $\bigoplus_{j \neq i} X_j$.*
b) *For the restriction $A_{|X_i}$ the identities*

$$\sigma\left(A|_{X_i}\right) = \{\lambda_i\} \quad and \quad \sigma\left(A|_{\bigoplus_{j \neq i} X_j}\right) = \sigma(A)\backslash\{\lambda_i\}$$

hold.

Proof. a) By Corollary 2.21, the spectrum of $A - \lambda_i$ restricted to X_i is zero. Hence, $\sigma(A|_{X_i}) = \{\lambda_i\}$ by the spectral mapping theorem, Theorem 2.20. Since all eigenvectors belonging to λ_i are contained in X_i, we have $\lambda_i \notin \sigma\left(A|_{\bigoplus_{j \neq i} X_j}\right)$. Again by Theorem 2.20, we obtain $0 \notin \sigma\left((A - \lambda_i)|_{\bigoplus_{j \neq i} X_j}\right)$. This means that $A - \lambda_i$ is bijective on $\bigoplus_{j \neq i} X_j$.

b) By the considerations above, we already have that $\sigma(A|_{X_i}) = \{\lambda_i\}$ and

$$\sigma\left(A|_{\bigoplus_{j \neq i} X_j}\right) \subseteq \sigma(A)\backslash\{\lambda_i\}.$$

Since $\lambda_k \in \sigma\left(A|_{\bigoplus_{j \neq i} X_j}\right)$ for $k \neq i$, the proof is complete. $\square$

The argument used in the second part of the proof of the Spectral Mapping Theorem 2.20 gives the following.

Corollary 2.23. *The eigenspace of $f(A)$ belonging to $f(\lambda_i)$ contains all eigenvectors of A that belong to the eigenvalues λ_j with $f(\lambda_j) = f(\lambda_i)$.*

It should be noted, however, that the eigenspace of $f(A)$ belonging to the eigenvalue $f(\lambda_i)$ need not be generated by eigenvectors of A, as the following simple example shows.

Example 2.24. The matrix $A = \left(\begin{smallmatrix} 0 & 0 \\ 1 & 0 \end{smallmatrix}\right)$ has spectrum $\sigma(A) = \{0\}$, the corresponding eigenspace has dimension 1. By contrast, since $A^2 = 0$, the eigenspace of A^2 belonging to 0 is two-dimensional.

The situation is much simpler for the spectral projections. First we need an auxiliary result.

Lemma 2.25. *For $g \in C^\infty_{f(A)}$ one has $g \circ f \in C^\infty_A$ and*

$$(g \circ f)(A) = g(f(A)).$$

Proof. The hypothesis means that g is a smooth function defined on a neighborhood of $\sigma(f(A)) = f(\sigma(A))$, hence $g \circ f$ is defined on a neighborhood of $\sigma(A)$. The rest is straightforward since we can now assume that f and g are polynomials. $\square$

Theorem 2.26. *The spectral projection of $f(A)$ belonging to $f(\lambda_i)$ for some $i \in \{1, \ldots, m\}$ is the sum of the spectral projections P_j of A with $f(\lambda_j) = f(\lambda_i)$.*

Proof. Let us denote by $P^f_i = \chi_{V_i}(f(A))$ the spectral projection of $f(A)$ belonging to $f(\lambda_i)$, where $V_i \subset \mathbb{C}$ is an open set such that $f(\lambda_i) \in V_i$. Using this notation, what we need to prove is that

$$P^f_i = \bigoplus_{f(\lambda_j) = f(\lambda_i)} P_j.$$

By Lemma 2.25, $P^f_i = (\chi_{V_i} \circ f)(A)$, so the statement follows since $\chi_{V_i} \circ f$ is a characteristic function of a neighborhood of exactly those eigenvalues λ_j for which $f(\lambda_j) = f(\lambda_i)$. $\square$

The question arises whether there are simple and general relations between the pole orders of $R(\cdot, A)$ at λ_i and $R(\cdot, f(A))$ at $f(\lambda_i)$. However, the following examples show that both an increase or a decrease can occur.

Examples 2.27.

a) Let us consider $A = \left(\begin{smallmatrix} 0 & 0 \\ 1 & 0 \end{smallmatrix}\right)$, $f(\lambda) = \lambda^2$. Then 0 is a pole of order 2 of $R(\cdot, A)$, but $0 = f(0)$ is a simple pole of $R(\cdot, f(A)) = R(\cdot, 0)$.

b) Considering

$$A = \begin{pmatrix} 2\pi i & 0 & 0 \\ 0 & 0 & 0 \\ 0 & 1 & 0 \end{pmatrix}, \text{ with } f(\lambda) = e^\lambda,$$

we see that $2\pi i$ is a pole of order 1 of $R(\cdot, A)$, but $1 = f(2\pi i)$ is a pole of order 2 of $R(\cdot, f(A))$ since $f(A) = e^A = \left(\begin{smallmatrix} 1 & 0 & 0 \\ 0 & 1 & 0 \\ 0 & 1 & 1 \end{smallmatrix}\right)$.

We close this section with a result that clarifies the relation between eigenvectors or pole orders with respect to A and $f(A)$ for the functions $f_t(\lambda) = e^{t\lambda}$, $t \in \mathbb{R}$.

Theorem 2.28. *Let us define e^{tA} as in Theorem 2.11, formula (2.9) for $t \neq 0$.*

a) *The eigenspace of e^{tA} belonging to $e^{t\lambda_i}$ is the sum of those eigenspaces of A that belong to eigenvalues λ_j with $e^{t\lambda_j} = e^{t\lambda_i}$.*

b) *The pole order of $e^{t\lambda_i}$ with respect to $R(\cdot, e^{tA})$ dominates the pole order of λ_i with respect to $R(\cdot, A)$.*

c) *If all λ_j with $e^{t\lambda_j} = e^{t\lambda_i}$ are first-order poles of $R(\cdot, A)$, then $e^{t\lambda_i}$ is a pole of order 1 of $R(\cdot, e^{tA})$.*

Proof. a) Let x be an eigenvector of e^{tA} for $e^{t\lambda_i}$ and some fixed i. Then by Theorem 2.26 we have $x \in \bigoplus_{e^{t\lambda_j} = e^{t\lambda_i}} X_j$. Let us denote by $i_1, \ldots, i_r$ those j with $e^{t\lambda_j} = e^{t\lambda_i}$, so

$$x = \sum_{s=1}^{r} y_s \quad \text{with } y_s \in X_{i_s}.$$

We will show that actually $y_s \in \ker(A - \lambda_{i_s})$ for all s. We first observe that each X_{i_s} is invariant under e^{tA}, hence

$$\sum_{s=1}^{r} e^{t\lambda_i} y_s = e^{t\lambda_i} x = e^{tA} x = \sum_{s=1}^{r} e^{tA} y_s$$

implies $e^{t\lambda_i} y_s = e^{tA} y_s$ for $s = 1, \ldots, r$. Now, for each fixed s, if $\nu_{i_s} = 1$, then we are done. Assume $\nu_{i_s} \geq 2$ and write

$$0 = e^{tA} y_s - e^{t\lambda_i} y_s = \sum_{\nu=1}^{\nu_{i_s}-1} \frac{e^{t\lambda_{i_s}} t^\nu}{\nu!} (A - \lambda_{i_s})^\nu y_s.$$

If $(A - \lambda_{i_s}) y_s \neq 0$, the right-hand side must be also nonzero by Lemma 2.16, a contradiction.

b) Generally speaking, for any operator B and any $\lambda_0 \in \sigma(B)$, the pole order of λ_0 with respect to $R(\cdot, B)$ is the order of nilpotency of $B - \lambda_0$ on X_0, the spectral subspace of B belonging to λ_0. Hence, for $z \in X_0$, this pole order is the maximal length of any chain

$$z, (B - \lambda_0)z, \ldots, (B - \lambda_0)^\nu z \neq 0.$$

Going now back to A, suppose $\nu_i \geq 2$ and take $z \in X_i$ with $(A - \lambda_i)^{\nu_i - 1} z \neq 0$. Then

$$\left(e^{tA} - e^{t\lambda_i} \right) z = \sum_{\nu=1}^{\nu_i - 1} \frac{e^{t\lambda_i} t^\nu}{\nu!} (A - \lambda_i)^\nu z =: z_1 \neq 0,$$

hence the pole order of $e^{t\lambda_i}$ with respect to $R(\cdot, e^{tA})$ is at least 2. If $\nu_i \geq 3$, we can apply $(A - \lambda_i)^{\nu_i - 2}$ to z_1 and obtain a vector $\neq 0$. Moreover,

$$\left(e^{tA} - e^{t\lambda_i} \right) z_1 = \sum_{\nu=1}^{\nu_i - 2} \frac{e^{t\lambda_i} t^\nu}{\nu!} (A - \lambda_i)^\nu z_1 \neq 0,$$

hence the pole order of $e^{t\lambda_i}$ is ≥ 3. If this argument is repeated $\nu_i - 1$ times, the conclusion follows.

c) This follows easily from Theorem 2.11, formula (2.9), and Theorem 2.26.
$\square$

2.4 Notes and Remarks

There are many ways to define a functional calculus for matrices. The first one applies the Jordan canonical form usually taught in a first linear algebra course. We have already mentioned the approach based on the power series representation (see Remark 2.5 and Corollary 2.14), which has the drawback that it only works nicely for entire functions f, making the analysis of spectral projections difficult. Using the Cauchy formula we can also show that our definition of $f(A)$ agrees with the Dunford's integral representation, meaning that for every $A \in \mathcal{L}(X)$ and $f \in C_A^\infty$ there is an open set $W \supset \sigma(A)$ such that

a) W is contained in the domain of f.

b) ∂W consists of a finite number of smooth closed curves oriented in the positive direction.

c) $f(A) = \dfrac{1}{2\pi i} \displaystyle\int_{\partial W} f(\lambda) R(\lambda, A) \, d\lambda$.

For a comprehensive survey on the functional calculus we refer to Gohberg, Goldberg, and Kaashoek [53, Section I.3].

2.5 Exercises

1. Find polynomials p with the following properties.
   a) $p(1) = 1$, $p(2) = 1$.
   b) $p(1) = 1$, $p'(1) = 2$.
   c) $p(1) = 1$, $p'(1) = 3$, $p(-1) = -1$.
   d) $p(1) = 1$, $p'(1) = 3$, $p(-1) = -1$, $p'(-1) = 3$.
   e) $p(1) = 1$, $p'(1) = 3$, $p(-1) = -1$, $p'(-1) = 2$.
   f) $p(1) = 1$, $p'(1) = 2$, $p''(1) = 3$.

   Discuss properties of $p(A)$ for some matrices $A \in \mathcal{L}(X)$.

2. Let A be a $n \times n$ diagonalisable matrix with m distinct eigenvalues $\lambda_1, \ldots, \lambda_m$. Prove that in this case its spectral projections are of the form

$$P_i = \prod_{j \neq i} \frac{A - \lambda_j}{\lambda_i - \lambda_j}, \quad i = 1, \ldots, m.$$

3. a) Determine the eigenvalues λ_i and the corresponding multiplicities ν_i for the matrices

$$A = \begin{pmatrix} 0 & 0 \\ 0 & 1 \end{pmatrix} \quad \text{and} \quad B = \begin{pmatrix} 1 & 0 & 0 \\ 0 & 1 & 0 \\ 0 & 1 & 1 \end{pmatrix}.$$

   b) Discuss further matrices you find interesting.

4. Calculate e^{tA}, A^n, and $\sin(tA)$ for $A = \begin{pmatrix} 6 & -1 \\ 3 & 2 \end{pmatrix}$. Discuss further matrices you find interesting.

5. Show that if $B = S^{-1}AS$, where S is an invertible matrix, and if $f \in C_A^\infty$, then $f \in C_B^\infty$ and $f(B) = S^{-1}f(A)S$.

6.   a) Show that Theorem 2.9 does not hold in the case of real scalars (i.e., in the situation of $X = \mathbb{R}^n$, $\mathcal{L}(X) = M_n(\mathbb{R})$, and $\mathbb{R}[x]$).

   b) Which one of the arguments leading to Theorem 2.9 does not hold in the real case?

7. Let $C \in \mathcal{L}(X)$ with $\|C\| < 1$. Show that

$$(I - C)^{-1} = \sum_{k=0}^\infty C^k. \tag{2.12}$$

8. Using the result of the previous exercise, show the Neumann series representation in (2.11) for $|\lambda| > \|A\|$.

9. Show, using the definition, the so-called resolvent equations

$$AR(\lambda, A) = \lambda R(\lambda, A) - I \tag{2.13}$$

for $\lambda \in \rho(A)$, and

$$R(\lambda, A) - R(\mu, A) = (\mu - \lambda)R(\lambda, A)R(\mu, A) \tag{2.14}$$

for $\lambda, \mu \in \rho(A)$.

10. Using the results of the previous two exercises, show that for $|\lambda - \mu| < \frac{1}{\|R(\mu,A)\|}$, we have

$$R(\lambda, A) = \sum_{k=0}^\infty (\mu - \lambda)^k R(\mu, A)^{k+1}. \tag{2.15}$$

11. Give an alternative proof of Theorem 2.28.b) using the formula

$$R(\mu, e^{tA}) = \sum_{i=1}^m \sum_{\nu=0}^{\nu_i - 1} \frac{g_\mu^{(\nu)}(\lambda_i)}{\nu!}(A - \lambda_i)^\nu P_i$$

with $g_\mu(\lambda) = (\mu - e^{t\lambda})^{-1}$ ($\mu \notin e^{t\sigma(A)}$). Can you determine the pole order in question?

12. Find conditions on f such that the assertions of Theorem 2.28 hold for $f(A)$ instead of e^{tA}.

Chapter 3

Powers of Matrices

In the previous chapter we presented known facts from the spectral theory of matrices in a coordinate-free way. We are, however, interested not simply in linear algebra, but mainly in the asymptotic behavior of dynamical systems, a central theme in this text.

We apply the knowledge we gained on the structure of linear operators on finite-dimensional vector spaces to investigate what happens to the sequence consisting of the powers of a matrix. Topics we cover include boundedness, convergence to zero, convergence, mean convergence (or Cesàro convergence), periodicity, and hyperbolic decomposition.

3.1 The Coordinate Sequences

Let X be a vector space with $\dim X = n < \infty$. We will employ the formulas obtained in Chapter 2 to study the asymptotic behavior of the powers of a given operator acting on X. In order to be consistent with the exponential function later on, we denote here the given operator by $T \in \mathcal{L}(X)$. We retain, however, the notation of Theorem 2.11 concerning the eigenvalues $\lambda_1, \ldots, \lambda_m$, the multiplicities $\nu_1, \ldots, \nu_m$ (as zeroes of the minimal polynomial of T), and the spectral projections $P_1, \ldots, P_m$.

Thus, by Theorem 2.11, formula (2.10), we have

$$T^k = \sum_{i=1}^{m} \sum_{\nu=0}^{\min\{\nu_i-1,k\}} \binom{k}{\nu} \lambda_i^{k-\nu} (T - \lambda_i)^\nu P_i \quad \text{for } k \in \mathbb{N}. \tag{3.1}$$

Example 3.1. Let (f_k) be the Fibonacci sequence considered as the motivating example in Section 1.1. We have seen that

$$\begin{pmatrix} f_{k+1} \\ f_k \end{pmatrix} = A^k \begin{pmatrix} 1 \\ 0 \end{pmatrix} \quad \text{for} \quad A = \begin{pmatrix} 1 & 1 \\ 1 & 0 \end{pmatrix}, \quad k \in \mathbb{N}_0.$$

It is not difficult to compute the eigenvalues of A,

$$\lambda_1 = \frac{1 + \sqrt{5}}{2} \quad \text{and} \quad \lambda_2 = \frac{1 - \sqrt{5}}{2},$$

and the corresponding spectral projections

$$P_1 = \frac{\sqrt{5}}{5} \begin{pmatrix} \frac{1+\sqrt{5}}{2} & 1 \\ 1 & \frac{-1+\sqrt{5}}{2} \end{pmatrix} \quad \text{and} \quad P_2 = \frac{\sqrt{5}}{5} \begin{pmatrix} \frac{-1+\sqrt{5}}{2} & -1 \\ -1 & \frac{1+\sqrt{5}}{2} \end{pmatrix}.$$

By formula (3.1) we obtain

$$A^k = \lambda_1^k P_1 + \lambda_2^k P_2$$
$$= \frac{\sqrt{5}}{5} \begin{pmatrix} \left(\frac{1+\sqrt{5}}{2}\right)^{k+1} - \left(\frac{1-\sqrt{5}}{2}\right)^{k+1} & \left(\frac{1+\sqrt{5}}{2}\right)^{k} - \left(\frac{1-\sqrt{5}}{2}\right)^{k} \\ \left(\frac{1+\sqrt{5}}{2}\right)^{k} - \left(\frac{1-\sqrt{5}}{2}\right)^{k} & \left(\frac{1-\sqrt{5}}{2}\right)^{k}\left(\frac{1+\sqrt{5}}{2}\right) - \left(\frac{1+\sqrt{5}}{2}\right)^{k}\left(\frac{1-\sqrt{5}}{2}\right) \end{pmatrix},$$

and thus the explicit formula for the entries of the Fibonacci sequence is

$$f_k = \frac{\sqrt{5}}{5} \left(\left(\frac{1+\sqrt{5}}{2}\right)^{k} - \left(\frac{1-\sqrt{5}}{2}\right)^{k} \right), \quad k \in \mathbb{N}_0.$$

Our main interest, however, lies in the asymptotic behavior. In order to understand what happens with T^k as $k \to \infty$, we use the linear independence of the set

$$B_T := \{(T - \lambda_i)^\nu P_i : i = 1, \ldots, m; \ \nu = 0, \ldots, \nu_i - 1\}$$

in the vector space $\mathcal{L}(X)$ (cf. Lemma 2.16). If we extend this set to a basis $\mathbb{B}_T$ of $\mathcal{L}(X)$, then (3.1) means that the (non-zero) coordinates of T^k with respect to $\mathbb{B}_T$ are

$$\left\{ \binom{k}{\nu} \lambda_i^{k-\nu} : i = 1, \ldots, m; \ \nu = 0, \ldots, \nu_i - 1 \right\}$$

(since from now on we consider $k \to \infty$, we allow ourselves to simplify the upper bound of ν from $\min\{\nu_i - 1, k\}$ to $\nu_i - 1$). Likewise, if we are interested in the "orbit" $\{T^k x : k \in \mathbb{N}\}$ of a single element $x \in X$ under the powers T^k of T, we may use a basis of X containing the set

$$\{(T - \lambda_i)^\nu P_i x : i = 1, \ldots, m; \ \nu = 0, \ldots, \nu_i - 1\} \setminus \{0\}.$$

Again, the corresponding coordinate sequences are among the ones obtained above. Thus, since convergence in a finite-dimensional vector space is convergence in every coordinate, no matter what basis is employed, the behavior of T^k (or of $T^k x$ for $x \in X$) as $k \to \infty$ is reflected by the behavior of the sequences

$$z_{\lambda,\nu}(k) := \binom{k}{\nu} \lambda^{k-\nu} \tag{3.2}$$

for $\lambda \in \sigma(T)$, $\nu = 0, \ldots, n-1$ (observe that $\nu_i \leq n$ for every i). In case all these sequences converge, (T^k) (or $T^k x$ for a given $x \in X$) converges and the coordinates of the limit are obtained as the limits of the corresponding sequences of coordinates.

The advantage of this approach is that the behavior of functions in (3.2) is easily understood and essentially depends on the modulus of λ.

- If $|\lambda| < 1$, then $z_{\lambda,\nu}(k) \to 0$ as $k \to \infty$ for all ν, since $\lim_{k \to \infty} k^\nu \lambda^k = 0$.
- If $|\lambda| > 1$, then $|z_{\lambda,\nu}(k)| \to \infty$ as $k \to \infty$ for all ν.
- If $|\lambda| = 1$ and $\nu = 0$, then $z_{\lambda,0}(k) = \lambda^k$.
- If $|\lambda| = 1$ and $\nu \geq 1$, then $|z_{\lambda,\nu}(k)| \to \infty$ as $k \to \infty$.

Using these facts, we are able to describe the asymptotics of $T^k x$ for $x \in X_i = \operatorname{im} P_i$, depending on λ_i and ν_i.

3.2 The Spectral Radius

We begin our investigations with estimates for $\|T^k\|$ related to the *spectral radius*

$$\mathrm{r}(T) := \max\left\{|\lambda| \,:\, \lambda \in \sigma(T)\right\}. \tag{3.3}$$

Lemma 3.2. *Let* $\|\cdot\|$ *be a norm on* $\mathcal{L}(X)$, $T \in \mathcal{L}(X)$, *and* $\mu > \mathrm{r}(T)$. *Then there exist constants* $N > 0$ *and* $M \geq 1$ *such that for all* $k \in \mathbb{N}$,

$$N \cdot \mathrm{r}(T)^k \leq \|T^k\| \leq M \cdot \mu^k.$$

If $\|\cdot\|$ *is an operator norm, we can choose* $N = 1$.

Proof. By the Spectral Mapping Theorem 2.20,

$$\mathrm{r}(T^k) = \mathrm{r}(T)^k,$$

hence the last part of the assertion is clear for any operator norm on $\mathcal{L}(X)$ by Corollary 1.10. Then the lower estimate

$$N \cdot \mathrm{r}(T)^k \leq \|T^k\|$$

(with suitable $N > 0$) follows by equivalence of norms for finite-dimensional spaces.

Turning to the upper estimate in the assertion, we note that the coordinates of T^k with respect to a basis of $\mathcal{L}(X)$ containing B_T are

$$z_{\lambda,\nu}(k) := \binom{k}{\nu} \lambda^{k-\nu},$$

where $1 \leq i \leq m$ and $0 \leq \nu \leq \nu_i$. Since

$$\binom{k}{\nu} \leq k^\nu,$$

and since for $|\lambda| < 1$

$$k^\nu \lambda^{k-\nu} \longrightarrow 0$$

as $k \to \infty$, for all $\nu \in \mathbb{N}$, the coordinate sequences of $\frac{1}{\mu^k} T^k$ remain bounded as $k \to \infty$. Thus,

$$\|T^k\|_\infty \leq C\mu^k, \qquad k \in \mathbb{N},$$

for a suitable constant C, where $\|\cdot\|_\infty$ denotes the usual maximum norm with respect to coordinates belonging to B_T. The desired estimate for $\|T^k\|$ again follows by the equivalence of norms.   □

It is extremely important that the spectral radius of T can be determined through the sequence $(\|T^k\|)$, regardless of the norm $\|\cdot\|$.

Proposition 3.3 (*Gelfand's formula*). *For an operator $T \in \mathcal{L}(X)$ the following holds.*

   a) $\mathrm{r}(T) = \lim_{k\to\infty} \|T^k\|^{1/k}$ *for any norm $\|\cdot\|$ on $\mathcal{L}(X)$.*
   b) *If $\|\cdot\|$ is an operator norm on $\mathcal{L}(X)$, then $\mathrm{r}(T) = \inf_{k>0} \|T^k\|^{1/k}$.*

Proof. The proof is an immediate consequence of the previous Lemma 3.2, which yields the estimate

$$N^{1/k}\mathrm{r}(T) \leq \|T^k\|^{1/k} \leq M^{1/k}\mu$$

whenever $\mu > \mathrm{r}(T)$ for suitable constants N, M. If $\|\cdot\|$ is an operator norm, then the (admissible) choice $N = 1$ in these estimates forces $\mathrm{r}(T) = \inf_{k>0} \|T^k\|^{1/k}$.   □

Let us now turn our attention back to estimates for $\|T^k x\|$ and suppose again that the same $\|\cdot\|$ denotes the corresponding operator norm. The special case when $x \in X_i$ for some spectral subspace $X_i = \mathrm{im}\, P_i$ can be treated analogously to our previous considerations.

Proposition 3.4. *Let us consider $T \in \mathcal{L}(X)$ and a norm $\|\cdot\|$ on X. Then for every $\mu > |\lambda_i|$ ($1 \leq i \leq m$) there exists a number $M \geq 1$ such that*

$$\|T^k x\| \leq M\mu^k \|x\|$$

for all $k \in \mathbb{N}$ and $x \in X_i$. Further, if $\lambda_i \neq 0$, then for every $0 < \rho < |\lambda_i|$ there exists $N > 0$ such that

$$N\rho^k \|x\| \leq \|T^k x\| \leq M\mu^k \|x\|$$

for all $k \in \mathbb{N}$ and $x \in X_i$.

Proof. Since $\|T^k x\| \leq \|T^k\| \cdot \|x\|$, and since $\sigma(T|_{X_i}) = \{\lambda_i\}$, we obtain from Lemma 3.2 the existence of a constant M for every $\mu > |\lambda_i|$ such that

$$\|T^k x\| \leq M\mu^k \|x\|$$

for all $k \in \mathbb{N}$ and $x \in X_i$.

On the other hand, if $\lambda_i \neq 0$, then $T|_{X_i}$ has an inverse S with $\sigma(S) = \{\lambda_i^{-1}\}$. So for every $0 < \rho < |\lambda_i|$, there exists $C \geq 1$ such that for all $z \in X_i$, we have that

$$\|S^k z\| \leq C\rho^{-k}\|z\|.$$

Choosing $z := T^k x$ for $k \in \mathbb{N}$ we have $S^k z = x$ and hence $\|x\| \leq C\rho^{-k}\|z\|$, that is

$$\|T^k x\| = \|z\| \geq \frac{1}{C}\rho^k\|x\|$$

yielding the desired inequality.   □

3.3 Asymptotics

After this intermezzo on the spectral radius $r(T)$, we now describe the long-time behavior of the sequence (T^k). We will consider various types of asymptotic behavior.

Definition 3.5. For $T \in \mathcal{L}(X)$ and any norm $\|\cdot\|$ on $\mathcal{L}(X)$ we say that the sequence (T^k) is

- *bounded*[3] if $\sup_{k \in \mathbb{N}} \|T^k\| < \infty$;
- *stable*[3] if $\lim_{k \to \infty} \|T^k\| = 0$;
- *convergent* if $\lim_{k \to \infty} T^k = P$ for some $P \in \mathcal{L}(X)$;
- *periodic* with *period* p if T is periodic with period p, i.e., $T^p = I$;
- *Cesàro summable* (or T is *mean-ergodic*) if the limit $\lim_{k \to \infty} \frac{1}{k}\sum_{l=0}^{k-1} T^l$ exists.

Remarks 3.6.

a) Note that boundedness of the sequence (T^k) is equivalent to boundedness of all coordinate sequences $z_{\lambda,\nu}(k)$ with respect to $\mathbb{B}_T$ defined in (3.2).

b) If the sequence (T^k) converges to P, then

$$TP = T \lim_{k \to \infty} T^k = \lim_{k \to \infty} T^{k+1} = P = PT = \lim_{k \to \infty} T^{2k} = P^2,$$

i.e., P is a projection commuting with T.

c) If (α_k) is a sequence in a vector space, then $\alpha^{(k)} := \frac{1}{k}\sum_{l=0}^{k-1} \alpha_l$ is called the corresponding sequence of *Cesàro means*. Analogously, we call

$$T^{(k)} := \frac{1}{k}\sum_{l=0}^{k-1} T^l, \quad k \in \mathbb{N},$$

appearing in the definition above, the *Cesàro means* of T.

---

[3]Note that working with ODEs or dynamical systems, a different terminology is also widely accepted: what we call "bounded" is often called "stable", and what we call "stable" is often called "asymptotically stable".

The asymptotic behavior of (T^k) will depend essentially on the size of $\mathrm{r}(T)$ compared to 1. By Γ we denote the unit circle in $\mathbb{C}$ and by

$$\Gamma_q := \{e^{2k\pi i/q} \colon k = 0, \ldots, q-1\}$$

the set of all qth roots of unity in $\mathbb{C}$. We call $\lambda_0 > 0$ a *radially dominant eigenvalue* if $\lambda_0 \in \sigma(T)$ and $|\lambda| < \lambda_0$ for all $\lambda \in \sigma(T)\setminus\{\lambda_0\}$.

Theorem 3.7. *For $T \in \mathcal{L}(X)$ the following assertions hold.*

a) (T^k) *is stable if and only if* $\mathrm{r}(T) < 1$.

b) (T^k) *is bounded if and only if* $\mathrm{r}(T) \leq 1$ *and all eigenvalues of modulus 1 are simple poles of* $R(\cdot, T)$.

c) (T^k) *is periodic with period p if and only if it is bounded and* $\sigma(T) \subseteq \Gamma_p$.

d) $1 \in \sigma(T)$ *and* $\lim_{k\to\infty} T^k = P_1$ *here (P_1 denotes the spectral projection of T belonging to 1) if and only if 1 is a radially dominant eigenvalue of T which is a simple pole of the resolvent* $R(\cdot, T)$.

e) $\sigma(T) \cap \Gamma = \emptyset$ *if and only if there exist T-invariant subspaces X_s and X_u such that $X = X_\mathrm{s} \oplus X_\mathrm{u}$ and*

$$\lim_{k\to\infty} \|T^k x\| = 0 \;\; \text{for } x \in X_\mathrm{s} \quad \text{and} \quad \lim_{k\to\infty} \|T^k x\| = \infty \;\; \text{for } x \in X_\mathrm{u}.$$

Proof. a) This is a direct consequence of Lemma 3.2.

b) Assume that (T^k) is bounded. Lemma 3.2 again yields $\mathrm{r}(T) \leq 1$. If an eigenvalue $\lambda \in \sigma(T)$ with $|\lambda| = 1$ would have multiplicity $\nu > 1$, then the corresponding coordinate sequence $z_{\lambda,\nu-1}(k)$ of T^k with respect to $\mathbb{B}_T$ would not be bounded, a contradiction.

On the other hand, if $\mathrm{r}(T) \leq 1$ and all eigenvalues of modulus 1 are simple poles of the resolvent, then all the coordinate sequences of T^k with respect to $\mathbb{B}_T$ are bounded.

c) Let (T^k) be bounded and $\sigma(T) \subseteq \Gamma_p$. By b), all nonzero eigenvalues have multiplicity 1, thus formula (3.1) is simplified as

$$T^k = \sum_{i=1}^m \lambda_i^k P_i \quad \text{for } k \in \mathbb{N}.$$

Since $\lambda_i^p = 1$ for all $i = 1, \ldots, m$, we have $T^p = I$.

If (T^k) is periodic, i.e., $T^p = I$ for some $p \in \mathbb{N}$, it is bounded and for every $\lambda \in \sigma(T)$ we have $\lambda^p = 1$.

d) Let $\lim_{k\to\infty} T^k = P_1$. From a) and b) we infer that in this case $\mathrm{r}(T) = 1$. If there exists an eigenvalue $\lambda \neq 1$ with $|\lambda| = 1$, then the corresponding coordinate sequence of T^k with respect to $\mathbb{B}_T$ contains $z_{\lambda,0}(k) = \lambda^k$, which does not converge as $k \to \infty$. Also, if 1 is not a simple pole, then T^k has coordinate $z_{\lambda,1}(k) = k$, which again does not converge as $k \to \infty$.

Conversely, if 1 is a radially dominant eigenvalue and a simple pole of the resolvent, then all coordinate sequences $z_{\lambda,0}(k)$ with respect to $\mathbb{B}_T$ converge. Since by a) the coordinate sequences belonging to eigenvalues with $|\lambda| < 1$ converge to 0, $T^k \to P_1$ as $k \to \infty$.

e) Let

$$X_{\mathrm{s}} := \bigoplus_{|\lambda_i|<1} X_i \quad \text{and} \quad X_{\mathrm{u}} := \bigoplus_{|\lambda_i|>1} X_i,$$

and define the operators $T_{\mathrm{s}} := T|_{X_{\mathrm{s}}}$ and $T_{\mathrm{u}} := T|_{X_{\mathrm{u}}}$. Since $\sigma(T) = \sigma(T_{\mathrm{s}}) \cup \sigma(T_{\mathrm{u}})$, the assertion follows using a) and b) for the operators T_{s} and T_{u}. $\qquad\square$

Summing up, we have seen that the limit $\lim_{k\to\infty} T^k$ exists only in one of the following two situations:

- if $\mathrm{r}(T) < 1$, then $T^k \to 0$ as $k \to \infty$;
- if $\mathrm{r}(T) = 1 = \lambda_1$ is a radially dominant eigenvalue with multiplicity $\nu_1 = 1$, then $T^k \to P_1$ as $k \to \infty$.

Example 3.8. As simple examples of different types of asymptotical behaviour consider the matrices

$$T_1 = \begin{pmatrix} 1 & 0 \\ 1 & 1 \end{pmatrix}, \quad T_2 = \frac{1}{2}\begin{pmatrix} 1 & 0 \\ 1 & 1 \end{pmatrix}, \quad T_3 = \frac{1}{2}\begin{pmatrix} 1 & 1 \\ 1 & 1 \end{pmatrix},$$

$$T_4 = \begin{pmatrix} 0 & 1 \\ 1 & 0 \end{pmatrix}, \quad T_5 = \begin{pmatrix} 1 & 1 \\ 1 & 1 \end{pmatrix}.$$

The spectra of these matrices are easy to compute and the considerations above yield that the sequences (T_1^k) and (T_5^k) are unbounded, the sequence (T_2^k) is stable, the sequence (T_3^k) converges to the spectral projection P_1, while the sequence (T_4^k) is periodic with period 2.

Let us take a closer look at two of the non-convergent sequences above. The eigenvalues of T_5 are $\lambda_1 = 0$ and $\lambda_2 = 2$, with corresponding spectral projections P_1 and P_2. Hence, we can decompose the space $X = \mathbb{C}_2$ into *stable* and *unstable* part as

$$X = P_1 X \oplus P_2 X = X_{\mathrm{s}} \oplus X_{\mathrm{u}}$$

such that $T_5|_{X_{\mathrm{s}}}$ is stable and $T_5|_{X_{\mathrm{u}}}$ is unbounded.

In contrast to the above, the sequence (T_4^k) is bounded. Moreover it is alternating, i.e., $T_4^{2l-1} = T_4$ and $T_4^{2l} = I$, $l \in \mathbb{N}$. So, it has convergent subsequences.

Cesàro Summability

In the remainder of this chapter we will discuss situations where a suitable subsequence of (T^k) converges, or where the Cesàro means of T converge.

First we show an auxiliary result.

Lemma 3.9 (Kronecker). *Let $\lambda_1, \ldots, \lambda_r \in \Gamma$. There exists a sequence $(s_k) \subset \mathbb{N}$, such that*

$$\lim_{k \to \infty} \lambda_i^{s_k} = 1$$

for all $1 \leq i \leq r$.

Proof. We give the proof for the case $r = 2$. The general case can be obtained in the same manner.

On the torus $\Gamma \times \Gamma = \{(e^{i\varphi_1}, e^{i\varphi_2}) : 0 \leq \varphi_{1,2} < 2\pi\}$, we define a metric d by

$$d\left((e^{i\varphi_1}, e^{i\varphi_2}), (e^{i\zeta_1}, e^{i\zeta_2})\right) := \max\{|\varphi_1 - \zeta_1|(\mathrm{mod}\ 2\pi), |\varphi_2 - \zeta_2|(\mathrm{mod}\ 2\pi)\}.$$

Let $(\lambda_1, \lambda_2) \in \Gamma \times \Gamma$. If there exists a $p \in \mathbb{N}$ such that $\lambda_1^p = 1 = \lambda_2^p$, then taking $s_k := p \cdot k$ for all $k \in \mathbb{N}$ we obtain

$$(\lambda_1, \lambda_2)^{s_k} := (\lambda_1^{s_k}, \lambda_2^{s_k}) = (1, 1),$$

and the assertion of the lemma holds.

Now assume that $\lambda_1 = e^{i\alpha}$ with $\frac{\alpha}{2\pi} \in \mathbb{R} \setminus \mathbb{Q}$. In this case all powers $(\lambda_1, \lambda_2)^p$, $p \in \mathbb{N}$, are pairwise distinct. Let $\xi_k := \frac{2\pi}{k}$ and consider all "boxes" of the form $\hat{\Gamma}_j \times \hat{\Gamma}_l$, where the sector $\hat{\Gamma}_j$ is defined as

$$\hat{\Gamma}_j := \left\{e^{i\xi} : (j-1)\xi_k \leq \xi \leq j\xi_k\right\}.$$

For $1 \leq j, l \leq n$ there are n^2 boxes of the form $\hat{\Gamma}_j \times \hat{\Gamma}_l$. Thus by Dirichlet's pigeonhole principle two of the $n^2 + 1$ powers $(\lambda_1, \lambda_2)^p$ with $0 \leq p \leq n^2$ are in the same box. If p and q are the exponents with $0 \leq p < q \leq n^2$ such that $(\lambda_1, \lambda_2)^p$ and $(\lambda_1, \lambda_2)^q$ are in the same box, then

$$d\left((\lambda_1, \lambda_2)^p, (\lambda_1, \lambda_2)^q\right) < \xi_k.$$

Observe that multiplication by a non-zero element of $\Gamma \times \Gamma$ is well defined and is an isometry for the metric d. Therefore,

$$d\left((1, 1), (\lambda_1, \lambda_2)^{q-p}\right) < \xi_k.$$

The statement of the lemma now follows upon letting $k \to \infty$. $\square$

Similar arguments as before together with Lemma 3.9 yield the following result on the convergence of a subsequence of (T^k).

Theorem 3.10. *Let $T \in \mathcal{L}(X)$. Then the following assertions are equivalent.*

(i) *There exists a subsequence of (T^k) which converges, as $k \to \infty$, to some limit $P \neq 0$.*

(ii) *$\mathrm{r}(T) = 1$ and all eigenvalues of modulus 1 are simple poles of the resolvent.*

If this is the case, the limit in question must be of the form $P = \sum_{|\lambda_i|=1} P_i$, where P_i are the spectral projections belonging to the eigenvalues of modulus 1.

Proof. (i) $\Longrightarrow$ (ii): This follows by the same reasoning as the proof of assertion b) of Theorem 3.7.

(ii) $\Longrightarrow$ (i): Let $\{\lambda_1, \ldots, \lambda_r\}$ be the (non-empty) set of eigenvalues of T of modulus 1. By Kronecker's Lemma 3.9, there exists a sequence $(s_k) \subset \mathbb{N}$, such that

$$\lim_{k \to \infty} \lambda_i^{s_k} = 1 \quad \text{for all } i = 1, \ldots, r.$$

For the subsequence (T^{s_k}) we use formula (3.1) and the fact that all λ_i are simple poles of the resolvent, hence

$$\lim_{k \to \infty} T^{s_k} = \sum_{i=1}^{r} \lambda_i^{s_k} P_i = \sum_{i=1}^{r} P_i \neq 0,$$

which proves (i) as well as the additional statement. $\qquad\square$

Definition 3.11. We call $T \in \mathcal{L}(X)$ a *spectral contraction* if the conditions in Theorem 3.10 are satisfied.

A simple example of a spectral contraction is the matrix $T_4 = \begin{pmatrix} 0 & 1 \\ 1 & 0 \end{pmatrix}$ considered in Example 3.8. The following is an alternative characterization of the spectral contraction that explains its name.

Theorem 3.12. *Let $T \in \mathcal{L}(X)$. Then the following assertions are equivalent.*

(i) *T is a spectral contraction.*
(ii) *There exists a norm $\|\cdot\|$ on X, such that for the corresponding operator norm on $\mathcal{L}(X)$, we have $\|T^k\| = 1$ for all $k \in \mathbb{N}$.*
(iii) *$\mathrm{r}(T) = 1$ and the sequence (T^k) is bounded.*

Proof. (i) $\Longrightarrow$ (ii): Let $\vertiii{\cdot}$ be a norm on $\mathbb{C}^n$. If (i) holds, then $\vertiii{T^k} \leq M, k \in \mathbb{N}$, for some $M < \infty$. Put

$$\|x\| := \sup_{k \in \mathbb{N}_0} \vertiii{T^k x}.$$

Then $\|\cdot\|$ is a norm on $\mathbb{C}^n$ such that for the corresponding operator norm $\|T^k\| = 1$ for all $k \in \mathbb{N}$.

(ii) $\Longrightarrow$ (iii): If (ii) holds, then $\mathrm{r}(T) = 1$ and, by the equivalence of norms, $(\|T^k\|)$ is bounded for any norm $\|\cdot\|$ on $\mathcal{L}(X)$.

(iii) $\Longrightarrow$ (i): Assuming (iii), Theorem 3.7.b) yields (ii) of Theorem 3.10 and the implication loop is closed. $\qquad\square$

Finally, we consider the convergence of Cesàro means $T^{(k)} = \frac{1}{k} \sum_{\nu=0}^{k-1} T^\nu$. We will see that the sequence (T^k) is Cesàro summable iff it is bounded, that is, if $\mathrm{r}(T) < 1$ or if T is a spectral contraction.

Theorem 3.13. *For $T \in \mathcal{L}(X)$ the following assertions are equivalent.*

(i) (T^k) *is Cesàro summable.*

(ii) $\lim_{k \to \infty} \left(k^{-1} T^k \right) = 0.$

(iii) (T^k) *is bounded.*

(iv) $\mathrm{r}(T) \leq 1$ *and each eigenvalue of modulus 1 is a simple pole of the resolvent.*

In any of these equivalent cases the sequence $\left(T^{(k)} \right)$ of Cesàro means of T converges to 0 if $1 \notin \sigma(T)$, and to the spectral projection P_1 belonging to 1 if $1 \in \sigma(T)$.

Proof. (i) $\implies$ (ii) follows since we can write $T^k = kT^{(k)} - (k-1)T^{(k-1)}$. Implication (ii) $\implies$ (iii) is clear and (iii) $\implies$ (iv) holds by Theorem 3.7.b). It remains to show (iv) $\implies$ (i) and the assertion on the limit of the Cesàro means.

The sequence $\left(T^{(k)} \right)$ has, as coordinate sequences with respect to $\mathbb{B}_T$, the Cesàro means of the coordinate sequences of (T^k):

$$z_{\lambda,\nu}^{(k)} := \frac{1}{k} \sum_{l=0}^{k-1} z_{\lambda,\nu}(l) = \frac{1}{k} \sum_{l=0}^{k-1} \binom{l}{\nu} \lambda^{l-\nu}.$$

These sequences clearly converge to 0 if $|\lambda| < 1$ and diverge if $|\lambda| > 1$ (see the computation below). Furthermore, for $|\lambda| = 1$ the sequence $z_{\lambda,\nu}^{(k)}$ is bounded only in the case $\nu = 0$, i.e., when λ is a simple pole of the resolvent.

If $\lambda \neq 1$, then compute

$$z_{\lambda,0}^{(k)} = \frac{1}{k} \sum_{l=0}^{k-1} \binom{l}{0} \lambda^l = \frac{1}{k} \frac{\lambda^k - 1}{\lambda - 1},$$

which tends to zero as $k \to \infty$, if $|\lambda| \leq 1, \lambda \neq 1$. On the other hand, for $\lambda = 1$ we have $z_{1,0}^{(k)} = 1$ for every $k \in \mathbb{N}$, and thus in this case $T^{(k)} \to P_1$ as $k \to \infty$. $\qquad\square$

This means that if the sequence (T^k) converges, its limit is the same as the Cesàro limit. On the other hand, there are Cesàro summable sequences which are not convergent, see the sequence (T_4^k) in Example 3.8.

3.4   Notes and Remarks

The question of convergence of (T^k) is extremely important for numerical methods. We just briefly mention a very simple method for computing a dominant eigenpair (λ_1, v_1) of a diagonalisable matrix $T \in M_n(\mathbb{R})$ with eigenvalues

$$|\lambda_1| > |\lambda_2| \geq \cdots \geq |\lambda_n|.$$

We know by Theorem 3.7 that

$$\lim_{k \to \infty} \left(\lambda_1^{-1} T \right)^k = P_1,$$

where P_1 is the spectral projection of $\lambda_1^{-1}T$ corresponding to 1, and consequently

$$\lim_{k \to \infty} \left(\lambda_1^{-1}T\right)^k x_0 = P_1 x_0$$

for any $x_0 \in \mathbb{R}^n$. Therefore, for $P_1 x_0 \neq 0$ the sequence $\|T^k x_0\|^{-1} T^k x_0$ converges to an eigenvector associated with λ_1. The iterative algorithm basing on these considerations is called *power method*:

$$x_{k+1} = \frac{T x_k}{\|T x_k\|}. \tag{3.4}$$

More information on this method as well as on other topics of this chapter can be found in the monograph by Meyer [94].

3.5 Exercises

1. For matrices $A = (a_{ij})$ and $B = (b_{ij})$ we write $A \leq B$ if $a_{ij} \leq b_{ij}$ for all i, j, and we denote $|A| := (|a_{ij}|)$. Prove that if $|A| \leq B$, then the following inequalities concerning spectral radii hold:

$$\mathrm{r}(A) \leq \mathrm{r}(|A|) \leq \mathrm{r}(B).$$

2. Show that if there exists an operator norm $\|\cdot\|$ on $\mathcal{L}(X)$ such that $\|T\| < 1$, then the sequence (T^k) is stable.

3. For the following matrices compute powers T^k and evaluate $\lim_{k \to \infty} T^k$ if it exists.

   a) $T = \begin{pmatrix} 1 & -1 \\ 0 & 1 \end{pmatrix}$,   b) $T = \frac{1}{2} \begin{pmatrix} -1 & 1 \\ 1 & 1 \end{pmatrix}$,   c) $T = \frac{1}{4} \begin{pmatrix} 2 & 2 \\ 1 & 3 \end{pmatrix}$.

4. Describe the asymptotic behavior of the sequence (T^k) for the following special classes of matrices $T \in M_n(\mathbb{R})$.

   a) T is idempotent (or involutary), i.e., $T^2 = I$.

   b) T is nilpotent, i.e., $T^q = 0$ for some $q \in \mathbb{N}$.

   c) T is unipotent, i.e., $T - I$ is nilpotent.

   d) T is orthogonal, i.e., $T^\top T = T T^\top = I$.

5.   a) Prove that if (a_k) converges in X, so does the sequence of Cesàro means $\left(a^{(k)}\right)$ in X and the limits coincide.

   b) Find an example of a non-convergent sequence whose associated sequence of Cesàro means converges.

6. For each of the following matrices determine whether its powers are convergent or Cesàro summable. Evaluate the limit of each convergent matrix and the Cesàro limit of each summable matrix.

$$A_1 = \begin{pmatrix} 0 & 1 & 0 \\ 1 & 0 & 1 \\ 0 & 1 & 0 \end{pmatrix}, \qquad A_2 = \begin{pmatrix} 0 & 1 & 0 \\ 0 & 0 & 1 \\ 1 & 0 & 0 \end{pmatrix}, \qquad A_3 = \frac{1}{2} \begin{pmatrix} -1 & 1 & -1 \\ 2 & 0 & -1 \\ 2 & -2 & 1 \end{pmatrix}.$$

Chapter 4

The Matrix Exponential Function

We continue our investigation of the asymptotic behavior of dynamical systems described by matrices, which was started in last chapter, now moving to the continuous time case. This means that we investigate the asymptotic properties of the matrix exponential function.

The importance of the topic should be clear for everyone reading this: the matrix exponential function always solves a corresponding system of ordinary differential equations, hence the asymptotic properties of matrix exponential functions provide information on the long-time behavior of solutions of ODEs. This subject has more than 100 years of history, with the famous Lyapunov stability theorem as its starting point.

Topics we cover include boundedness, convergence to zero, convergence, mean convergence (or Cesàro convergence), periodicity, hyperbolic decomposition, and are presented in analogy to the results achieved in the previous chapter.

4.1 Main Properties

Let X be a n-dimensional vector space. The *exponential function of a complex matrix* $A \in \mathcal{L}(X)$ is the mapping

$$\exp : \mathbb{R} \longrightarrow \mathcal{L}(X), \quad t \longmapsto \exp(tA) = e^{tA}.$$

Here, as explained in Section 2.2, $\exp(tA) = e^{tA}$ stands for the matrix $f_t(A)$ with $f_t(\lambda) := e^{t\lambda}$. Therefore, Theorem 2.11, formula (2.9), says that e^{tA} can be written as

$$e^{tA} = \sum_{i=1}^{m} \sum_{\nu=0}^{\nu_i - 1} \frac{e^{t\lambda_i} t^{\nu}}{\nu!} (A - \lambda_i)^{\nu} P_i, \tag{4.1}$$

where $\lambda_1, \ldots, \lambda_m$ are the eigenvalues of A with corresponding multiplicities ν_1, $\ldots, \nu_m$ (as roots of the minimal polynomial) and spectral projections $P_1, \ldots, P_m$.

Alternatively, according to Corollary 2.14, the matrix e^{tA} is represented by the exponential series

$$e^{tA} = \sum_{k=0}^{\infty} \frac{t^k A^k}{k!}. \tag{4.2}$$

Formula (4.1) gives an easy access to the following properties of e^{tA}. Firstly, since $f_0(\lambda) = 1$, we have

$$f_0(A) = e^{0A} = I.$$

By the multiplicativity of the functional calculus and the fact that

$$f_{s+t}(\lambda) = f_s(\lambda) \cdot f_t(\lambda), \qquad \lambda \in \mathbb{C},$$

we infer that

$$e^{(s+t)A} = e^{sA} \cdot e^{tA} \tag{4.3}$$

for $s, t \in \mathbb{R}$. Hence, $(e^{tA})_{t \in \mathbb{R}}$ is a subgroup of the multiplicative semigroup $\mathcal{L}(X)$, and the mapping $t \mapsto e^{tA}$ is a homomorphism of the additive group $(\mathbb{R}, +)$ into $\mathcal{L}(X)$.

Remark 4.1. It is usual to refer to these properties of $t \mapsto e^{tA}$ by saying that $(e^{tA})_{t \in \mathbb{R}}$ is the *matrix group generated by* A. If we consider only $t \geq 0$, we call $(e^{tA})_{t \geq 0}$ the *matrix semigroup* generated by A

Furthermore, the function $t \mapsto e^{tA}$ has nice analytic properties.

Theorem 4.2. *The matrix exponential function* $t \mapsto e^{tA}$ *is differentiable on* $\mathbb{R}$ *with derivative*

$$\frac{d}{dt} e^{tA} = A e^{tA} = e^{tA} A, \quad t \in \mathbb{R}. \tag{4.4}$$

Proof. Let $f(\lambda) := \lambda e^{t\lambda}$ and observe that

$$\left(\lambda e^{t\lambda}\right)^{(\nu)} (\lambda) = \left(\lambda t^{\nu} + \nu t^{\nu-1}\right) e^{t\lambda}.$$

Hence applying Theorem 2.11 for $f(\lambda) = \lambda e^{t\lambda}$ we obtain

$$
\begin{aligned}
A e^{tA} &= \sum_{i=1}^{m} \sum_{\nu=0}^{\nu_i-1} \left(\lambda_i t^{\nu} e^{t\lambda_i} + \nu t^{\nu-1} e^{t\lambda_i}\right) \frac{(A - \lambda_i)^{\nu}}{\nu!} P_i \\
&= \sum_{i=1}^{m} \sum_{\nu=0}^{\nu_i-1} \left[\frac{d}{dt}(t^{\nu} e^{t\lambda_i})\right] \frac{(A - \lambda_i)^{\nu}}{\nu!} P_i \\
&= \frac{d}{dt} \left(\sum_{i=1}^{m} \sum_{\nu=0}^{\nu_i-1} t^{\nu} e^{t\lambda_i} \frac{(A - \lambda_i)^{\nu}}{\nu!} P_i\right) \\
&= \frac{d}{dt} e^{tA}.
\end{aligned}
$$

Notice that since $\lambda e^{t\lambda} = e^{t\lambda} \lambda$, by the properties of the functional calculus we see that $A e^{tA} = e^{tA} A$. □

The following consequence of formula (4.4) motivates our interest in the behavior of the function $t \mapsto e^{tA}$ as $t \to \infty$.

Corollary 4.3. *Let $A = (a_{ij}) \in \mathcal{L}(X)$. Then for each $x = (x_1, \ldots, x_n)^\top \in \mathbb{C}^n$ the function*

$$t \longmapsto e^{tA}x =: (x_1(t), x_2(t), \ldots, x_n(t))^\top$$

is the unique solution of the system of differential equations

$$\frac{\mathrm{d}}{\mathrm{d}t}x_1(t) = a_{11}x_1(t) + \cdots + a_{1n}x_n(t)$$

$$\frac{\mathrm{d}}{\mathrm{d}t}x_2(t) = a_{21}x_1(t) + \cdots + a_{2n}x_n(t)$$

$$\vdots$$

$$\frac{\mathrm{d}}{\mathrm{d}t}x_n(t) = a_{n1}x_1(t) + \cdots + a_{nn}x_n(t),$$

with the initial condition

$$(x_1(0), x_2(0), \ldots, x_n(0)) = (x_1, x_2, \ldots, x_n).$$

Proof. A look at the differential quotient defining the derivative $\frac{\mathrm{d}}{\mathrm{d}t}(e^{tA}x)$ and Theorem 4.2 convinces us that

$$\frac{\mathrm{d}}{\mathrm{d}t}(e^{tA}x) = \left(\frac{\mathrm{d}}{\mathrm{d}t}e^{tA}\right)x = (Ae^{tA})x = A(e^{tA}x)$$

for all $t \in \mathbb{R}$. Since $e^{0A} = I$, we infer that $e^{tA}x$ is a solution of the above initial value problem. Now, let $x(t)$ be any solution and define $y(t) := e^{-tA}x(t)$. Then

$$\frac{\mathrm{d}}{\mathrm{d}t}y(t) = \left(\frac{\mathrm{d}}{\mathrm{d}t}e^{-tA}\right)x(t) + e^{-tA}\frac{\mathrm{d}}{\mathrm{d}t}x(t)$$

$$= -Ae^{-tA}x(t) + e^{-tA}Ax(t)$$

$$= 0.$$

Therefore, $t \mapsto y(t) = e^{-tA}x(t)$ is constant. Since for $t = 0$ we have $y(0) = x$, we conclude that $x(t) = e^{tA}x$ for all $t \in \mathbb{R}$. $\qquad\square$

Remark 4.4. In short, Corollary 4.3 tells us that the matrix semigroup generated by $A = (a_{ij})$ solves the initial value problem

$$\begin{cases} \dot{x}(t) = Ax(t), & t \geq 0, \\ x(0) = x_0, \end{cases}$$

in the sense that the orbit $\{e^{tA}x_0 : t \in \mathbb{R}\} \geq 0$ of the initial value $x_0 \in \mathbb{C}^n$ is the unique solution of the problem. At present, we note that Theorem 4.2 remains true if t is allowed to run through $\mathbb{C}$, hence

$$z \longmapsto e^{zA}$$

is a holomorphic function on $\mathbb{C}$.

4.2   Coordinate Functions

We intend to study the behavior of the function $t \mapsto e^{tA}$ (or of $t \mapsto e^{tA}x$ for a given $x \in X$), as $t \to \infty$, following the same pattern as in the previous Chapter 3 for the matrix powers. Nevertheless, a few comments seem to be appropriate.

While in Chapter 3 we studied the sequence (T^k) and based our considerations on the characterization of convergence of the coordinate sequences given in Section 3.1, we will now have to deal with a function $t \mapsto e^{tA}$ of the real variable t. We will formulate, without going into a detailed discussion, the following versions of the convergence properties discussed in Section 1.2.

If $t \mapsto y(t)$ is a real function with values in a finite-dimensional vector space X with a basis $\{y_1, \ldots, y_n\}$, then

$$y(t) = \sum_{i=1}^{n} \eta_i(t) y_i$$

with uniquely determined values of $\eta_i(t)$ for each $t \in \mathbb{R}$. We call the functions $t \mapsto \eta_i(t)$ the *coordinate functions* of $y(t)$ with respect to $\{y_1, \ldots, y_n\}$. Convergence of $y(t)$ as $t \to \infty$ in X is equivalent to the convergence of all coordinate functions $\eta_i(t)$, no matter what basis is employed, the coordinates of the limit being the limits of the respective coordinate functions.

In order to discuss the function $t \mapsto e^{tA}$, in analogy to Lemma 2.16, we use a basis $\mathcal{B}_A$ of $\mathcal{L}(X)$ containing the set

$$B_A := \left\{ \frac{(A - \lambda_i)^\nu}{\nu!} P_i : i = 1, \ldots, m;\ \nu = 0, \ldots, \nu_i - 1 \right\}.$$

By (4.1), the non-zero coordinate functions with respect to this basis are

$$g_{\nu, \lambda_i}(t) := t^\nu e^{t\lambda_i} \tag{4.5}$$

for $i = 1, \ldots, m$ and $\nu = 0, \ldots, \nu_i - 1$. Likewise, if we wish to study $e^{tA}x$ for a given $x \in X$, we use a basis $\mathcal{B}_{A,x}$ of X containing the non-zero elements of

$$B_{A,x} := \left\{ \frac{(A - \lambda_i)^\nu x}{\nu!} : i = 1, \ldots, m;\ \nu = 0, \ldots, \nu_i - 1 \right\}.$$

Again, the coordinate functions of $e^{tA}x$ with respect to this basis are among the functions $g_{\nu, \lambda_i}(t)$ defined in (4.5).

The behavior of a function $g_{\nu, \lambda}(t) := t^\nu e^{t\lambda}$ is easy to understand and essentially depends on the real part of λ. The following cases are possible.

- $\underline{\operatorname{Re} \lambda < 0}$. Then, for each fixed value of ν, $e^{t\lambda} t^\nu \to 0$ as $t \to \infty$, where the decay is exponential in the following sense:
  for any $0 < \delta < -\operatorname{Re} \lambda$ there is $M_\delta \geq 1$ such that

$$\left| e^{\lambda t} t^\nu \right| = t^\nu e^{t \operatorname{Re} \lambda} \leq M_\delta e^{-\delta t} \quad \text{for all } t \geq 0.$$

- <u>Re $\lambda > 0$</u>. Then, for each fixed value of ν, $\left|e^{t\lambda}t^{\nu}\right| \to \infty$ as $t \to \infty$, but it remains exponentially bounded in the following sense: for each $w > \mathrm{Re}\,\lambda > \delta > 0$ there is $M_w \geq 1$ such that

$$\left|e^{\lambda t}t^{\nu}\right| \leq M_w e^{wt} \quad \text{for all } t \geq 0.$$

- <u>Re $\lambda = 0$ and $\nu = 0$</u>. Then $e^{t\lambda}$ is constant (for $\lambda = 0$) or periodic of period $\frac{2\pi i}{\lambda}$ (for $\lambda \neq 0$).

- <u>Re $\lambda = 0$ and $\nu \geq 1$</u>. Then $\left|e^{\lambda t}t^{\nu}\right| = t^{\nu} \to \infty$ as $t \to \infty$.

After these preparations, we now look at the behavior of $e^{tA}x$ on the spectral subspaces $X_i = \mathrm{im}\,P_i$ of X.

Theorem 4.5. *Let $A \in \mathcal{L}(X)$, let $\|\cdot\|$ be a norm on X, and fix an $i \in \{1, \ldots, m\}$. Then the following assertions hold.*

a) *For every $\rho < \mathrm{Re}\,\lambda_i < \omega$ there exist $M \geq 1$ and $N > 0$ such that*

$$N e^{\rho t}\|x\| \leq \left\|e^{tA}x\right\| \leq M e^{\omega t}\|x\|$$

for all $t \geq 0$ and all $x \in X_i$.

b) *If $\mathrm{Re}\,\lambda_i = 0$, then*

$$\left\{e^{tA}x : t \geq 0\right\}$$

is bounded for every $x \in X_i$ if and only if λ_i is a simple pole of $R(\cdot, A)$, i.e., if $\nu_i = 1$. In this case, $e^{tA}x = e^{t\lambda_i}x$ for every $x \in X_i$ and $t \geq 0$.

Proof. a) By formula (4.1), we have for $x \in X_i$ that

$$\left\|e^{tA}x\right\| = \left\|\sum_{\nu=0}^{\nu_i-1} e^{t\lambda_i}t^{\nu}\frac{(A-\lambda_i)^{\nu}}{\nu!}x\right\| \leq \sum_{\nu=0}^{\nu_i-1}\left\|\frac{(A-\lambda_i)^{\nu}}{\nu!}\right\|\left|e^{t\lambda_i}t^{\nu}\right| \cdot \|x\|$$

$$\leq M e^{\omega t}\|x\|,$$

for all $\omega > \mathrm{Re}\,\lambda_i$ and some $M \geq 1$.

Now, we apply the above estimate to $-A$ which has $-\lambda_i$ as an eigenvalue with the same spectral projection P_i and spectral subspace X_i as before. Hence,

$$\left\|e^{-tA}y\right\| \leq M e^{-\rho t}\|y\|$$

for all $y \in X_i$, $t \geq 0$, and some $M \geq 1$. Since $e^{-tA}e^{tA} = I$, we find for every $x \in X_i$ an element $y \in X_i$ such that $x = e^{-tA}y$. This implies

$$\left\|e^{tA}x\right\| = \|y\| \geq \frac{1}{M}e^{\rho t}\|x\|,$$

for all $t \geq 0$.

b) If $\nu_i = 1$, then $e^{tA}x = e^{t\lambda_i}x$ for all $x \in X_i$. On the other hand, if $\nu_i \geq 2$, then $\ker(A-\lambda_i) \subsetneqq X_i$, hence there is an $x \in X_i$ with $(A-\lambda_i)x \neq 0$. The coordinate function of $(A-\lambda_i)x$ with respect to the basis element $x \in \mathcal{B}_{A,x}$ equals $te^{t\lambda_i}$, which is unbounded as $t \to \infty$. $\qquad\square$

4.3   The Spectral Bound

Now we introduce the following constant which plays the same role for the exponential function $t \mapsto e^{tA}$ as the spectral radius $r(T)$ does for the powers $k \mapsto T^k$ (see Section 3.2).

Definition 4.6. For $A \in \mathcal{L}(X)$ the number

$$s(A) := \sup\{\operatorname{Re}\lambda : \lambda \in \sigma(A)\}$$

is called the *spectral bound* of A.

We note that the spectral bound of A can be determined from $\|e^{tA}\|$ in the following way (compare with Proposition 3.3).

Proposition 4.7. *If* $\|\cdot\|$ *is any norm on* $\mathcal{L}(X)$, *then*

$$s(A) = \lim_{t \to \infty} \frac{1}{t} \log \|e^{tA}\|. \tag{4.6}$$

If $\|\cdot\|$ *is an operator norm, then*

$$s(A) = \inf_{t>0} \frac{1}{t} \log \|e^{tA}\|. \tag{4.7}$$

Proof. By the equivalence of norms on $\mathcal{L}(X)$, the limit

$$\lim_{t \to \infty} \frac{1}{t} \log \|e^{tA}\|,$$

if it exists, does not depend on the specific norm. Hence, we can use the supremum norm $\|\!|\cdot|\!\|$ with respect to the basis $\mathcal{B}_A$ above. Then we have

$$\|\!|e^{tA}|\!\| = \left|t^\nu e^{t\lambda_i}\right| = t^\nu e^{t\operatorname{Re}\lambda_i}$$

for some $i \in \{1, \ldots, m\}$ and some $0 \le \nu \le n-1$. Hence,

$$\frac{1}{t} \log \|\!|e^{tA}|\!\| = \frac{\nu}{t} \log t + \operatorname{Re}\lambda_i$$

for all $t > 0$ and i and ν as before. Since the function $t \mapsto \frac{\nu}{t} \log t$ tends to zero as $t \to \infty$, we obtain

$$\frac{\nu}{t} \log t + \operatorname{Re}\lambda_i \longrightarrow \operatorname{Re}\lambda_i$$

as $t \to \infty$, yielding Formula (4.6).

Now let $\|\cdot\|$ be an operator norm. The Spectral Mapping Theorem 2.20 and Corollary 1.10 imply

$$\log \|e^{tA}\| \ge t|\lambda_i| \ge t\operatorname{Re}\lambda_i,$$

and (4.7) follows. $\square$

A repeated application of Theorem 4.5 yields the following.

Corollary 4.8. *Let $A \in \mathcal{L}(X)$ and let $\| \cdot \|$ be any norm on X. Then for every $w > s(A)$ there is a constant $M \geq 1$ such that*

$$\|e^{tA}x\| \leq Me^{wt}\|x\|$$

for all $t \geq 0$ and $x \in X$. Furthermore,

$$s(A) = \omega_0(T)$$

where

$$\omega_0(T) := \inf\{w \in \mathbb{R} : \exists\, M \geq 1 \text{ such that } \|e^{tA}\| \leq Me^{wt} \text{ for } t \geq 0\}. \qquad (4.8)$$

Remark 4.9. The number $\omega_0(T)$ defined in (4.8) is known as the *growth bound* of the matrix semigroup $T(t) := e^{tA}$. Note that if X is an infinite-dimensional vector space, the equality $s(A) = \omega_0(T)$ need no longer hold in general.

4.4 Asymptotics

Now we put all the information together to describe the action of e^{tA} on all of X. As in Section 3.3, we first define different types of long-time behavior of e^{tA}.

Definition 4.10. For $A \in \mathcal{L}(X)$ and any norm $\| \cdot \|$ on X we say that the semigroup $(e^{tA})_{t \geq 0}$ is

- *bounded*[4] if $\sup_{t \geq 0} \|e^{tA}\| < \infty$;
- *stable* if $\lim_{t \to \infty} \|e^{tA}\| = 0$;
- *exponentially stable* if there exist $M \geq 1$ and $\varepsilon > 0$ such that $\|e^{tA}\| \leq Me^{-\varepsilon t}$ for all $t \geq 0$;
- *convergent* if $\lim_{t \to \infty} e^{tA} = P$ for some $P \in \mathcal{L}(X)$;
- *periodic* if $e^{t_0 A} = I$ for some $t_0 > 0$; in this case the smallest such t_0 is called the *period* of e^{tA};
- *hyperbolic* if there exist A-invariant subspaces X_s and X_u, such that $X = X_s \oplus X_u$ and

$$\|e^{tA}x\| \leq Me^{-\varepsilon t}\|x\| \quad \text{for } x \in X_s, \qquad (4.9)$$

$$\|e^{tA}x\| \geq \frac{1}{M}e^{\varepsilon t}\|x\| \quad \text{for } x \in X_u, \qquad (4.10)$$

for all $t \geq 0$ and some constants $M \geq 1$, $\varepsilon > 0$; X_s and X_u are called the *stable* and *unstable* subspaces, respectively.

---

[4]Note that working with ODEs or dynamical systems, a different terminology is also widely accepted: what we call "bounded" is often called "stable", what we call "stable" is often called "asymptotically stable", and what we call "hyperbolic" is often called "exponential dichotomy".

Remarks 4.11.

a) Since pointwise and norm convergence on $\mathcal{L}(X)$ coincide, a statement about the long-time behavior of $\|e^{tA}x\|$ for all $x \in X$ is equivalent to the same statement regarding $\|e^{tA}\|$ for the appropriate operator norm.

b) Note that stability of a matrix semigroup is equivalent to exponential stability, see Exercise 3. We will prove this in a more general form later (cf. Proposition 12.4).

c) Using (4.8), we see that

$$(e^{tA})_{t \geq 0} \text{ is (exponentially) stable} \iff \omega_0(T) < 0. \tag{4.11}$$

We now classify the asymptotic behavior of e^{tA} in terms of spectral properties of the matrix A.

Theorem 4.12. *Let $A \in \mathcal{L}(X)$ and take any norm $\|\cdot\|$ on X.*

a) $(e^{tA})_{t \geq 0}$ *is (exponentially) stable if and only if* $\mathrm{s}(A) < 0$.

b) $(e^{tA})_{t \geq 0}$ *is bounded if and only if* $\mathrm{s}(A) \leq 0$ *and all eigenvalues of A with real part equal to 0 are simple poles of the resolvent $R(\cdot, A)$.*

c) $(e^{tA})_{t \geq 0}$ *is periodic with period t_0 if and only if it is bounded and $\sigma(A) \subset \frac{2\pi i}{t_0}\mathbb{Z}$ for some $t_0 > 0$.*

d) $\lim_{t \to \infty} e^{tA} = P_1$ *(P_1 denotes the spectral projection of A belonging to the eigenvalue 0) if and only if $\mathrm{s}(A) = 0$ is a simple pole of the resolvent $R(\cdot, A)$ and $\sigma(A) \cap i\mathbb{R} = \{0\}$.*

e) $(e^{tA})_{t \geq 0}$ *is hyperbolic if and only if $\sigma(A) \cap i\mathbb{R} = \emptyset$.*

Proof. a) This is a consequence of relation (4.11) and Corollary 4.8.

b) $(e^{tA})_{t \geq 0}$ is bounded iff the same is true for the coordinate functions in formula (4.5), which holds iff for each i, either $\operatorname{Re}\lambda_i < 0$ or $\operatorname{Re}\lambda_i = 0$ and $\nu_i = 1$, that is, if and only if $\mathrm{s}(A) \leq 0$ and every eigenvalue λ_i with $\operatorname{Re}\lambda_i = 0$ is a simple pole of the resolvent.

c) Again we use coordinate functions and observe that $e^{t_0 A} = I$ for some $t_0 > 0$ iff $\lambda_i \in \frac{2\pi i}{t_0}\mathbb{Z}$, with $\nu_i = 1$, which by b) holds iff $(e^{tA})_{t \geq 0}$ is bounded and $\sigma(A) \subset \frac{2\pi i}{t_0}\mathbb{Z}$.

d) $\lim_{t \to \infty} e^{tA} = P_1$ iff all coordinate functions converge, that is, iff either $\operatorname{Re}\lambda_i < 0$, or $\lambda_i = 0$ and $\nu_i = 1$. This is true iff $\mathrm{s}(A) \leq 0$, which is a simple pole of the resolvent and the only eigenvalue on the imaginary axis. Moreover, the semigroup converges to the corresponding spectral projection.

e) $(e^{tA})_{t \geq 0}$ is hyperbolic iff there exist A-invariant subspaces X_s and X_u such that $X = X_s \oplus X_u$ and inequality (4.9) holds, that is, iff $e^{tA}|_{X_s}$ and $e^{-tA}|_{X_u}$ are both exponentially stable. By a), this is equivalent to

$$\mathrm{s}(A|_{X_s}) < 0 \text{ and } \mathrm{s}(-A|_{X_u}) < 0 \iff \sigma(A) \cap i\mathbb{R} = \emptyset. \qquad \square$$

Thus, in complete analogy to the situation in Section 3, convergence of e^{tA} as $t \to \infty$ is restricted to one of the following situations.

- $\lim_{t\to\infty} e^{tA} = 0$: this is the case if and only if $s(A) < 0$;
- $\lim_{t\to\infty} e^{tA} = P_1$, where P_1 is the spectral projection belonging to $\lambda_1 = 0$: this is the case if and only if $s(A) = 0$, $\sigma(A) \cap i\mathbb{R} = \{0\}$, and 0 is a simple pole of the resolvent $R(\cdot, A)$.

Example 4.13. Analyzing the spectral properties of the matrices

$$A_1 = \begin{pmatrix} 0 & 1 \\ -1 & 0 \end{pmatrix}, \quad A_2 = \begin{pmatrix} 0 & 1 \\ 1 & 0 \end{pmatrix}, \quad A_3 = \begin{pmatrix} 1 & 1 \\ -1 & -1 \end{pmatrix},$$

$$A_4 = \begin{pmatrix} -1 & 1 \\ 0 & 0 \end{pmatrix}, \quad A_5 = \begin{pmatrix} -1 & 0 \\ 1 & -1 \end{pmatrix},$$

one obtains that the semigroup $\left(e^{tA_1}\right)_{t\geq 0}$ is periodic with period 2π, $\left(e^{tA_2}\right)_{t\geq 0}$ is hyperbolic, $\left(e^{tA_3}\right)_{t\geq 0}$ is unbounded, $\left(e^{tA_4}\right)_{t\geq 0}$ converges to $P_1 = \begin{pmatrix} 0 & 1 \\ 0 & 1 \end{pmatrix}$, and $\left(e^{tA_5}\right)_{t\geq 0}$ is exponentially stable.

Decomposing the space we can study stability concepts more in detail. One example of this approach is the definition of hyperbolicity of a matrix semigroup. Let us use this approach to obtain another asymptotic property.

Definition 4.14. For $A \in \mathcal{L}(X)$ we call the semigroup $(e^{tA})_{t\geq 0}$ *asymptotically periodic* if there is a direct sum decomposition

$$X = X_0 \oplus X_1$$

into A-invariant subspaces X_0 and X_1 such that

a) $e^{tA}|_{X_0}$ is stable, i.e., $\lim_{t\to\infty} e^{tA}x = 0$ for all $x \in X_0$, and

b) $e^{tA}|_{X_1}$ is periodic, i.e., there exists $t_0 > 0$ such that $e^{t_0 A}y = y$ for all $y \in X_1$.

Again, this property can be described by spectral properties of A.

Theorem 4.15. *For $A \in \mathcal{L}(X)$ the following assertions are equivalent.*

(i) $(e^{tA})_{t\geq 0}$ *is asymptotically periodic.*

(ii) $(e^{tA})_{t\geq 0}$ *is bounded and $\sigma(A) \cap i\mathbb{R} \subset 2\pi i\alpha\mathbb{Z}$ for some $\alpha \in \mathbb{R}$.*

(iii) *$s(A) \leq 0$, the set $\sigma(A) \cap i\mathbb{R}$ consists of simple poles of the resolvent $R(\cdot, A)$ and is contained in $2\pi i\alpha\mathbb{Z}$ for some $\alpha \in \mathbb{R}$.*

Proof. (i) $\implies$ (ii): The boundedness of $(e^{tA})_{t\geq 0}$ follows directly. Let $X = X_0 \oplus X_1$ be the corresponding decomposition. Then

$$\sigma(A) = \sigma(A|_{X_0}) \cup \sigma(A|_{X_1}),$$

where $\sigma(A|_{X_0}) \subset \{\lambda \in \mathbb{C} : \operatorname{Re}\lambda < 0\}$ by Theorem 4.12.a) and $\sigma(A|_{X_1}) \subset 2\pi i\alpha\mathbb{Z}$ for some $\alpha \in \mathbb{R}$ by Theorem 4.12.c).

(ii) $\Longrightarrow$ (iii): This follows by Theorem 4.12.b).

(iii) $\Longrightarrow$ (i): Define

$$X_0 := \bigoplus_{\operatorname{Re} \lambda_i < 0} X_i \qquad \text{and} \qquad X_1 := \bigoplus_{\operatorname{Re} \lambda_i = 0} X_i$$

and apply Theorem 4.12.a) and c). $\square$

Finally, we ask under which conditions a subsequence or the Cesàro means of $(e^{tA})_{t \geq 0}$ converge. First recall the concept of a spectral contraction in Definition 3.11.

Theorem 4.16. *The following assertions are equivalent for $A \in \mathcal{L}(X)$.*

(i) e^{tA} *is spectral contraction for one/all $t > 0$.*

(ii) $s(A) = 0$ *and all eigenvalues of A with real part equal to 0 are simple poles of the resolvent $R(\lambda, A)$.*

(iii) *There is an operator norm $\|\cdot\|$ on $\mathcal{L}(X)$ such that $\|e^{ktA}\| = 1$ for all $k \in \mathbb{N}$ and one/all $t > 0$.*

(iv) *There exists a sequence (t_m) of the form $t_m := tk_m$, where (k_m) is a subsequence of (k), such that $(e^{t_m A})$ converges to some limit $P \neq 0$ for one/all $t > 0$.*

Proof. The equivalence (i) $\Longleftrightarrow$ (ii) follows by Theorem 3.10 combined with the Spectral Mapping Theorem 2.20 and Theorem 2.28, the equivalence (i) $\Longleftrightarrow$ (iv) again by Theorem 3.10, while (i) $\Longleftrightarrow$ (iii) holds by Theorem 3.12. $\square$

Definition 4.17. We say that $(e^{tA})_{t \geq 0}$ is a *spectral contraction semigroup*, if any of the equivalent assertions of Theorem 4.16 is true.

The following is the continuous-time analogue of the Cesàro means introduced in Chapter 3.

Definition 4.18. Let $A \in \mathcal{L}(X)$. The matrices

$$C(r) := \frac{1}{r} \int_0^r e^{sA} \, ds, \qquad r > 0,$$

are called the *Cesàro means* of the semigroup $(e^{tA})_{t \geq 0}$. The semigroup $(e^{tA})_{t \geq 0}$ is *mean ergodic* (or *Cesàro summable*), if $\lim_{r \to \infty} C(r)$ exists.

Theorem 4.19. *For $A \in \mathcal{L}(X)$ the semigroup $(e^{tA})_{t \geq 0}$ is mean ergodic if and only if either $s(A) < 0$, or $(e^{tA})_{t \geq 0}$ is a spectral contraction semigroup.*

In the case $0 \in \sigma(A)$, the Cesàro means $C(r)$ converge to the spectral projection of A belonging to 0, in all other cases of mean ergodic semigroups $C(r)$ converge to 0.

Proof. First note that the coordinate functions of $C(r)$ with respect to $\mathcal{B}_A$ are of the form

$$g_{\nu,\lambda_i}^{(r)} := \frac{1}{r} \int_0^r g_{\nu,\lambda_i}(s)\, \mathrm{d}s = \frac{1}{r} \int_0^r e^{s\lambda_i} s^\nu\, \mathrm{d}s. \tag{4.12}$$

Following the discussion on page 46–47, we see that $g_{\nu,\lambda}^{(r)}$ converges only in two cases: either for $\operatorname{Re}\lambda < 0$, or for $\operatorname{Re}\lambda = 0 = \nu$. This proves the first assertion of the theorem.

Since for $\operatorname{Re}\lambda < 0$ we have $g_{\nu,\lambda}^{(r)} \to 0$ as $r \to \infty$, in the case when $s(A) < 0$ we obtain $C(r) \to 0$. On the other hand, in the case of a spectral contraction semigroup, the only nonzero limit of the coordinate functions as $r \to \infty$ is $g_{0,0}^{(r)} = 1$. By (4.1), the Cesàro means $C(r)$ then converge towards the spectral projection of A belonging to $\lambda = 0$. $\qquad\square$

4.5 Notes and Remarks

There are many ways how to compute the exponential function of a matrix numerically and we refer here to an excellent survey paper by Moler and van Loan [99].

Theorem 4.12.a) is Lyapunov's Stability Theorem proved in 1892 (see [88]). The results of this chapter are presented in many books on ordinary differential equations, like for example Amann [3] or Teschl [139, Ch. 3].

4.6 Exercises

1. Show that if $A, B \in \mathcal{L}(X)$ commute, then $e^{t(A+B)} = e^{tA}e^{tB}$. Find an example to show that the commutativity assumption is necessary.

2. Let $B \in \mathcal{L}(X)$. Under which conditions is there an $A \in \mathcal{L}(X)$ such that $e^{kA} = B^k$ for all $k \in \mathbb{N}$?

3. Prove that for $A \in \mathcal{L}(X)$ the matrix semigroup e^{tA} is stable if and only if it is exponentially stable.

4. Show that $\{e^{tA}x : t \in \mathbb{R}\}$ is bounded for every $x \in X$ if and only if $\sigma(A) \subset i\mathbb{R}$ and all eigenvalues are simple poles of the resolvent $R(\cdot, A)$.

5. Show that the semigroup $(e^{tA})_{t\geq 0}$ is hyperbolic if and only if $\sigma(e^{tA}) \cap \Gamma = \emptyset$ for some/all $t > 0$, where Γ denotes the unit circle in $\mathbb{C}$.

6. Compute the matrix exponential e^{tA} for

$$A = \begin{pmatrix} -a & b \\ a & -b \end{pmatrix} \quad \text{with } a + b \neq 0.$$

7. For every one of the matrices in Example 4.13 compute the spectrum and the corresponding semigroup, and then describe its asymptotic behavior.

Chapter 5

Positive Matrices

We call a real matrix positive if its entries are greater or equal to zero. Positivity naturally occurs in many applications and it turns out to have deep consequences on the spectral properties of the matrix. In this chapter we discuss spectral properties of positive matrices, incorporating seminal work by Perron, Frobenius, and Wielandt.

5.1 Positivity

Let $x = (\xi_1, \ldots, \xi_n)$, $y = (\eta_1, \ldots, \eta_n) \in \mathbb{R}^n$. We say that

$$x \leq y \quad \text{if } \xi_i \leq \eta_i \quad \text{for all } 1 \leq i \leq n.$$

Similarly, for real $n \times n$ matrices $T = (\tau_{ij})$ and $S = (\sigma_{ij})$ we say that

$$T \leq S \quad \text{if } \tau_{ij} \leq \sigma_{ij} \quad \text{for all } 1 \leq i, j \leq n.$$

The symbol $x < y$ $(T < S)$ means that $\xi_i \leq \eta_i$ for all i, j and there are indeces i, j such that $\xi_j < \eta_j$.

Definition 5.1. A vector $x = (\xi_1, \ldots, \xi_n)$ (a matrix $T = (\tau_{ij})$) is called *positive*, if $\xi_i \geq 0$ for all i ($\tau_{ij} \geq 0$ for all i, j). In this case we write $x \geq 0$ ($T \geq 0$). Next, $x > 0$ ($T > 0$) if $x \geq 0$ ($T \geq 0$) and there is at least one nonzero coordinate (entry).

We point out that, in our terminology, a positive vector need not to have *all* coordinates larger than zero and a positive matrix need not have *all* entries larger than zero. Likewise, a vector $x > 0$ (matrix $T > 0$) may have many coordinates (entries) equal to zero, but at least one coordinate (entry) must be larger than zero.

Definition 5.2. A vector $x = (\xi_1, \ldots, \xi_n)$ (a matrix $T = (\tau_{ij})$) is called *strictly positive*[5], if $\xi_i > 0$ for all its coordinates ($\tau_{ij} > 0$ for all its entries). In this case, we write $x \gg 0$ ($T \gg 0$).

By the *absolute value* of a vector $x = (\xi_1, \ldots, \xi_n) \in X = \mathbb{C}^n$ we mean the vector

$$|x| := (|\xi_1|, \ldots, |\xi_n|).$$

Similarly, for a matrix $T = (\tau_{ij}) \in \mathcal{L}(X)$ we call

$$|T| := (|\tau_{ij}|)$$

the *absolute value* of T.

We start with some basic observations.

Lemma 5.3. *Let $S, T \in \mathcal{L}(X)$. Then the following properties hold.*

a) $T \geq 0$ *if and only if $Tx \geq 0$ for all $x \geq 0$.*

b) $T \leq S$ *if and only if $Tx \leq Sx$ for all $x \geq 0$.*

c) $|Tx| \leq |T| \, |x|$, *hence $|Tx| \leq T \, |x|$ if $T \geq 0$.*

Proof. a) For positive T and x, the product Tx is obviously positive. Conversely, if $Tx \geq 0$ for every standard basis vector $x = u_j$, $j = 1, \ldots, n$, then all the columns of T are positive, and so is T.

b) Since $T \leq S$ iff $T - S \geq 0$, this is a consequence of a).

c) This follows by the triangle inequality, since for every coordinate we have

$$(|Tx|)_i = \left| \sum_{j=1}^n \tau_{ij} x_j \right| \leq \sum_{j=1}^n |\tau_{ij}| |x_j| = (|T| |x|)_i, \quad i = 1, \ldots, n. \qquad \square$$

In the following, we will always use the maximum norm on $X = \mathbb{C}^n$,

$$\|x\| := \|x\|_\infty = \max_{1 \leq i \leq n} |\xi_i|,$$

while on $\mathcal{L}(X)$ we will use the corresponding operator norm (see Example 1.11), given by

$$\|T\| := \|T\|_\infty = \max_{\|x\| \leq 1} \|Tx\| = \max_{1 \leq i \leq n} \sum_{j=1}^n |\tau_{ij}|.$$

The reason for this special choice becomes clear from the following lemma.

---

[5]There is a different terminology that has its followers, calling vectors or matrices that are positive in our sense *non-negative* and reserving the term positive for what we call strictly positive. This may of course lead to misunderstandings and confusion, but the coexistence of both terminologies is a fact.

Lemma 5.4. *For $x, y \in X$ and $S, T \in \mathcal{L}(X)$ the following assertions hold.*

a) *The inequality $|x| \leq |y|$ implies $\|x\| \leq \|y\|$, in particular $\| |x| \| = \|x\|$ for all x.*

b) *The inequality $|S| \leq |T|$ implies $\|S\| \leq \|T\|$. In particular,*

$$\| |T| \| = \|T\|$$

for all T, and $|S| \leq T$ implies $\|S\| \leq \|T\|$.

c) *If $|S| \leq T$, then $r(S) \leq r(|S|) \leq r(T)$.*

d) *If $T \geq 0$, then*

$$\|T\| = \|T\mathbb{1}\|,$$

where $\mathbb{1} := (1, \ldots, 1)^{\top}$.

Proof. a) For $x = (\xi_1, \ldots, \xi_n)$ and $y = (\eta_1, \ldots, \eta_n)$, inequality $|x| \leq |y|$ implies, for every i, that

$$|\xi_i| \leq |\eta_i| \leq \max_j |\eta_j| \leq \|y\|,$$

and taking maximum also on the left side one obtains $\|x\| \leq \|y\|$.

b) This follows in the same way as above.

c) From $|S| \leq T$ follows $|S^k| \leq |S|^k \leq T^k$ for every $k \in \mathbb{N}$. Now assertion b) and Gelfand's formula for the spectral radius imply the desired inequality.

d) Observe that $(T\mathbb{1})_i = \sum_{j=1}^n |\tau_{ij}|$ for all i and use the definition of the maximum norm and the corresponding operator norm. $\qquad\square$

The following observation on the resolvent has remarkable consequences.

Proposition 5.5. *Let T be a positive matrix with spectral radius $r(T)$ and $\mu \in \rho(T)$.*

a) *The resolvent $R(\mu, T)$ is positive whenever $\mu > r(T)$.*

b) *If $|\mu| > r(T)$, then*

$$|R(\mu, T)| \leq R(|\mu|, T).$$

Proof. We use the Neumann series representation

$$R(\mu, T) = \sum_{k=0}^{\infty} \frac{T^k}{\mu^{k+1}} \tag{5.1}$$

for the resolvent, which is valid for $|\mu| > r(T)$ by Corollary 2.15.

a) If $T \geq 0$, then $T^k \geq 0$ for all k, hence for $\mu > r(T)$, we have

$$R(\mu, T) = \lim_{N \to \infty} \sum_{k=0}^{N} \frac{T^k}{\mu^{k+1}} \geq 0$$

since the finite sums are positive and convergence holds in every entry.

b) We have for $|\mu| > \mathrm{r}(T)$ that

$$|R(\mu, T)| = \left| \lim_{N \to \infty} \sum_{k=0}^{N} \frac{T^k}{\mu^{k+1}} \right| \le \lim_{N \to \infty} \sum_{k=0}^{N} \left| \frac{T^k}{\mu^{k+1}} \right|$$

$$= \lim_{N \to \infty} \sum_{k=0}^{N} \frac{T^k}{|\mu|^{k+1}} = R(|\mu|, T). \qquad \square$$

The following result, a fundamental property of positive matrices, was discovered by O. Perron in 1907 and can be considered as the first major result in the theory of positive matrices.

Theorem 5.6 (Perron). *If T is a positive matrix, then $\mathrm{r}(T)$ is an eigenvalue of T with positive eigenvector.*

Proof. Assertion b) of Proposition 5.5 and Lemma 5.4 imply

$$\|R(\mu, T)\| \le \|R(|\mu|, T)\| \qquad \text{for } |\mu| > \mathrm{r}(T).$$

Recall the formula for the resolvent proved in Theorem 2.11:

$$R(\mu, T) = (\mu - T)^{-1} = \sum_{i=1}^{m} \sum_{\nu=0}^{\nu_i - 1} \frac{(T - \lambda_i)^\nu}{(\mu - \lambda_i)^{\nu+1}} P_i \qquad \text{for } \mu \notin \{\lambda_1, \ldots, \lambda_m\}. \tag{5.2}$$

Here λ_i are the eigenvalues of T with respective multiplicities of the minimal polynomial ν_i and spectral projections P_i, $i = 1, \ldots, m$.

Let now $\lambda_j \in \sigma(T)$ such that $|\lambda_j| = \mathrm{r}(T)$. Then $\|R(\mu, T)\| \to \infty$ whenever μ approaches λ_j. This is obtained immediately from (5.2) by looking at the sup-norm on $\mathcal{L}(X)$ with respect to a basis containing the set B_T in the sense of Lemma 2.16.b). Putting $\mu = s\lambda_j$ with $s > 1$ the above estimate yields

$$\|R(s\mathrm{r}(T), T)\| \ge \|R(s\lambda_j, T)\| \to \infty \quad \text{as } s \downarrow 1,$$

hence $\mathrm{r}(T)$ must be an eigenvalue of T.

Finally, again by (5.2),

$$\lim_{\mu \downarrow \mathrm{r}(T)} R(\mu, T) (\mu - \mathrm{r}(T))^{\nu_1} = (T - \mathrm{r}(T))^{\nu_1 - 1} P_1,$$

where P_1 denotes the spectral projection corresponding to $\lambda_1 = \mathrm{r}(T)$ and ν_1 is the pole order of $R(\cdot, T)$ at $\mathrm{r}(T)$. Hence $(T - \mathrm{r}(T))^{\nu_1 - 1} P_1 \ge 0$ by Proposition 5.5.a). Since $(T - \mathrm{r}(T))^{\nu_1 - 1} P_1 \ne 0$, there necessarily exists a *positive* vector y_1 with

$$x_1 := (T - \lambda_1)^{\nu_1 - 1} P_1 y_1 \ne 0.$$

Any such x_1 is a positive eigenvector of T belonging to $\mathrm{r}(T)$. $\qquad \square$

Before continuing, let us state here an immediate corollary of Perron's result.

Corollary 5.7. *Let $T \geq 0$ and $\mu \in \rho(T)$. Then $R(\mu, T) \geq 0$ implies $\mu > \mathrm{r}(T)$.*

Proof. Let $x \geq 0$ be a positive eigenvector of T belonging to the eigenvalue $\mathrm{r}(T)$. Then $R(\mu, T)x = (\mu - \mathrm{r}(T))^{-1}x \geq 0$ and we are done. □

Originally, Perron studied strictly positive matrices and thus obtained stronger results. Under the assumption of strict positivity of T he proved that $r = \mathrm{r}(T) > 0$, it is a first-order pole of the resolvent $R(\cdot, T)$, and is the only eigenvalue of T of modulus r. The corresponding eigenspace is one-dimensional and spanned by a strictly positive vector.

However, as already Frobenius noticed, it is not the non-existence of zero entries in a given matrix, but the positions of them, that implies all these nice spectral properties. Frobenius defined the term irreducibility of a matrix, which we discuss in the next section.

5.2 Irreducibility

We now turn to the question under which conditions on a positive matrix T (other than strict positivity) the spectral radius $\mathrm{r}(T)$ is a first-order pole of the resolvent, a property with important consequences for the behavior of the powers T^k as $k \to \infty$. The main consequences we have in mind will, however, be discussed in the next chapter and concern the situation where $T = e^{tA}$.

The following property of T, again relatively easy to recognize from the matrix entries, will turn out to be sufficient.

Definition 5.8. A matrix $T \in \mathcal{L}(X)$ is called *reducible* if there exists a subspace

$$J_M := \left\{ (\xi_1, \ldots, \xi_n)^\top : \xi_i = 0 \text{ for } i \in M \right\} \subset X \tag{5.3}$$

for some $\emptyset \neq M \subsetneq \{1, \ldots, n\}$ which is invariant under T. If T is not reducible, it is called *irreducible*.

Remark 5.9. According to the definition, for $n = 1$, i.e., $X = \mathbb{R}$, any 1×1 matrix (a) is irreducible, including the case $a = 0$.

It is important to note that arbitrary coordinate transformations may destroy or produce irreducibility of a given matrix. However, a permutation of the canonical basis vectors of X does not affect it. So, T is reducible if and only if, after a reordering of the canonical basis vectors of X, there is $1 \leq k < n$ such that

$$J_{M_k} := \left\{ (\xi_1, \ldots, \xi_n)^\top : \xi_{k+1} = \cdots = \xi_n = 0 \right\} \tag{5.4}$$

is invariant under T. This leads to the following characterization, which can be applied easily to concrete matrices.

Lemma 5.10. *A matrix $T \in \mathcal{L}(X)$ is reducible if and only if there exists a permutation matrix P such that*

$$S := PTP^{-1}$$

has block-triangular form

$$S = \begin{pmatrix} A & B \\ 0 & C \end{pmatrix},$$

with square matrices A and C.

We will apply the notion of irreducibility mainly to positive matrices and list now two most important examples.

Examples 5.11.

a) A matrix $T = (\tau_{ij})$ with all off-diagonal entries $\tau_{ij} > 0$ $(i \neq j)$ is irreducible, since obviously there is no permutation matrix P for which PTP^{-1} would be block triangular with square diagonal blocks.

b) If $T \geq 0$ is irreducible and $T \leq S$, then S is irreducible. Indeed, if S would be reducible, then there would exist an appropriate S-invariant subspace J_M defined as in (5.3) which would also be T-invariant, since for every $(\xi_1, \ldots, \xi_n)^{\top} \in J_M$ we have $T\xi_i \leq S\xi_i = 0$ for $i \in M$: we reached a contradiction.

The following result shows that every matrix can, in a certain way, be decomposed into irreducible blocks.

Proposition 5.12. *For every matrix $T \in \mathcal{L}(X)$ there exists a permutation matrix P such that $S := PTP^{-1}$ has block-triangular form*

$$\begin{pmatrix} T_{11} & * & \cdots & * \\ & T_{22} & \ddots & \vdots \\ 0 & & \ddots & * \\ & & & T_{mm} \end{pmatrix},$$

where the (square) diagonal blocks T_{ii} are all irreducible.

Proof. We prove the result by induction on n. The case $n = 1$ is clear. Let $n > 1$ and suppose that the result holds for matrices of size $\leq n - 1$. If T is irreducible, there is nothing to prove. If T is reducible, a reordering of the canonical basis produces the form

$$\begin{pmatrix} T_{11} & T_{12} \\ 0 & T_{22} \end{pmatrix},$$

where T_{11} is a $k \times k$ block for suitable $1 \leq k < n$. Since $k < n$, we can, by assumption, rearrange the first k basis vectors in such a way that T_{11} has block triangular form with irreducible diagonal blocks. Since T_{22} can be treated in the same way, we obtain the assertion. $\square$

The following is our main result on *irreducible positive* matrices. It was proved by F.G. Frobenius in 1912.

Theorem 5.13 (Perron–Frobenius). *Let $T \in \mathcal{L}(X)$ be an irreducible and positive matrix. If $n = \dim X > 1$, then the spectral radius $r := \mathrm{r}(T)$ satisfies $r > 0$, and r is a first-order pole of the resolvent $R(\cdot, T)$. The corresponding eigenspace is one-dimensional and spanned by a strictly positive vector $z = (\zeta_1, \ldots, \zeta_n)^\top$, i.e., with $\zeta_i > 0$ for all i.*

Proof. Suppose that $n = \dim X > 1$. By Theorem 5.6, there exists $0 < z = (\zeta_1, \ldots, \zeta_n)^\top$ such that $Tz = rz$. Suppose now that z is not strictly positive. After a reordering of the coordinates we may assume $\zeta_i > 0$ for $i = 1, \ldots, k$ and $\zeta_i = 0$ for $i = k+1, \ldots, n$. Note that $0 \leq k < n$, hence $J_{M_k} \neq \{0\}, X$ (see (5.4)). Now for every $y \in J_{M_k}$ there is a $c > 0$ such that $|y| \leq c \cdot z$ holds. Thus

$$|Ty| \leq T|y| \leq cTz = cr \cdot z,$$

which shows that $Ty \in J_{M_k}$, i.e., J_{M_k} is T-invariant. Since T was supposed to be irreducible, this is impossible. Therefore, z must be strictly positive.

For the next step assume $r = 0$, hence $Tz = 0$ for a strictly positive vector z. As before, we conclude that

$$|Ty| = 0$$

for all $y \in X$. Thus $T = 0$, which is not irreducible since $n = \dim X > 1$.

Now we show that the eigenspace belonging to r is one-dimensional. Let

$$Ty = ry$$

for some $0 \neq y \in X$, $y \neq dz$ for $d \in \mathbb{R}$. Since T is a positive, hence a real matrix, we infer that the real and imaginary parts of y are eigenvectors belonging to r. Therefore, we can assume $0 \neq y \in \mathbb{R}^n$. Since the eigenvector z found above is strictly positive, there exists a $c \in \mathbb{R}$ such that

$$x := z - cy$$

is positive, but not strictly positive. The identity $Tx = rx$ implies that the subspace J_M corresponding to the zero coordinates of x is invariant under T, hence must be $\{0\}$. This implies $z = cy$.

Finally, we determine the pole order of r. Since $r > 0$, we may, after rescaling, take $r = 1$, hence

$$Tz = z$$

for some strictly positive $z = (\zeta_1, \ldots, \zeta_n)^\top$. Define $D_z := \mathrm{diag}(\zeta_1, \ldots, \zeta_n)$ and

$$S := D_z^{-1} T D_z.$$

Then $S \geq 0$ and $S\mathbb{1} = \mathbb{1} = (1, \ldots, 1)^\top$, hence $\|S\| = 1$ by Lemma 5.4.d). This implies $\|S^k\| \leq 1$ and

$$\|T^k\| = \|D_z S^k D_z^{-1}\| \leq \|D_z\| \cdot \|D_z^{-1}\|$$

for all $k \in \mathbb{N}$, hence (T^k) is bounded. By Theorem 3.7.b), the number 1 is a simple pole of the resolvent. $\qquad \square$

Example 5.14. For the positive irreducible matrix

$$T = \begin{pmatrix} 0 & 1 & 1 \\ 1 & 0 & 1 \\ 1 & 1 & 0 \end{pmatrix}$$

one has $\mathrm{r}(T) = 2 \in \sigma(A) = \{-1, 2\}$ and the corresponding strictly positive eigenvector is $\mathbb{1}$.

The strictly positive vector z appearing in Theorem 5.13 is called the *Perron vector* for T and is unique up to multiplication by positive scalars (assuming $\|z\| = 1$ for a norm $\| \cdot \|$ on X, we have uniqueness).

If T is irreducible, then so is its transpose, and consequently $T^\top$ has a strictly positive eigenvector w that spans the one-dimensional eigenspace corresponding to $r = \mathrm{r}(T) = \mathrm{r}(T^\top)$,

$$T^\top w = rw \quad \text{or, equivalently,} \quad w^\top T = rw^\top.$$

Using this we immediately obtain the following result.

Lemma 5.15. *Suppose that T is a positive irreducible matrix with $r = \mathrm{r}(T)$ and $rx \leq Tx$ for some nonzero $x \geq 0$. Then $Tx = rx$ and $x \gg 0$.*

Proof. Assume $Tx - rx > 0$. Taking the Perron vector $w \gg 0$ for $T^\top$ we have $w^\top(Tx - rx) = (w^\top T - rw^\top)x > 0$, which is not possible. Hence $Tx = rx$ and by the Perron–Frobenius Theorem, x is strictly positive. $\qquad \square$

Let us now use the obtained result to describe the long-term behavior of the powers of a positive irreducible matrix.

Corollary 5.16. *Let T be a positive irreducible matrix with $\mathrm{r}(T) = 1$ radially dominant[6]. Then*

$$\lim_{k \to \infty} T^k = P_1 \gg 0,$$

where P_1 denotes the spectral projection belonging to the eigenvalue 1. Moreover, P_1 is of the form

$$P_1 x = \langle x, y \rangle z, \quad x \in X,$$

where $z \gg 0$ and $y \gg 0$, and $\langle z, y \rangle = 1$.

---

[6]For the definition of a radially dominant eigenvalue, see page 36.

Proof. The convergence of the powers T^k to the spectral projection P_1 is an immediate consequence of Theorems 3.7 and 5.13.

Now let $z \gg 0$ and $w \gg 0$ be the respective Perron vectors for T and $T^\top$. Put $y = \frac{w}{\langle z, w \rangle}$ and define $Px := \langle x, y \rangle z$ for all $x \in X$. It is clear that $P^2 = P \gg 0$ and $\operatorname{im} P = \ker(T - I)$ is one-dimensional. Since

$$\langle Tx - x, y \rangle z = \langle x, T^\top y \rangle z - \langle x, y \rangle z = 0, \quad x \in X,$$

we have $\operatorname{im}(T - I) \subseteq \ker P$. By the above, the dimensions of both subspaces equal $n - 1$, hence they are equal. This means that P is a projection on $\ker(T - I)$ along $\operatorname{im}(T - I)$, therefore $P = P_1$. $\qquad\square$

We showed almost the same properties for irreducible positive matrices as Perron has obtained for strictly positive matrices. However, we were not able to show that $r = \mathrm{r}(T)$ is the only eigenvalue of modulus r.

5.3 Imprimitivity

The number of eigenvalues on the spectral circle has interesting impact on the asymptotic behavior of T^k for a positive irreducible matrix T.

Definition 5.17. *The boundary spectrum of a matrix T with spectral radius $r = \mathrm{r}(T)$ is the set*

$$\sigma_{\mathrm{b}}(T) := \{\lambda \in \mathbb{C} \colon |\lambda| = r\} \cap \sigma(T).$$

A positive irreducible matrix T with $\sigma_{\mathrm{b}}(T) = \{r\}$ is called a *primitive matrix*. If a positive irreducible matrix has exactly $h > 1$ eigenvalues in the set $\sigma_{\mathrm{b}}(T)$, it is called *imprimitive* and h is referred to as the *index of imprimitivity*.

The next result was proved by H. Wielandt in 1950.

Lemma 5.18 (Wielandt)**.** *Let S be any complex matrix and T a positive irreducible matrix such that $|S| \leq T$. Let $r = \mathrm{r}(T)$. Then for any $\lambda \in \sigma(S)$ we have $|\lambda| \leq r$. Moreover,*

$$|\lambda| = r \quad \text{if and only if} \quad S = \mathrm{e}^{\mathrm{i}\varphi} D T D^{-1},$$

where $\mathrm{e}^{\mathrm{i}\varphi} = \lambda/r$ and D is a diagonal matrix with $|D| = I$. If we set $d_{11} = 1$, the matrix $D = \operatorname{diag}(d_{ii})$ is uniquely determined.

Proof. From Lemma 5.4.c) we know that $\mathrm{r}(S) \leq \mathrm{r}(T)$.

Assume that $|\lambda| = r$ and let x be an eigenvector of S corresponding to the eigenvalue λ, i.e., $Sx = \lambda x$. Then

$$r|x| = |\lambda||x| = |\lambda x| = |Sx| \leq |S||x| \leq T|x|.$$

By Lemma 5.15, we have $T|x| = r|x|$, $|x| \gg 0$, and as we have seen also $|S||x| = r|x|$, therefore $(T - |S|)|x| = 0$. Since $T - |S| \geq 0$ and $|x| \gg 0$, we see that $T = |S|$.

Let now $\lambda = |\lambda|e^{i\varphi} = re^{i\varphi}$ and $x_k = |x_k|e^{i\theta_k}$ for some $\varphi, \theta_1, \ldots, \theta_n \in \mathbb{R}$ and define $D := \mathrm{diag}(e^{i\theta_1} \ldots, e^{i\theta_n})$. Then $x = D|x|$ and taking $V := e^{-i\varphi}D^{-1}SD$ we have

$$V|x| = r|x| = T|x|.$$

Since $|V| = |S| = T$, we obtain $(V - |V|)|x| = 0$. Taking only the real part of this equation and noting that $|V| \geq \mathrm{Re}(V)$ and $|x| \gg 0$, we see that $\mathrm{Re}(V) = |V|$, which further implies $V = \mathrm{Re}(V) = |V| = T$, i.e., $S = e^{i\varphi}DTD^{-1}$.

The converse implication is obvious.   □

Using Wielandt's lemma we see that the eigenvalues on the spectral boundary of an imprimitive matrix are exactly the hth roots of the spectral radius. The following can be regarded as a continuation of the Perron–Frobenius Theorem 5.13.

Theorem 5.19. *Let T be an imprimitive matrix with index of imprimitivity h and spectral radius $r = \mathrm{r}(T)$. Then the following holds.*

a) *All eigenvalues of T of modulus r are simple poles of the resolvent and the corresponding eigenspaces are one-dimensional.*

b) *$\sigma_b(T) = \{r, r\omega, r\omega^2, \ldots, r\omega^{h-1}\}$, where $\omega = e^{2\pi i/h}$.*

c) *The whole spectrum $\sigma(T)$ is invariant under rotation about the origin through an angle $2\pi/h$, but not through any other positive smaller angle.*

d) *There exists a permutation matrix P such that*

$$PTP^{-1} = \begin{pmatrix} 0 & T_{12} & 0 & \cdots & 0 \\ 0 & 0 & T_{23} & \cdots & 0 \\ \vdots & \vdots & \ddots & \ddots & \vdots \\ 0 & 0 & \ddots & 0 & T_{h-1,h} \\ T_{h1} & 0 & \cdots & 0 & 0 \end{pmatrix}, \tag{5.5}$$

where the blocks on the main diagonal are square.

Proof. a) Let $\sigma_b(T) = \{\lambda_1, \ldots, \lambda_h\}$ where $\lambda_k = re^{i\varphi_k}$, $k = 1, \ldots, h$. Wielandt's lemma (Lemma 5.18) with $S = T$ and $\lambda = \lambda_k$ yields

$$T = e^{i\varphi_k}D_kTD_k^{-1}, \quad k = 1, \ldots, h, \tag{5.6}$$

showing that T and $e^{i\varphi_k}T$ are similar. Let $Tz = rz$, where $z \gg 0$ is the Perron vector for T. Then for $z_k := D_k z$ we have $Tz_k = \lambda_k z_k$ and z_k is an eigenvector corresponding to the simple eigenvalue λ_k, which is unique up to multiplication by scalars. This proves a).

b) By relation (5.6) we also have

$$\begin{aligned} T &= e^{i\varphi_{k_1}}D_{k_1}TD_{k_1}^{-1} = e^{i\varphi_{k_1}}D_{k_1}\left(e^{i\varphi_{k_2}}D_{k_2}TD_{k_2}^{-1}\right)D_{k_1}^{-1} \\ &= e^{i(\varphi_{k_1}+\varphi_{k_2})}\left(D_{k_1}D_{k_2}\right)T\left(D_{k_1}D_{k_2}\right)^{-1}. \end{aligned} \tag{5.7}$$

Consequently, $e^{i(\varphi_{k_1}+\varphi_{k_2})} \in \sigma_b(T)$ for any pair $k_1, k_2 \in \{1, \ldots, h\}$. Thus $\sigma_b(T)$ is a multiplicative abelian group of order h, yielding b).

c) Now let $\sigma(T) = \{\lambda_1, \ldots, \lambda_n\}$ and note that multiplying by $\omega = e^{2\pi i/h}$ we have

$$\sigma(\omega T) = \{\omega\lambda_1, \ldots, \omega\lambda_n\}.$$

As above we obtain that ωT and T are similar, hence $\sigma(\omega T) = \sigma(T)$. On the other hand no rotation by less than $2\pi/h$ keeps $\sigma_b(T)$ invariant, therefore the same holds for the whole spectrum $\sigma(T)$.

d) By b), the eigenvalues on the boundary are of the form $\lambda_k = r\omega^k$, $k = 0, \ldots, h-1$. For the matrices D_k from (5.6) and the Perron eigenvector z of T, from (5.7) it follows that $D_{k_1} D_{k_2} z$ is an eigenvector of T corresponding to the eigenvalue $r\omega^{k_1+k_2}$. We may assume that the upper left entry of each diagonal matrix D_k is 1 and is therefore uniquely determined. Hence also the diagonal matrices D_k form a multiplicative abelian group of order h. In particular, $D_1^h = I_n$, so its main diagonal consists of hth roots of unity.

Let P be a permutation matrix such that

$$PD_1P^{-1} = \operatorname{diag}(\omega^{m_1} I_{n_1}, \omega^{m_2} I_{n_2}, \ldots, \omega^{m_s} I_{n_s}),$$

where I_{n_j} are identity matrices of size $n_j \times n_j$, $\sum_{j=1}^s n_j = n$ and $0 = m_1 < m_2 < \cdots < m_s \leq h - 1$. Using the same permutation matrix P we obtain the block matrix

$$PTP^{-1} = \begin{pmatrix} T_{11} & T_{12} & \cdots & T_{1s} \\ T_{21} & T_{22} & \cdots & T_{2s} \\ \vdots & \vdots & \ddots & \vdots \\ T_{s1} & T_{s2} & \cdots & T_{ss} \end{pmatrix},$$

where each block T_{pq} is of size $n_p \times n_q$. Now, equating the (p,q)-blocks on both sides of the matrix equation

$$PTP^{-1}\omega(PD_1P^{-1})(PTP^{-1})(PD_1^{-1}P^{-1})$$

we obtain a system of s^2 equations

$$T_{pq} = \omega^{1+m_p-m_q} T_{pq}, \quad p, q = 1, \ldots, s.$$

Therefore, $T_{pq} \neq 0$ if and only if

$$m_q = m_p + 1 \mod h. \tag{5.8}$$

T is an irreducible matrix, hence for every p there is a q such that $m_q = m_p + 1 \mod h$. Since m_i, $i = 1, \ldots, s$, are strictly ordered numbers from the set $\{0, \ldots, h-1\}$, the only possibility is that $s = h$ and $m_k = k - 1$, $k = 1, \ldots, h$. So, the block matrix PTP^{-1} has exactly h nonzero blocks: $T_{pq} \neq 0$ iff $q = p + 1 \mod h$. $\square$

Let us verify all these properties on a given matrix.

Example 5.20. The matrix

$$T = \begin{pmatrix} 0 & 2 & 0 & 0 \\ 1 & 0 & 1 & 0 \\ 0 & 1 & 0 & 1 \\ 0 & 0 & 2 & 0 \end{pmatrix}$$

is positive and irreducible (there is no permutation matrix P such that $P^{-1}TP$ is block triangular with square diagonal blocks). Computing the spectrum we obtain $\sigma(T) = \{\pm 1, \pm 2\}$, so, $\sigma_{\mathrm{b}}(T) = \{\pm 2\}$ and T is imprimitive with index of imprimitivity $h = 2$, $\sigma(T)$ is invariant under rotation through the angle π, and

$$PTP^{-1} = \begin{pmatrix} 0 & 0 & 2 & 0 \\ 0 & 0 & 1 & 1 \\ 1 & 1 & 0 & 0 \\ 0 & 2 & 0 & 0 \end{pmatrix} \quad \text{for} \quad P = \begin{pmatrix} 1 & 0 & 0 & 0 \\ 0 & 0 & 1 & 0 \\ 0 & 1 & 0 & 0 \\ 0 & 0 & 0 & 1 \end{pmatrix}.$$

We give some alternative characterizations of primitivity.

Proposition 5.21. *For a positive irreducible matrix $T \neq 0$ with spectral radius $r = \mathrm{r}(T)$ the following assertions are equivalent.*

 (i) *T is primitive.*
 (ii) *$\lim_{k \to \infty}(T/r)^k = P_1$, where $P_1 \gg 0$ denotes the spectral projection belonging to the eigenvalue r.*
 (iii) *$T^m \gg 0$ for some m enough large.*
 (iv) *T^k is irreducible for all $k \geq 1$.*

Moreover, if any of the above assertions holds, then

$$P_1 x = \langle x, y \rangle z, \quad x \in X,$$

where $z \gg 0$ and $y \gg 0$ are Perron vectors for T and $T^\top$, respectively, normalized to satisfy $\langle z, y \rangle = 1$.

Proof. (i) $\Longrightarrow$ (ii): Observe that T is primitive if and only if T/r is primitive, which is true if and only if $1 = \mathrm{r}(T/r)$ is radially dominant. Now the implication as well as the additional assertion follow by Corollary 5.16.

(ii) $\Longrightarrow$ (iii): Since $\lim_{k \to \infty}(T/r)^k = P_1 \gg 0$, one has that $T^k \gg 0$ for k sufficiently large.

(iii) $\Longrightarrow$ (iv): If T^k would be reducible for some $k \geq 0$, then so would be T^{kl} for all $l \in \mathbb{N}$, which conflicts with $T^m \gg 0$ for large m.

(iv) $\Longrightarrow$ (i): Suppose T has index of imprimitivity $h > 1$. Then there exists an eigenvalue $r\alpha \in \sigma_{\mathrm{b}}(T)$ with $|\alpha| = 1$ and $\alpha \neq 1$. By Theorem 5.19, $\alpha^h = 1$. Therefore, T^h must have at least two independent eigenvectors corresponding to $r = \mathrm{r}(T)$ which, by the Perron–Frobenius theorem, contradicts the irreducibility of T. $\square$

We have seen that the powers of a primitive matrix converge to a strictly positive projection. We conclude the description of the asymptotic behavior of imprimitive matrices.

Definition 5.22. For $T \in \mathcal{L}(X)$ we call sequence (T^k) *asymptotically periodic with period p* if there is a direct sum decomposition

$$X = X_s \oplus X_u$$

into T-invariant subspaces X_s and X_u such that

a) $T|_{X_s}$ is stable, i.e., $\lim_{k \to \infty} \|T^k x\| = 0$ for all $x \in X_s$, and

b) $T|X_u$ is periodic with period p, i.e., $T^p y = y$ for all $y \in X_u$ and $p \in \mathbb{N}$ is the smallest natural number with this property.

We close with a consequence of Proposition 5.19.

Corollary 5.23. *Let T be an imprimitive matrix with index of imprimitivity h and spectral radius $r = \mathrm{r}(T)$. Then the sequence $((T/r)^k)$ is asymptotically periodic with period h.*

Proof. Let

$$X_s := \bigoplus_{|\lambda_i| < r} X_i \qquad \text{and} \qquad X_u := \bigoplus_{|\lambda_i| = r} X_i$$

and use Theorems 3.7 and 5.19 for the matrix T/r restricted to its invariant subspaces X_s and X_u. $\qquad\square$

5.4 Notes and Remarks

Theorem 5.6 can be found in the paper by O. Perron [111], which was the beginning of the general theory of positive matrices. Theorem 5.13 goes back to F.G. Frobenius [48]. Almost 40 years later, H. Wielandt [155] proved Theorem 5.18. The form of an imprimitive matrix presented in Theorem 5.19 was originally found by Frobenius [48] and is known as the *Frobenius Form*.

The literature on Perron–Frobenius theory is vast. We refer to some classical monographs by Berman and Plemmons [18], Ding and Zhou [32], Meyer [94], Minc [96], and Schaefer [126].

Let us make a few remarks on terminology because we have witnessed bitter disputes between communities about it. These matrices were first called "matrices with non-negative entries", which is a rather long name, and was shortened to "non-negative matrices", as it turned out they were important and the term was used frequently. Independently, starting with the work of F. Riesz and L. Kantorovich, the theory of positive operators on ordered Banach spaces was developed. We decided to accept the terminology of Riesz spaces and Banach lattices to make our presentation consistent with the infinite-dimensional part following later on, and hope that nobody gets confused by this.

5.5   Exercises

1. For the matrix $T = \begin{pmatrix} 1-a & b \\ a & 1-b \end{pmatrix}$, where $a, b > 0$ and $a + b = 1$, verify that $r(T)$ is an eigenvalue of T, and find the appropriate eigenvector.

2. For $T \geq 0$, prove the existence of a positive eigenvector x_0 belonging to the eigenvalue $r := r(T)$ through the following steps.

   a) There exists $y \geq 0$ such that $\left\| R(r + \frac{1}{k}, T)y \right\| \to \infty$ as $k \to \infty$.

   b) The sequence
   $$y_k := \frac{R(r + \frac{1}{k}, T)y}{\left\| R(r + \frac{1}{k}, T)y \right\|}$$
   has a convergent subsequence (y_{k_l}).

   c) The sequence $(r - T)y_{k_l} \to 0$ as $l \to \infty$.

   d) The limit $x_0 := \lim_{k \to \infty} y_{k_l}$ is a positive eigenvector of T belonging to r.

3. Show that, if $a_1, \ldots, a_n \in \mathbb{C}$ are all non-zero, then
   $$\begin{pmatrix} 0 & a_1 & 0 & \cdots & 0 \\ 0 & 0 & a_2 & \cdots & 0 \\ \vdots & \vdots & \ddots & \ddots & \vdots \\ 0 & 0 & \ddots & 0 & a_{n-1} \\ a_n & 0 & \cdots & 0 & 0 \end{pmatrix}$$
   is irreducible.

4. Prove that for a positive matrix T the following conditions are equivalent.

   (i) T is irreducible.

   (ii) $R(\mu, T)x \gg 0$ for some $\mu > r(T)$ and all $x > 0$.

   (iii) $R(\mu, T)x \gg 0$ for all $\mu > r(T)$ and all $x > 0$.

5. Show that $0 \neq T \geq 0$ is irreducible if and only if the eigenspaces of T and of $T^\top$ belonging to $r(T) = r(T^\top)$ are one-dimensional and spanned by a strictly positive vector.

6. If $0 \leq T < S$ and T is irreducible, then $r(T) < r(S)$. In other words: The spectral radius is a strictly monotone function on the set of irreducible positive matrices.

7. Let T be a positive irreducible matrix. Prove that, if the trace $\operatorname{tr} T > 0$, then T is primitive.

8. Verify irreducibility and imprimitivity of the matrices T_i, $i = 1, 2$, below and discuss the asymptotic behavior of the sequence $\left((T_i / r(T_i))^k \right)$.

$$T_1 = \begin{pmatrix} 0 & 1 & 0 \\ 0 & 0 & 1 \\ 1 & 0 & 0 \end{pmatrix}, \qquad T_2 = \begin{pmatrix} 0 & 1 & 0 \\ 1 & 0 & 1 \\ 0 & 1 & 0 \end{pmatrix}.$$

Chapter 6

Applications of Positive Matrices

We have now accumulated enough material to pause for a while to discuss its consequences in concrete situations. We have revised linear algebra facts from a functional analytic perspective and obtained a construction to get functions of matrices in a coordinate-free manner, without the use of the Jordan normal form. This was useful when we considered positive matrices, and enabled us to see important and deep spectral consequences of positivity.

The applications of the developed theory are numerous and we have selected just a few representing our taste: graph matrices, the Google matrix, and age-structured population models.

6.1 Motivating Examples Revisited

We start by revisiting our motivating examples from Section 1.1.

Graphs

Let $G = (V, E)$ be a directed graph with n vertices $V = \{v_1, \ldots, v_n\}$ and a set of directed edges E. The graph G is called *strongly connected* if for every $v_i \in V$ and every $v_j \in V$ there is walk in G from v_i to v_j. This property of the graph can be read from its adjacency matrix.

Proposition 6.1. *A graph G is strongly connected if and only if its adjacency matrix is irreducible.*

Proof. Let $A = (a_{ij})$ be the adjacency matrix of G. By Lemma 5.10, A is reducible iff we can partition the sets of vertices $V = V_1 \cup V_2$ into two disjoint subsets such that after relabeling the vertices we obtain a block-triangular form for A,

$$A = \begin{pmatrix} A_{11} & A_{12} \\ 0 & A_{22} \end{pmatrix}, \tag{6.1}$$

where the block $A_{k\ell}$ for each $k, \ell \in \{1, 2\}$ corresponds to connections from the set of vertices V_k to the set V_ℓ. Note that $A_{21} = 0$ is equivalent to the fact that there are no direct edges from a vertex in V_2 to a vertex in V_1.

Let $v_i \in V_2$ and $v_j \in V_1$ and assume there exists a walk in G from v_i to v_j. Then

$$a_{ii_1} a_{i_1 i_2} \cdots a_{i_s j} \neq 0$$

for some $i_1, \ldots, i_s \in \{1, \ldots, n\}$. Observe that in this product there must be a nonzero entry with "mixed" indices, i.e., $a_{i_k i_\ell} \neq 0$ with $v_{i_k} \in V_2$ and $v_{i_\ell} \in V_1$, which contradicts (6.1). So, if G is strongly connected, A must be irreducible.

For the converse assume that G is not strongly connected. Hence there exist vertices $v_i, v_j \in V$ such that there is no walk starting in v_i and ending in v_j. Let V_1 be the set of all initial vertices of walks which end in v_j, and let $V_2 = V \backslash V_1$. The sets V_1 and V_2 are disjoint and nonempty. According to the partition $V = V_1 \cup V_2$ the adjacency matrix has block-triangular form given in (6.1), so A is reducible. $\square$

As a corollary we obtain a combinatorial characterization of positive irreducible matrices. Note that every positive matrix can be seen as the adjacency matrix of a graph.

Corollary 6.2. *A positive $n \times n$ matrix A, $n \geq 2$, is irreducible if and only if for every $i, j \in \{1, \ldots, n\}$ there exists an $s \in \mathbb{N}$ such that $(A^s)_{ij} > 0$.*

We will illustrate another property of the adjacency matrix A in terms of the structure of the graph G. Recall from Theorem 5.19 that any imprimitive matrix with index of imprimitivity h can be written in Frobenius form as follows:

$$PAP^{-1} = \begin{pmatrix} 0 & A_{12} & 0 & \ldots & 0 \\ 0 & 0 & A_{23} & \ldots & 0 \\ \vdots & \vdots & \ddots & \ddots & \vdots \\ 0 & 0 & \ddots & 0 & A_{h-1,h} \\ A_{h1} & 0 & \ldots & 0 & 0 \end{pmatrix} \tag{6.2}$$

with square blocks on the main diagonal.

Lemma 6.3. *Let A be an imprimitive matrix with index of imprimitivity h and Frobenius form (6.2). Then $A_{12} A_{23} \cdots A_{h1}$ is a primitive matrix.*

Proof. First introduce the matrices

$$\tilde{A}_1 := A_{12} A_{23} \cdots A_{h1}, \quad \tilde{A}_2 := A_{23} A_{34} \cdots A_{12}, \quad \ldots \quad \tilde{A}_h := A_{h1} A_{12} \cdots A_{h-1,h}.$$

Observe that all of them are positive matrices and their spectra coincide.

Using the Frobenius form (6.2), one sees that

$$\left(PAP^{-1}\right)^{sh} = \begin{pmatrix} \tilde{A}_1^s & 0 & \cdots & 0 \\ 0 & \tilde{A}_2^s & \cdots & 0 \\ \vdots & \vdots & \ddots & \vdots \\ 0 & 0 & \cdots & \tilde{A}_h^s \end{pmatrix} \qquad \text{for all } s \in \mathbb{N}.$$

Since A is irreducible, so is PAP^{-1}. Combining Corollary 6.2 and the above block diagonal form yields the irreducibility of $\tilde{A}_1$.

By Theorem 5.19, the boundary spectrum of A equals

$$\sigma_{\mathrm{b}}(A) = \left\{ r, r\omega, r\omega^2, \ldots, r\omega^{h-1} \right\},$$

where $r = \mathrm{r}(A)$ and $\omega = \mathrm{e}^{2\pi\mathrm{i}/h}$. Hence,

$$\{r^h\} = \sigma_{\mathrm{b}}(A^h) = \sigma_{\mathrm{b}}\left(PA^hP^{-1}\right) = \sigma_{\mathrm{b}}(\tilde{A}_1),$$

and the matrix $\tilde{A}_1$ is indeed primitive. $\qquad\square$

Proposition 6.4. *Let G be a strongly connected graph whose adjacency matrix A is imprimitive with index of imprimitivity h. Then h equals the greatest common divisor*

- d_i *of lengths of all closed walks through a vertex v_i in G,*
- d_{W} *of lengths of all closed walks in G, and*
- d_{C} *of lengths of all cycles in G.*

Proof. Let us first show that $d_{\mathrm{C}} = d_{\mathrm{W}}$. Clearly, $d_{\mathrm{W}} | d_{\mathrm{C}}$, as every cycle is also a closed walk. Now observe that every closed walk can be partitioned into cycles and the length of the closed walk is the sum of the lengths of these cycles, hence divisible by d_{C}.

Now fix a vertex v_i of G. By definition, $d := d_{\mathrm{C}} = d_{\mathrm{W}}$ divides d_i. Choose an arbitrary closed walk C in G. If it contains v_i, then its length $\ell(C)$ is divisible by d_i. Otherwise, take a vertex $v_j \in C$. Since G is strongly connected, there exist a walk W_{ij} from v_i to v_j and a walk W_{ji} from v_j to v_i. Now $W_{ij}CW_{ji}$ is a closed walk in G that contains v_i, hence its length

$$\ell(W_{ij}CW_{ji}) = \ell(W_{ij}) + \ell(C) + \ell(W_{ji})$$

is divisible by d_i. But also $W_{ij}W_{ji}$ is a closed walk in G that contains v_i and thus also $\ell(W_{ij}W_{ji}) = \ell(W_{ij}) + \ell(W_{ji})$ is divisible by d. Therefore d_i divides $\ell(C)$ and since W was arbitrary, it divides d. We conclude that $d_i = d$.

Again take a vertex v_i of G. It remains to show that $d_i = h$. The existence of a closed walk in G of length ℓ through a vertex v_i is equivalent to the condition

$(A^{\ell})_{ii} > 0$, see Proposition 1.1. Therefore $(A^{kd_i})_{ii} > 0$ for all sufficiently large $k \in \mathbb{N}$ and $(A^s)_{ii} = 0$ if s is not a multiple of d_i.

On the other hand, we may assume that A is in Frobenius form (6.2). Then only powers of A^h can have nonzero diagonal elements. Note that, by Lemma 6.3, the square diagonal blocks of A^h consist of primitive matrices, hence $A^{mh} \gg 0$ for some $m \in \mathbb{N}$, see Proposition 5.21. Therefore $(A^{mh})_{ii} > 0$ for all sufficiently large $m \in \mathbb{N}$ and $(A^s)_{ii} = 0$ if s is not a multiple of h.

Altogether we thus have that $h = d_i = d$. $\square$

Remark 6.5. Observe that in the case when A is a primitive matrix the same proof yields $d_i = d_{\mathrm{W}} = d_{\mathrm{C}} = 1$.

Markov chains

Now let a positive stochastic $n \times n$ matrix $P = (p_{ij})$ be the transition matrix of a discrete finite homogeneous Markov chain with the state space $V = \{v_1, \ldots, v_n\}$. The *kth step probability distribution vector* $p(k) = (p_1(k), p_2(k), \ldots, p_n(k))^{\top}$ is defined as a positive stochastic vector, i.e.,

$$0 \leq p_i(k) \leq 1, \quad \sum_{i=1}^{n} p_i(k) = 1,$$

where $p_i(k)$ is the probability of Markov process being in the state v_i after k steps. By the Markov property and Remark 1.2, the kth step distribution is determined from the initial distribution $p(0)$ by means of the transition matrix:

$$p(k) = (P^k)^{\top} p(0), \quad k \in \mathbb{N}.$$

Therefore the long-run (or limiting) probability distribution depends on the behavior of P^k for $k \to \infty$. Using our results from Chapters 3 and 5 we can describe it in terms of spectral properties of P.

Let us first state some spectral properties of P.

Lemma 6.6. *For the transition matrix P the following holds.*

a) $\mathrm{r}(P) = 1$ *is an eigenvalue of P with corresponding eigenvector*

$$\mathbb{1} = (1, 1, \ldots, 1)^{\top}.$$

b) *All eigenvalues of P with modulus 1 are simple poles of the resolvent.*

Proof. a) Since P is row-stochastic, $P^k \mathbb{1} = \mathbb{1}$ holds for all $k \geq 1$. Hence, by Gelfand's formula, $\mathrm{r}(P) = 1$ and 1 is an eigenvalue with eigenvector $\mathbb{1}$.

b) Since $\|P^k\|_{\infty} = 1$ for all $k \in \mathbb{N}$, the sequence (P^k) is bounded, and by Theorem 3.13, all eigenvalues with modulus 1 are simple poles of the resolvent. $\square$

As a consequence, P is always Cesàro summable with Cesàro means converging to the spectral projection of P belonging to 1 (cf. Theorem 3.13). The sequence P^k, however, does not converge as $k \to \infty$ unless 1 is radially dominant (see Theorem 3.7).

The Cesàro means of $p(k)$ have an illustrative interpretation in the context of Markov chains. Pick a state v_j and define a sequence of random variables $(X_i)_{i=0}^{\infty}$ by

$$X_i = \begin{cases} 1 & \text{if the chain is in the state } v_j \text{ after } i \text{ steps,} \\ 0 & \text{otherwise.} \end{cases}$$

Then $\frac{1}{k}\sum_{i=0}^{k-1} X_i$ represents the fraction of time that the state v_j is visited in $k-1$ steps. Since the expected value of each X_i is $E(X_i) = p_j(i)$, we have

$$E\left(\frac{1}{k}\sum_{i=0}^{k-1} X_i\right) = \left(\frac{1}{k}\sum_{i=0}^{k-1} p(i)\right)_j.$$

This means that the jth component of the Cesàro limit vector represents the fraction of time that the chain spends in the state v_j in the long-run.

Assume now, that the matrix P is irreducible (i.e., all states v_i are reachable from each other in a finite number of steps). In this case we have two possibilities.

- If P is a primitive matrix, then

$$\lim_{k\to\infty} P^k = P_1 \text{ with } P_1 x = \langle x, y \rangle \mathbb{1} \quad \text{and} \quad \lim_{k\to\infty} p(k) = y, \qquad (6.3)$$

  where y is the stochastic Perron vector for $P^{\top}$, see Proposition 5.21.

- If P is an imprimitive matrix, then the above limits do not exist. However, for the corresponding Cesàro means,

$$\lim_{k\to\infty} P^{(k)} = P_1 \text{ with } P_1 x = \langle x, y \rangle \mathbb{1} \quad \text{and} \quad \lim_{k\to\infty} \frac{1}{k}\sum_{i=0}^{k-1} p(i) = y, \qquad (6.4)$$

  where again y is the stochastic Perron vector for $P^{\top}$, see Theorem 3.13.

A Markov chain with an irreducible and imprimitive transition matrix is called *periodic*. In such a chain all states are visited periodically, with the period equal to the index of imprimitivity of P, see Corollary 5.23.

Note that the value of the (Cesàro) limit is independent of the initial distribution $p(0)$. The vector y in equations (6.3) and (6.4) is called the *stationary distribution vector* for the Markov chain. It is the unique stochastic vector satisfying $P^{\top} y = y$. Its components represent the long-run fraction of time that the chain spends in the corresponding state.

6.2   The Google Matrix

We shall demonstrate now that we encounter positive matrices and their Perron vectors on an everyday basis. We will look at the mathematics behind Google[7], currently the world biggest web search engine.

Every web search engine must build its web-page repository and index the pages stored there in the best possible way. For this purpose they use crawler software that creates virtual robots, called spiders, that constantly travel the web. The spiders number each page, collect important data from it (such as title, key words, link names, anchors, etc.) and create an index of all visited pages. Now the pages have to be ranked according to their importance. When the user does an internet search it is desired that more relevant pages are placed at the beginning of the produced list. This is actually the most important and delicate step for a search engine. It is because of intelligent ranking that Google got ahead its competitors when it appeared on the market. The core of Google is the ranking algorithm *PageRank*, developed in 1998 by Larry Page and Sergey Brin, then PhD students at Stanford University, California.

PageRank

Assume we have n web pages $W = \{W_k \mid k = 1, \dots, n\}$. For a page W_k we denote by $I_k := \{i \mid W_i \to W_k\}$ the set of indices of all *inlinks* to W_k, by $O_k := \{j \mid W_k \to W_j\}$ the set of indices of all *outlinks* of W_k, and by $x_k \geq 0$ the *rank of the page* W_k. Now the question is, how to define x_k properly?

The answer of Page and Brin is: *A page is important if it is pointed to by other important pages*. Their formula for the rank is thus recursive and it is not clear at this point whether it admits a solution:

$$x_k := \sum_{i \in I_k} \frac{x_i}{|O_i|}, \quad k = 1, \dots, n. \tag{6.5}$$

Here it is assumed that a link from a page to itself does not count.

The internet can be viewed as a huge directed graph with n vertices (= web pages) whose edges are hyperlinks. Let H be the transposed adjacency matrix of this graph, called also the *hyperlink matrix*, with entries

$$H_{ij} = 1/|O_j| \quad \text{iff } W_j \to W_i \quad \text{and} \quad H_{ij} = 0 \text{ otherwise.}$$

We can interpret the values H_{ij} as probabilities of accessing page W_i from page W_j. Collecting single ranks into a *ranking vector* $x := (x_1, \dots, x_n)^\top$, we can now write the recursive relation (6.5) as a matrix equation

$$x = Hx. \tag{6.6}$$

---

[7]The name comes from the misspelled number googol $= 10^{100}$.

The solution vector, if it exists, is thus the fixed vector of the hyperlink matrix H. To assure uniqueness, we impose from now on that the ranking vector x is stochastic, i.e., $\|x\|_1 = 1$.

Note that H is a positive matrix, thus by Perron's theorem (see Theorem 5.6), its spectral radius $r(H)$ is an eigenvalue of H with positive eigenvector. Matrix H is also substochastic, i.e., $\sum_{i=1}^{n} H_{ij} \leq 1$ for all j, hence $r(H) \leq 1$. Having equation (6.6) in mind, we would like that $r(H) = 1$. Observe that the sum of non-zero columns actually equals 1, but H might have some zero columns which represent the so-called *dangling nodes*, that is, pages without outlinks. Brin and Page therefore suggested to adjust the matrix H: replace all zero columns with $(1/n, \ldots, 1/n)^\top$. The adjusted matrix becomes stochastic and thus equation (6.6) with the modified matrix H has a solution. We can also interpret this adjustment. Imagine a random surfer traveling the web using hyperlinks, which he chooses randomly. At some point he might find himself at a dangling node. His way out is to randomly type an url and thus jump to any page with probability $1/n$.

In order to assure the uniqueness of the solution to equation (6.6), we would like H to be irreducible. By Proposition 6.1, H is irreducible if and only if the web is strongly connected, which is clearly a nonrealistic assumption. However, Brin and Page overcame also this problem with a new adjustment: they replaced the matrix H by the *Google matrix*

$$G := \alpha H + (1 - \alpha)S, \tag{6.7}$$

where $S = (1/n)_{n \times n}$ and $\alpha \in [0, 1]$ is some fixed number. The interpretation of this adjustment is a continuation of the one above: a random surfer sometimes decides to jump to some other page directly by typing an url instead of following some hyperlink, even if he is not at the dangling node. The role of the parameter α is to balance between the original web structure given by H and a fully connected web represented by S. We would of course like to weight the original hyperlink structure heavily and take α close to 1.

For any $\alpha \in [0, 1)$, the Google matrix G is positive, irreducible, and column stochastic, hence Frobenius Theorem 5.13 guarantees that the equation $Gx = x$ has a unique strictly positive stochastic solution. Thus the desired ranking vector is nothing but the Perron vector for G!

Computation of the Perron vector

To compute the Perron vector for G we can use a very simple numerical method called the *power method* that was already mentioned at the end of Chapter 3. It is an iterative method defined by

$$x^{(k+1)} = Gx^{(k)}.$$

From this we infer that $x^{(k+1)} = G^k x^{(0)}$, thus convergence of this process is assured by Corollary 5.16, independent of the choice of the initial vector $x^{(0)} \neq 0$. Here it is important that 1 is a strictly dominant eigenvalue of the positive irreducible matrix G.

It is well known that the rate of convergence of the power method is governed by the magnitude of the second eigenvalue $|\lambda_2|$ of the matrix. For the Google matrix it can be shown that $|\lambda_2| \leq \alpha$. This means that the convergence is faster for smaller α. Since we argued above that α should be close to 1, one has to accept a compromise here. It is reported that Google uses $\alpha = 0.85$, the value set already by Brin and Page in 1998.

6.3 Age-structured Population Models

Plant, animal, and human population models are typical examples for positive dynamical systems in which the state variables represent biomass, density, or the number of individuals in the population. Many of these models, in particular those describing predation, competition, and symbiosis among species, are nonlinear and therefore deemed to investigation by other means. An important and still widely used exception is the well-known *Leslie model*, which describes the time evolution of a population in which fertility and survival rates of individuals strongly depend on their age. For this reason, such populations are called age-structured populations. In the Leslie model, the time is discrete and represents the reproduction season (typically the year in case of mammals), while the variables $x_1(t), x_2(t), \ldots, x_n(t)$ represent the number of females (or individuals, or couples) of age $1, 2, \ldots, n$ at the beginning of year t.

In the simplest possible case one can describe the aging process by means of the equations

$$x_{i+1}(t+1) = s_i x_i(t), \quad i = 1, 2, \ldots, n-1,$$

where $s_i > 0$ is the survival coefficient at age i, that is, the fraction of females of age i that survive at least for 1 year. The first state equation takes into account the reproduction process, and is

$$x_1(t+1) = s_0(f_1 x_1(t) + f_2 x_2(t) + \cdots + f_n x_n(t)),$$

where $s_0 > 0$ is the survival coefficient during the first year of life and $f_i \geq 0$ is the fertility rate of females of age i, that is, the mean number of females born from each female of age i. These equations, originally proposed by Leslie, lead to a positive linear autonomous model

$$x(t+1) = Ax(t),$$

where the matrix A, called the *Leslie matrix*, is given as

$$A = \begin{pmatrix} s_0 f_1 & s_0 f_2 & \cdots & s_0 f_{n-1} & s_0 f_n \\ s_1 & 0 & \cdots & 0 & 0 \\ 0 & s_2 & \cdots & 0 & 0 \\ \vdots & \vdots & & \vdots & \vdots \\ 0 & 0 & \cdots & s_{n-1} & 0 \end{pmatrix}. \tag{6.8}$$

Though Leslie models appear to be quite coarse at first sight, they are extensively used for making demographic projections, i.e., forecasting

$$x(k) = A^k x(0)$$

given $x(0)$.

Let us comment on the usefulness of these models first. In Leslie models, survival and fertility rates depend exclusively on age. In reality, this is more or less true provided the individuals in each age class are not too many. In fact, as soon as the density of the individuals increases, some phenomena show up, which may reduce fertility and/or survival rates. For example, finding appropriate niches for reproduction becomes more difficult if the number of fertile individuals increases; the spreading of epidemics is favoured by high population densities; the search for food becomes more and more difficult as a population increases, and so on. This means that Leslie models are well suited for describing the dynamics of populations doomed to extinction, that is, characterized by small densities $x_i(t)$ for which we can suppose that survival and fertility rates are constant as time evolves. Leslie models are also extremely effective yielding short term forecasts in growing populations.

Investigating the properties of the Leslie matrix, we see that it is positive and, if $f_n > 0$, it is also irreducible. Looking at the directed weighted graph whose adjacency matrix is given by equation (6.8) and using Proposition 6.4 one easily obtains that the index of imprimitivity of the Leslie matrix equals

$$h = \gcd \{k \in \{1, \ldots, n\} : f_k > 0\}.$$

Hence, if there are two consecutive ages with strictly positive fertility age, then the Leslie matrix is primitive.

The (normalized) Perron eigenvector of the Leslie matrix is called the *stable age structure*, which is roughly the asymptotic age distribution as time evolves. More precisely, we have the following result as a consequence of Proposition 5.21.

Proposition 6.7. *Consider the Leslie matrix A given in (6.8) with $f_n > 0$ and assume that A is a primitive matrix. Denote the Perron eigenvalue by $\lambda_1 = \mathrm{r}(A)$ and the corresponding eigenvector by $x_1 \gg 0$. Then*

$$\lambda_1^{-k} A^k - P_1 \longrightarrow 0$$

as $k \to \infty$, where P_1 is the projection to the one-dimensional subspace spanned by x_1.

Age Class	Average Reprod./ Year	Low Reprod./ Year	High Reprod./ Year	Average Annual Survival	Low Annual Survival	High Annual Survival
Cub	0.00	0.00	0.00	0.80	0.41	0.99
1-year-old	0.00	0.00	0.00	0.75	0.41	0.99
2-year-old	0.00	0.00	0.00	0.71	0.41	0.90
3-year-old	0.28	0.00	0.50	0.84	0.69	0.93
Adult	0.58	0.23	0.82	0.84	0.69	0.93

Table 6.1: Input parameters for Leslie Matrix population model (based on females only) of Virginias hunted black bear populations as estimated between 1994–1999.

Let us note that in many applications it is better to structure the population not in age groups, but in so-called *stage groups*. As an example, we consider Virginias hunted black bear populations. A statistical analysis, the details of which we omit, leads to the following table, which is only reproduced here to show the complexity of such problems.

In this case, as we see, it is better to investigate the so-called stage-based Leslie model. Stage-based models are frequently used for long-lived species because data on specific ages are not available, demographic variables within age classes are not different, and individual age classes for a species that lives, for example, up to 30 years (like black bear), would result in matrices of sizes up to 30×30. Analysis of this table can lead to the following Leslie matrix, where various other effects have been taken into account, and which was used successfully in analysis done by biologists:

$$
A = \begin{pmatrix}
0 & 0 & 0 & 0.275 & 0.575 \\
0.80 & 0 & 0 & 0 & 0 \\
0 & 0.75 & 0 & 0 & 0 \\
0 & 0 & 0.71 & 0 & 0 \\
0 & 0 & 0 & 0.84 & 0.84
\end{pmatrix}. \tag{6.9}
$$

Here the last row stands for the whole adult stage, the element in the lower right corner of the matrix representing the rate of the adult population remaining alive after the year.

We now consider a second model, which is famous in the literature. The eastern wild turkey (Meleagris gallopavo silvestris) inhabits more or less the eastern part of the United States. Turkey hunting has a substantial economic effect in many rural communities. It is not only important because of the actual turkey hunting, but it also takes part in the development of the related industries of turkey-hunting clothes and equipment. Improvement of the knowledge of turkey population dynamics is important for formulating hunting regulations and other turkey management practices. A Leslie matrix model can be developed for the population dynamics of eastern wild turkeys in Iowa based on local studies. Here a three-stage model is chosen in order to simplify the modeling procedure. The first category is "poults", aged from 0 to 1, the second category is "yearlings", aged from 1 to 2, and the last category is "adults", aged 2 and older. Reproduction occurs from yearlings onwards. The time unit is one year. The Leslie matrix obtained is

$$A = \begin{pmatrix} 0 & 0.880 & 1.860 \\ 0.445 & 0 & 0 \\ 0 & 0.616 & 0.610 \end{pmatrix}. \tag{6.10}$$

This grouping makes sense for example if there are regulations allowing only the adult population to be hunted, see Exercise 6.

6.4 Notes and Remarks

For further reading on search engines and the PageRank algorithm we recommend the excellent monograph by Langville and Meyer [82]. The modeling and investigation of age-structured populations was initiated by Leslie in 1945 [87], and extended to stage structured populations by Lefkovitch [86]. Virginia's hunted black bear populations is discussed in the PhD dissertation by Klenzendorf [75]. Much research about the rates of reproduction, mortality, and survival, and the movement of wild turkeys has been done by Dickson [30].

6.5 Exercises

1. Verify that the matrix

$$A = \begin{pmatrix} 0 & 1 & 0 & 0 & 0 & 0 \\ 1 & 0 & 0 & 0 & 1 & 0 \\ 0 & 1 & 0 & 0 & 0 & 0 \\ 1 & 0 & 1 & 0 & 0 & 0 \\ 0 & 0 & 0 & 1 & 0 & 1 \\ 0 & 0 & 1 & 0 & 1 & 0 \end{pmatrix}$$

   is irreducible and imprimitive using graph-theoretical interpretations. Compute also its index of imprimitivity.
2. Explain the statements given in (6.3) and (6.4). Why is the limiting distribution independent of $p(0)$?
3. Find the limiting distribution for the Markov chain given by the transition matrix

$$P = \begin{pmatrix} 0 & 1/2 & 1/2 \\ 1/3 & 0 & 2/3 \\ 1/3 & 2/3 & 0 \end{pmatrix}.$$

4. Translate the PageRank algorithm into the language of Markov chains.

5. Compute the ranking vector for the web depicted in Figure 6.1. Choose sev-

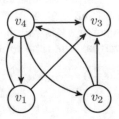

Figure 6.1: The web graph for Exercise 5.

eral values for α and observe how this choice does affect the ranking and the computation time.

6. Consider the Leslie matrix in (6.10) corresponding to the turkey population in Iowa. Use an appropriate computer software if necessary.

   a) Calculate the Perron eigenvalue and the corresponding stable age structure. Is the population growing?

   b) Assume we can change the survival rate of the adult population. How should we change the survival rate of the adult population to ensure that the Perron eigenvalue equals 1, meaning that the population remains balanced?

   c) Using a 1977 survey, the age structure in a region in Iowa was estimated as $x_1(0) = 580$, $x_2(0) = 123$, $x_3(0) = 156$. How many adults should be hunted down at the end of the first year to ensure this decrease in the survival rate of adults?

7. What is the Perron eigenvalue and the corresponding stable age distribution of the Leslie matrix in (6.9) corresponding to the bear population? Is the population growing, balanced, or dying out? Use an appropriate computer software if necessary.

8. To connect two topics of this chapter, google further Leslie matrix models, for example for the annual bluegrass (poa annua) or the brown rat (rattus norvegicus) populations.

Chapter 7

Positive Matrix Semigroups and Applications

Now we investigate positive one-parameter matrix semigroups, or, using a more common name, positive matrix exponentials. As expected, positivity and irreducibility in this case also lead to remarkable spectral and asymptotic properties.

Some applications of the theory are also presented to emphasize the importance of the subject.

7.1 Positive Semigroups

In this section we combine the matrix exponential from Chapter 4 with the positivity from Chapter 5. More precisely, we consider *positive matrix semigroups* $(e^{tA})_{t\geq 0}$, i.e., we assume that each e^{tA}, $t \geq 0$, is a positive matrix. As a first step, we characterize this property by the entries of A. In particular, we show that positivity of A is sufficient, but not necessary for this.

Let $A = (\alpha_{ij})$ be given. By Theorem 4.2,

$$A = \lim_{t \downarrow 0} \frac{e^{tA} - I}{t},$$

which can be rewritten coordinatewise as

$$\alpha_{ij} = \lim_{t \downarrow 0} \left\langle \frac{e^{tA} u_j - u_j}{t}, u_i \right\rangle, \tag{7.1}$$

for $i, j = 1, \ldots, n$ and u_i the ith standard unit vector in $\mathbb{C}^n$. If we denote the (i, j)th entry of e^{tA} by $\tau_{ij}(t)$, then (7.1) implies that

$$\alpha_{ij} = \begin{cases} \lim_{t \downarrow 0} \frac{\tau_{ij}(t)}{t} & \text{for } i \neq j, \\ \lim_{t \downarrow 0} \frac{\tau_{ii}(t) - 1}{t} & \text{for } i = j. \end{cases} \tag{7.2}$$

If $(e^{tA})_{t \geq 0}$ is positive, i.e., $\tau_{ij}(t) \geq 0$ for all t, i and j, then

$$\alpha_{ij} \geq 0 \quad \text{for } i \neq j, \text{ and}$$
$$\alpha_{ii} \in \mathbb{R} \quad \text{for } i = j.$$

We call such matrices *positive off-diagonal*. Thus, we have shown the necessity part of the following characterization.

Theorem 7.1. *The matrix* $A = (\alpha_{ij}) \in \mathcal{L}(X)$ *generates a positive semigroup if and only if it is real and positive off-diagonal.*

Proof. It remains to show the sufficiency of the condition.

Since A is real and positive off-diagonal, we can find $\rho \in \mathbb{R}$ such that

$$B_\rho := A + \rho I \geq 0 \tag{7.3}$$

(e.g., take $\rho := \max_{1 \leq i \leq n} |\alpha_{ii}|$). Note that also $e^{tB_\rho} \geq 0$ for all $t \geq 0$. Applying the functional calculus introduced in Section 2.2 to the function $f(\lambda) := e^{t\lambda - t\rho}$, we obtain

$$e^{tA} = e^{[t(A + \rho I) - t\rho I]}$$
$$= f(B_\rho)$$
$$= e^{-t\rho} \cdot e^{tB_\rho} \geq 0$$

for all $t \geq 0$. $\qquad\square$

Let us mention another terminology here. A real and positive off-diagonal matrix A is also called a *Metzler matrix* and $-A$ is called a *Z-matrix*.

Since $e^{tA} \geq 0$ does not imply $A \geq 0$, Perron's theorem (see Theorem 5.6) is not directly applicable and $r(A)$ may not be an eigenvalue of A. However, considering the positive matrix B_ρ defined in (7.3) we obtain an important property of the spectral bound $s(A)$ of A.

Theorem 7.2. *If A generates a positive semigroup* $(e^{tA})_{t \geq 0}$, *then* $s(A)$ *is a strictly dominant eigenvalue in the lateral sense, i.e.,* $s(A) \in \sigma(A)$ *and*

$$\text{Re } \lambda < s(A)$$

for all other eigenvalues λ *of A.*

Proof. As already noticed in the proof of Theorem 7.1, $B_\rho := A + \rho I \geq 0$ for $\rho := \max_{1 \leq i \leq n} |\alpha_{ii}|$. Perron's theorem (see Theorem 5.6) yields $r(B_\rho) \in \sigma(B_\rho)$. Evidently, $r(B_\rho) = s(B_\rho)$, which is strictly dominant in $\sigma(B_\rho)$. Since

$$\sigma(B_\rho) = \sigma(A + \rho I) = \sigma(A) + \rho,$$

and thus $s(B_\rho) = s(A + \rho I) = s(A) + \rho$, we obtain that $s(A) \in \sigma(A)$ and is strictly dominant. $\qquad\square$

We state here another auxiliary result.

Lemma 7.3. *Supppose A generates a positive semigroup $(e^{tA})_{t\geq 0}$ and $\mu \in \rho(A)$. Then*
$$R(\mu, A) \geq 0 \iff \mu > s(A).$$

Proof. Again take $\rho \geq 0$ such that $B_\rho := A + \rho I \geq 0$. By Perron's theorem (see Theorem 5.6), $r(B_\rho) = s(B_\rho)$, and as in the proof of Theorem 7.2 we see that $s(B_\rho) = s(A) + \rho$. By Proposition 5.5 and Corollary 5.7 we also know that
$$R(\lambda, B_\rho) \geq 0 \iff \lambda > r(B_\rho) = s(A) + \rho,$$
which, by taking $\mu := \lambda - \rho$, yields $R(\mu, A) \geq 0 \iff \mu > s(A)$. $\square$

As we have seen in Theorem 4.12, $s(A)$ determines the asymptotic behavior of e^{tA} as $t \to \infty$. The case $s(A) < 0$, yielding stability of the semigroup, is of particular importance. In the case of positive semigroups, we thus obtain the following characterization of stability.

Corollary 7.4. *If A generates a positive semigroup $(e^{tA})_{t\geq 0}$, then the following assertions are equivalent.*

(i) $s(A) < 0$.
(ii) *The characteristic polynomial of A has no real root ≥ 0.*
(iii) *The matrix $-A^{-1}$ exists and is positive.*
(iv) *There exists $x \geq 0$ such that $Ax = -\mathbb{1}$.*
(v) *The semigroup $(e^{tA})_{t\geq 0}$ is exponentially stable.*

Proof. The equivalence of (v) and (i) follows directly from Theorem 4.12 and the equivalence (i) $\iff$ (ii) from Theorem 7.2. Since $-A^{-1} = R(0, A)$, we have by Lemma 7.3 the equivalence (iii) $\iff$ (i). Next, (iii) $\implies$ (iv) follows by taking $x := -A^{-1}\mathbb{1}$.

We close the implications loop by showing (iv) $\implies$ (i). If $Ax = -\mathbb{1}$, then
$$x \geq x - e^{tA}x = -A \int_0^t e^{sA}x \, ds = \int_0^t e^{sA}\mathbb{1} \, ds \quad \text{for all } t > 0.$$

Since e^{tA} is positive for all $t \geq 0$, we infer that the function $t \mapsto (\int_0^t e^{sA}\mathbb{1} \, ds)$ is increasing and satisfies
$$0 \leq \int_0^t e^{sA}\mathbb{1} \, ds \leq x.$$

Hence, $\int_0^\infty e^{sA}\mathbb{1} \, ds$ exists and since
$$A \int_0^t e^{sA}\mathbb{1} \, ds = e^{tA}\mathbb{1} - \mathbb{1},$$

we obtain the existence of $\lim_{t\to+\infty} e^{tA}\mathbb{1}$. This implies $r(e^{tA}) \leq 1$, since $\|e^{tA}\|_\infty = \|e^{tA}\mathbb{1}\|_\infty$. Thus $s(A) \leq 0$. Take now $\varepsilon > 0$. Then, by Lemma 7.3,

$$0 \leq R(\varepsilon, A)\mathbb{1} = -R(\varepsilon, A)Ax$$
$$= x - \varepsilon R(\varepsilon, A)x \leq x.$$

So, we have the existence of $\lim_{\varepsilon\to 0} R(\varepsilon, A)\mathbb{1}$ since the function $\varepsilon \to R(\varepsilon, A)$ is decreasing. Therefore,

$$\lim_{\varepsilon\to 0} \|R(\varepsilon, A)\|_\infty$$

exists, and hence $s(A) < 0$.   □

If $-A$ has the above properties, A is also called a *nonsingular M-matrix*.

In the case $s(A) = 0$ the following is a consequence of Theorem 4.12.

Corollary 7.5. *Let A generate a positive semigroup $(e^{tA})_{t\geq 0}$ and assume $s(A) = 0$. Then $\lim_{t\to\infty} e^{tA}$ exists if and only if 0 is a first-order pole of the resolvent $R(\cdot, A)$. In this case, $\lim_{t\to\infty} e^{tA}$ is the spectral projection of A belonging to 0, and its range is the kernel of A.*

Again, it is important to assure that $s(A)$ is a simple pole. As shown in Theorem 5.13, irreducibility can help for this purpose.

Theorem 7.6. *Let A generate a positive semigroup $(e^{tA})_{t\geq 0}$. Then any of the following conditions implies*

$$\lim_{t\to\infty} e^{tA} = P_1$$

for P_1 a projection of the form

$$P_1 x = \langle x, y\rangle z, \quad x \in \mathbb{C}^n,$$

with strictly positive vectors $y \gg 0$ and $z \gg 0$ such that $\langle z, y\rangle = 1$.

a) *A is irreducible with $s(A) = 0$.*
b) *A is irreducible, $(e^{tA})_{t\geq 0}$ is bounded, and $0 \in \sigma(A)$.*
c) *$e^{t_0 A}$ is irreducible for some $t_0 > 0$ and $s(A) = 0$.*
d) *$e^{t_0 A}$ is irreducible for some $t_0 > 0$, $(e^{tA})_{t\geq 0}$ is bounded, and $0 \in \sigma(A)$.*

Proof. First note that by formula (2.9), the invariant subspaces of A and e^{tA} coincide. Hence, irreducibility of A is equivalent to irreducibility of e^{tA} for some/all $t > 0$.

Next, if A is irreducible, then $B_\rho = A + \rho I$ given by (7.3) is positive and irreducible, hence by Theorem 5.13 its spectral radius $r(B_\rho)$ is a first-order pole of the resolvent $R(\cdot, B_\rho)$. Therefore, also $s(A)$ is a first-order pole of the resolvent $R(\cdot, A)$.

All assertions now follow by Corollary 7.5 and Theorem 4.12.b). The formula for $P_1 x$ can be verified as in the proof of Corollary 5.16.   □

7.2   The Competitive Market Model

As a first application let us revise the competitive market model presented in Chapter 1. Recall that the dynamics of the prices $p(t)$ in this model is given by

$$p(t) = p^0 + e^{tKA}c, \quad t \geq 0, \quad \text{where} \quad c = p(0) - p^0. \tag{7.4}$$

Here p^0 are equilibrium prices, $p(0)$ initial prices, $K = \text{diag}(k_1, \ldots, k_n)$ a diagonal matrix of positive adjustment speeds. The coefficients of the matrix $A = (a_{ij})$ satisfy

$$a_{ij} \geq 0 \text{ for } i \neq j \text{ and } a_{ii} < 0.$$

Using the theory we developed so far we are able to study the behavior of the prices depending on spectral properties of the matrix KA.

Let us first determine in which case the prices eventually return to the equilibrium p^0. Assuming $p(0) \neq p^0$, by Theorem 4.12 this happens if and only if $s(KA) < 0$. Moreover, since KA is real and positive off-diagonal, it generates a positive semigroup and $s(KA)$ is the largest real eigenvalue of KA (cf. Theorems 7.1 and 7.2). Furthermore, by Corollary 7.4 we have

$$
\begin{aligned}
s(KA) < 0 \quad &\Longleftrightarrow \quad (KA)^{-1} \leq 0 \\
&\Longleftrightarrow \quad A^{-1} \leq 0 \\
&\Longleftrightarrow \quad s(A) < 0 \\
&\Longleftrightarrow \quad \text{there exists } x \geq 0 \text{ such that } Ax = -\mathbb{1}.
\end{aligned}
$$

Hence automatic return to the equilibrium p^0 can be checked by determining A^{-1} (if it exists; one positive entry in A^{-1} implies $s(KA) \geq 0$), or by solving the equation $Ax = -\mathbb{1}$.

If $s(A) > 0$, the prices will unboundedly rise. For $s(A) = 0$ we have two possibilities: the prices may either converge to a new equilibrium or rise unboundedly.

Let us consider the case when A is strictly positive off-diagonal and $s(A) = 0$. By Theorem 7.6 (see also Exercise 5.5.3) in this case e^{tKA} converges to a projection of the form $P = u \otimes v$ with $u, v \gg 0$. Hence $p(t)$ converges, as $t \to \infty$, to $p^0 + d$ where

$$d = \lim_{t \to \infty} e^{tKA}c = \langle c, u \rangle v.$$

This produces the strange effect that for $\langle c, u \rangle > 0$ a new equilibrium $\tilde{p}^0 = p^0 + d$ develops with $d \gg 0$. On the other hand, $\langle c, u \rangle < 0$ produces a new equilibrium with $d \ll 0$.

7.3   Queueing Models

Systems, in which an operation on some objects or individuals is performed, are frequently encountered in applications. Such systems are generally composed of two parts, the queue line and the service:

Figure 7.1: Structure of a queueing system

An airport with a queue of airplanes waiting for landing, an office where the papers for driving licence renewal are processed, or the waiting room of a service center (telephone, gas, electricity, etc.) are typical examples characterized by a waiting time followed by a service. Such systems can be modeled when the statistics of the arrivals, the rule for selecting the next user, and the statistics of the service times are specified.

We will assume that the arrivals and the departures are random processes characterized by the property that the probability of one arrival or departure during a small time Δt is proportional to Δt itself, meaning that they are Poisson processes. The proportionality coefficients, denoted by η and μ, respectively, depend on the total number of people in the system.

To build the equations governing the system, we make the following assumption. At time $t + \Delta t$ there are no users in the system if one of the following happens: either there are no users in the system at time t and no user arrives during this time interval, or there is one user in the system at time t, which leaves the system, and no other user arrives. We can derive the equations

$$y_0(t + \Delta t) = y_0(t)(1 - \eta_0 \Delta t) + y_1(t)\mu_1 \Delta t (1 - \eta_1 \Delta t),$$

where $y_0(t)$ is the probability that no user is in the system at time t, and $y_1(t)$ is the probability that exactly one user is in the system at time t.

Taking the limit as $\Delta t \to 0$, we arrive at the equation

$$\dot{y}_0 = -\eta_0 y_0(t) + \mu_1 y_1(t), \tag{7.5}$$

which is the first state equation of the system. With a similar reasoning, we obtain for $i > 0$ that

$$\dot{y}_i(t) = \eta_{i-1} y_{i-1}(t) - (\eta_i + \mu_i) y_i(t) + \mu_{i+1} y_{i+1}(t), \tag{7.6}$$

where $y_i(t)$ is the probability that i users are in the system at time t. This means that there are four possibilities for the system to change into the state i:

- there were $i - 1$ users and someone arrived;
- there were i users and someone arrived;
- there were i users and someone left;
- there were $i + 1$ users and someone left.

Such problems are naturally modeled by infinite-dimensional systems. In many applications, however, we have a natural bound on the number of possible users entering the system on the whole, hence we have a finite-dimensional system, and this is the case we are investigating now. Assuming that all the proportionality constants η_j and μ_j are non-zero, this leads to a system of differential equations

$$\dot{y}(t) = Ay(t), \tag{7.7}$$

where $y = (y_0, y_1, \ldots, y_n)^\top \in \mathbb{R}^{n+1}$ and

$$A = \begin{pmatrix} -\eta_0 & \mu_1 & 0 & 0 & \ldots & 0 & 0 \\ \eta_0 & -\eta_1 - \mu_1 & \mu_2 & 0 & \ldots & 0 & 0 \\ 0 & \eta_1 & -\eta_2 - \mu_2 & \mu_3 & \ldots & 0 & 0 \\ \vdots & \vdots & \vdots & \vdots & & \vdots & \vdots \\ 0 & 0 & 0 & 0 & \ldots & \eta_{n-1} & -\mu_n \end{pmatrix}. \tag{7.8}$$

The matrix A is a so-called *band matrix*, has negative elements on the diagonal, positive elements below and above the diagonal, and zeros elsewhere. We infer that the matrix is irreducible and hence the positive and irreducible matrix $B_\rho = A + \rho I$ defined in (7.3) has a unique normalized strictly positive Perron eigenvector x_P. Note that x_P is also an eigenvector of A with the corresponding strictly dominant eigenvalue $\lambda_1 \in \mathbb{R}$. Moreover, since all the columns of A have a zero sum, $\lambda_1 = 0$. Since all the eigenvalues of A, except λ_1, have negative real part, Theorem 7.6 yields

$$e^{tA} \longrightarrow P_1 \qquad \text{as } t \longrightarrow \infty.$$

Hence, the system will always converge to a stationary probability distribution given by the Perron eigenvector x_P, and the expected value of the clients in the queue (and the expected waiting time) can be estimated using this on the long run. This asymptotic distribution can be easily evaluated because A is a tridiagonal matrix. In fact, we can see that

$$\bar{x}_{i+1} = \frac{\eta_i}{\mu_{i+1}} \bar{x}_i,$$

where $x_P = (\bar{x}_0, \bar{x}_1, \ldots, \bar{x}_n)^\top$. Hence,

$$\bar{x}_i = \varphi_i \bar{x}_0,$$

where

$$\varphi_i = \frac{\eta_0 \eta_1 \cdots \eta_{i-1}}{\mu_1 \mu_2 \cdots \mu_i}.$$

Since we normalize x_P to represent a probability distribution, its coordinates sum up to 1, hence

$$\bar{x}_0 = \frac{1}{1 + \sum_{i=1}^n \varphi_i}.$$

We also assume that there are s servers and that c people can get in the queue, i.e., $n = c + s$. Then the average number of people in the system is

$$\bar{n} = \sum_{i=0}^{n} i\bar{x}_i,$$

and we have

$$\bar{c} = \sum_{i=s+1}^{n} (i - s)\bar{x}_i$$

persons waiting in the queue.

7.4   Disease Transition Models

The spread and persistence of infectious diseases is a result of the complex interaction between individual epidemiological units (e.g., individual, city, county, etc.), disease characteristics, and various control programs that are aimed at halting disease transmission or bringing infection prevalence to a level as low as possible. The aim of many models is to gain insight into how diseases transmit and to identify the most effective strategies for prevention and control. The early work by Kermack and McKendrick from 1927 provides the basis of differential-equation-based models which lie at the heart of modern quantitative epidemiology. Traditionally, mathematical epidemiology is based on differential equation models and these operate on the basis of some strong simplifying assumptions about the behaviour of the individuals and the biology of the disease. A key component in any disease transmission model is the population contact structure, and in most cases, this is highly heterogeneous with strong correlations and nontrivial large scale structure.

We consider here a really elementary, so-called "SIS" model, where individuals can have two possible states: susceptible (but healthy), or infected. Infected individuals can infect susceptible ones at a certain rate, and infected individuals recover at a certain rate, and can be infected afterwards again. Usually there are natural time delays (like incubation) and non-linear dependencies in the model, or other stages (like immunized individuals), but we consider here a rather simple case.

Assume that there are n individuals in the model, and that there is a graph describing the connections between them where the infection can spread. Usually the graph is huge and some random graph models have to be used, but let us restrict ourselves to the case where we have a rather small group of individuals with a clear social network. Specifically, the assumptions on the model are the following.

Suppose the individuals are connected by a weighted undirected graph, with adjacency matrix $G = (w_{ij})$, where the weights $0 \le w_{ij} \le 1$ describe the strength of the connection. Each individual can recover at a rate μ and infect a connected person at a rate η. Denoting with $y_i(t)$ the probability of the ith individual to be infected, a simple model for the change of this probability after a small time Δt is

$$y_i(t + \Delta t) = (1 - \mu \Delta t)y_i(t) + \eta \Delta t \sum_{j=1}^{n} w_{ij} y_j(t).$$

This leads to the differential equation

$$\dot{y}(t) = (\eta G - \mu I)y(t). \tag{7.9}$$

It is usual to assume that the graph is connected and has no loops. Hence, the matrix $A := \eta G - \mu I$ is irreducible. Theorem 7.6 is applicable if $\mu = \eta s(G)$.

Convergence to zero here is implied by the condition $s(G) < \frac{\mu}{\eta}$, and means that everyone recovers from the illness eventually, and convergence to a projection means that there is a stationary distribution of probabilities and the system tries to achieve it.

7.5 Discrete Maximum Principles

Maximum principles play an important part in many mathematical subdisciplines and it is known that there is a deep connection between maximum principles and positive semigroups. Discrete maximum principles are particularly useful in numerical analysis: when you discretize a differential operator, you not only want to obtain some good approximation result, but it is also important to preserve some qualitative properties of the underlying differential equation. Hence, if you discretize an elliptic problem, it is good to know whether the discretization also satisfies some kind of a maximum principle. The literature is vast even in the matrix case and we only mention here one illustrative example.

Let $A = (a_{ij})$ be a real $n \times n$ matrix and for a vector $y = (y_1, \ldots, y_n)^\top$ let us use the notation $N^+(y) = \{i \in \{1, \ldots, n\} : y_i > 0\}$. We say that the matrix A satisfies the *discrete maximum principle* (DMP) if

$$-Ax = y \tag{7.10}$$

with $y \geq 0$ implies $x \geq 0$ and

$$\max_{1 \leq i \leq n} x_i = \max_{i \in N^+(y)} x_i. \tag{7.11}$$

We present a simple sufficient condition to ensure the discrete maximum principle.

Theorem 7.7. *Suppose that A generates an exponentially stable positive semigroup and that $A\mathbb{1} \ll 0$, where $\mathbb{1} = (1, \ldots, 1)^\top$. Then the discrete maximum principle holds.*

Proof. Let $y > 0$ and $-Ax = y$. Then $x = -A^{-1}y > 0$ since $-A^{-1}$ exists and is positive by Corollary 7.4. Let $x_k = \max_{1 \leq i \leq n} x_i$. By assumption, we have $\sum_{l=1}^{n} a_{kl} = (A\mathbb{1})_k < 0$ and using Theorem 7.1 we obtain

$$y_k = -\sum_{l=1}^{n} a_{kl}x_l = -a_{kk}x_k - \sum_{l \neq k} a_{kl}x_l \geq -x_k \left(a_{kk} + \sum_{l \neq k} a_{kl} \right) \geq 0, \qquad (7.12)$$

hence $k \in N^+(y)$. □

7.6 Notes and Remarks

For nonsingular M-matrices we refer to the monograph by Berman and Plemmons [18]. For queueing systems see the excellent exposition in Feller [46, Section XVII.7]. For population equations see the monograph by Diekmann and Heesterbeek [31], where many of the problems we omitted here are discussed.

The investigation of discrete maximum principles was initiated in the late sixties and early seventies, see Varga [146] and Ciarlet [24]. We follow here Stoyan [134], where a much more general statement is formulated. In the infinite-dimensional setting, we shall discuss the connection between analogous minimum principles and generation of positive semigroups. Here we recommend the paper by Kalauch [71] for a direct generalization to Banach lattices.

7.7 Exercises

1. Let $A \in \mathcal{L}(X)$. If $\operatorname{Re} \lambda > \mathrm{s}(A)$, then

$$R(\lambda, A) = \int_0^\infty \mathrm{e}^{-\lambda t} \mathrm{e}^{tA} \, \mathrm{d}t.$$

2. Let $A \in \mathcal{L}(X)$ generate a positive semigroup. Show that the following are equivalent.

    (i) A is irreducible.

    (ii) $\mathrm{e}^{t_0 A}$ is irreducible for some/all $t_0 > 0$.

    (iii) $\mathrm{e}^{tA} \gg 0$ for all $t > 0$.

3. Let A be positive off-diagonal and irreducible, $\mathrm{s}(A) = 0$. Show that e^{tA} converges to a projection P of the form

$$P = u \otimes v$$

    with $u, v \gg 0$, $u \in \ker A^\top$, $v \in \ker A$.

4. a) Characterize those matrices $A \in \mathcal{L}(X)$ for which e^{tA} is positive for all $t \in \mathbb{R}$.

b) Find all positive periodic semigroups, and all positive, periodic, and irreducible semigroups.

5. Consider the Competitive Market Model given by equation (7.4).

   a) List the conditions for the matrix A under which the prices will behave in a periodic manner.

   b) The price average at time $T > 0$ can be expressed by the Cesàro mean

   $$C(T) := \frac{1}{T} \int_0^T e^{tKA} \, dt.$$

   What can we say about the long-time behavior of the price averages depending on the properties of matrix A?

6. Consider the queuing system and give formulas for the average frequency of arrivals, the fraction of time during which the system is not used, and the probability that a user cannot be served on arrival.

7. Consider a telephone system of a large company where calls are accepted if at least one line is free and rejected otherwise. Thus, the frequency of arrivals η is independent of the number of busy lines if at least one is free. Each accepted call engages the line for an average time $1/\mu$. Hence, $c = 0$, $s = n$, $\eta_i = \eta$ for $i = 0, 1, \ldots, n-1$, $\eta_n = 0$, and $\mu_i = i\mu$ for $i = 1, 2, \ldots, n$. Develop formulas for this system. Discuss some concrete examples.

8. Assume that we have a group of individuals who are arranged in the vertices of a graph and the disease can spread along the edges. Each individual can recover from the illness with a rate of $\mu = 1/4$. Discuss the role of the parameter η if the graph is

   a) a complete graph with 4 vertices,

   b) a cycle of length 5 (regular pentagon),

   c) a cube (8 vertices),

   d) the graph from Exercise 1.4.2.

9. It is also possible to speak about maximum principles in connection with difference equations and boundary value problems. Let $a_i, b_i > 0$ and $c_i \geq a_i + b_i$ for $i = 0, 1, \ldots, n+1$. Define the difference operator

   $$Ly_i = a_i y_{i-1} - c_i y_i + b_i y_{i+1} \quad \text{for } i = 1, \ldots, n.$$

   Let $\alpha, \beta \geq 0$ and consider the vector $0 \leq f \in \mathbb{R}^n$. Let $y \in \mathbb{R}^{n+2}$ be such that

   $$Ly_i = f_i \quad \text{for } i = 1, \ldots, n, \quad y_0 = \alpha, \ y_{n+1} = \beta$$

   holds, which we call a non-homogeneous boundary value problem. Show that

   $$\max_{i=0,\ldots,n+1} y_i = \max\{\alpha, \beta\}.$$

Chapter 8

Positive Linear Systems

We present one important large field of applications to the theory developed so far: control theory. More specifically, we present an elementary introduction to positive linear systems.

We cover some very special aspects of linear time-invariant systems, like controllability or stabilizability. Many of these problems can be naturally posed with additional positivity assumptions: we have a positive system, we would like to apply positive controls, or we would like to steer our system into positive states.

We discuss only continuous-time systems, but the definitions and most results can be modified for the discrete-time case in a straightforward way.

8.1 Externally and Internally Positive Systems

First, we set the stage and present the relevant notation and terminology. For the sake of simplicity, we only refer to the case of time-invariant, finite-dimensional input-output systems, which are described by state equations of the form

$$\begin{cases} \dot{x}(t) = Ax(t) + Bu(t), \\ x(0) = x_0, \\ y(t) = Cx(t), \end{cases} \tag{8.1}$$

where the objects involved are the following:

- $X = \mathbb{C}^n$ is the *state space*, $Y = \mathbb{C}^q$ is the *observability space*, and $U = \mathbb{C}^p$ is the *control space*.
- The function $x : \mathbb{R}^+ \to X$ is the *state vector*, the operator $A \in \mathcal{L}(X)$ is the *state (or system) operator*.
- The function $u : \mathbb{R}^+ \to U$ is the *control*, the operator $B \in \mathcal{L}(U, X)$ is the *input (or control) operator*.
- The function $y : \mathbb{R}^+ \to Y$ is the *output (or observation)*, the operator $C \in \mathcal{L}(X, Y)$ is the *output (or observation) operator*.
- The vector $x_0 \in X$ is the *initial value*.

System (8.1) is often referred to as $\Sigma(A, B, C)$. The interpretation of this set
of equations is the following. There is a system described by a set of n equations
and governed by the operator A. This is also referred to as the "free system",
the system without intervention. The function u is the control we apply from the
outside, and the operator B represents the action of u on the system. Finally, the
function y is the set of parameters we are able to measure, and the measurement
process is described by the observation operator C.

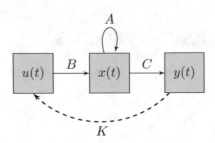

Figure 8.1: An input-output system (with feedback).

If we need to stress the dependence of the solution x on the initial value x_0,
then we shall write $x(t) = x(t; x_0)$.

Before turning our attention to controllability concepts, let us make the following crucial observation and present a representation formula. Suppose that the
control u is locally integrable and set

$$z(t) = e^{tA}x_0 + \int_0^t e^{(t-s)A}Bu(s)\,\mathrm{d}s.$$

Then, $\dot{z}(t) = Az(t) + Bu(t)$ and $z(0) = x_0$. Since x is the solution of $\Sigma(A, B, C)$
in (8.1), we infer that $\dot{z}(t) - \dot{x}(t) = A(z(t) - x(t))$ and $z(0) - x(0) = 0$. Hence, by
uniqueness,

$$x(t) = z(t) = e^{tA}x_0 + \int_0^t e^{(t-s)A}Bu(s)\,\mathrm{d}s. \tag{8.2}$$

This yields the formula

$$y(t) = Ce^{tA}x_0 + \int_0^t Ce^{(t-s)A}Bu(s)\,\mathrm{d}s. \tag{8.3}$$

The function $h(t) = Ce^{tA}B$ is sometimes called the *impulse response*.

Often the control u is designed depending on the observation, and such systems are called *feedback systems*. If $u(t) = Ky(t)$, then the operator $K \in \mathcal{L}(Y, U)$
is called the feedback operator. Note that in this case we have the representation
formula

$$x(t) = e^{t(A+BKC)}x_0. \tag{8.4}$$

We list now a few properties that a time-invariant linear system can have, and which are important in view of applications.

Definition 8.1. The linear system $\Sigma(A, B, C)$ in (8.1) is said to be *externally positive*, if the output corresponding to the zero initial state is positive for every positive input function. In other words, $u(t) \geq 0$ implies $y(t) \geq 0$ if $x_0 = 0$.

In the following we characterize externally positive linear systems.

Proposition 8.2. *A linear system is externally positive if and only if its impulse response is positive.*

Proof. By the representation formula (8.3), the sufficiency is clear. Suppose now that there is $t_0 > 0$ such that $h(t_0) = Ce^{t_0 A}B$ is not positive. Then, by continuity, at least one entry of $h(t)$ would be negative on a whole nondegenerate interval $[t_1, t_2]$. Thus, the appropriate entry of the output would be negative for every input function which is strictly positive in $[t - t_2, t - t_1]$ and zero elsewhere. Hence, the system cannot be externally positive. $\qquad\square$

Let us give a simple example of an externally positive linear system.

Example 8.3. Here we suppose that $U = Y = \mathbb{C}$ and $X = \mathbb{C}^2$. Let us consider

$$A = \begin{pmatrix} -a & -a \\ 1 & -1 \end{pmatrix}, \quad B = \begin{pmatrix} a \\ 0 \end{pmatrix}, \quad C = \begin{pmatrix} 0 & 1 \end{pmatrix},$$

for a parameter $a > 0$. For which values of a will the system $\Sigma(A, B, C)$ in (8.1) be externally positive?

Using the above proposition one has to check for which values of a the corresponding impulse response $h(t) = Ce^{tA}B$ is positive. The special forms of B and C imply that $h(t) = a(e^{tA})_{2,1}$, where $(e^{tA})_{2,1}$ is the $(2,1)$th entry of e^{tA}. The eigenvalues λ_i, $i = 1, 2$, of A are the roots of $\lambda^2 + (a + 1)\lambda + 2a$. So, the following holds:

$$\lambda_{1,2} = \begin{cases} \frac{-1-a\pm\sqrt{a^2-6a+1}}{2} & \text{if } 0 < a < 3 - 2\sqrt{2} \text{ or } a > 3 + 2\sqrt{2}, \\ -\frac{1+a}{2} =: \lambda_0 & \text{if } a = 3 - 2\sqrt{2} \text{ or } a = 3 + 2\sqrt{2}, \\ \in \mathbb{C} \setminus \mathbb{R} & \text{if } 3 - 2\sqrt{2} < a < 3 + 2\sqrt{2}. \end{cases}$$

We have to investigate only the first two cases, since the third one corresponds to an oscillating e^{tA}, which can have negative values. From the Theorem 2.11 one has $e^{tA} = \alpha_0 I + \alpha_1 A$, where α_0 and α_1 satisfy the system of equations

$$e^{t\lambda_1} = \alpha_0 + \alpha_1\lambda_1,$$
$$e^{t\lambda_2} = \alpha_0 + \alpha_1\lambda_2$$

in the first case above, and

$$e^{t\lambda_0} = \alpha_0 + \alpha_1\lambda_0,$$
$$te^{t\lambda_0} = \alpha_1$$

in the second one. Thus, by a simple computation, we obtain

$$h(t) = \begin{cases} \dfrac{a}{\sqrt{a^2-6a+1}} e^{t\lambda_2} \left(e^{2t\sqrt{a^2-6a+1}} - 1 \right) & \text{if } 0 < a < 3 - 2\sqrt{2} \text{ or } a > 3 + 2\sqrt{2}, \\ ate^{t\lambda_0} & \text{if } a = 3 - 2\sqrt{2} \text{ or } a = 3 + 2\sqrt{2}. \end{cases}$$

Therefore, for $a > 0$, the system $\Sigma(A, B, C)$ is externally positive, iff $0 < a \le 3 - 2\sqrt{2}$ or $a \ge 3 + 2\sqrt{2}$.

The concept of (internal) positivity and irreducibility is defined as follows.

Definition 8.4. The linear system $\Sigma(A, B, C)$ in (8.1) is said to be *positive* (or internally positive), if the state and the output corresponding to a positive initial state are positive for every positive input function. In other words, $u(t) \ge 0$ and $x_0 \ge 0$ implies $x(t) \ge 0$ and $y(t) \ge 0$. The system is said to be *reducible*, if the matrix A is reducible, and *irreducible* otherwise.

The positivity of linear systems can be characterized in terms of the positivity of B, C, and e^{tA}.

Proposition 8.5. *The linear system $\Sigma(A, B, C)$ is positive if and only if $B \ge 0$, $C \ge 0$, and A generates a positive matrix semigroup.*

Proof. Assume that the system is positive. Then, letting $x_0 = 0$ and $u(t) = u$, a nonnegative constant, we see that

$$0 \le \frac{x(t)}{t} = \frac{1}{t} \int_0^t e^{(t-s)A} Bu \; ds = \left(\frac{1}{t} \int_0^t e^{sA} \; ds \right) Bu.$$

So, by letting $t \to 0$, we obtain the positivity of B.

Since $Cx_0 = Cx(0) = y(0) \ge 0$ for every $x_0 \ge 0$, the operator C has to be positive too. Finally, applying the zero control, we see that $x(t) = e^{tA}x_0 \ge 0$ for every $x_0 \ge 0$.

To prove the converse implication, suppose that $B \ge 0$, $C \ge 0$, and that A generates a positive matrix semigroup. Taking $x_0 \ge 0$ and $u \ge 0$, we see that

$$e^{tA}x_0 \ge 0$$

and that $Bu(s) \ge 0$ for each $s \in [0, t]$, hence $e^{(t-s)A}Bu(s) \ge 0$, implying

$$\int_0^t e^{(t-s)A} Bu(s) \; ds \ge 0,$$

which in view of (8.2) and (8.3) proves the statement. □

Let us define now excitable positive linear systems.

Definition 8.6. A positive system is said to be *excitable*, if each state variable can be made strictly positive by applying an appropriate positive input to the system initially at rest. In other words, for each $i = 1, 2, \ldots, n$ there are a control $u_i \ge 0$ and a time t_i such that $x_i(t_i) > 0$ if $x_0 = 0$.

Excitable systems enjoy some remarkable properties. To be able to present some of them, we introduce some new concepts. To keep the presentation as simple as possible, we restrict ourselves for the rest of this section to the case $Y = U = \mathbb{C}$, i.e., we only consider one-dimensional control and observation spaces.

The *influence graph* of the system $\Sigma(A, B, C)$ in (8.1) is a directed graph $G = (V, E)$ with $n+2$ vertices $V = \{v_0, v_1, \ldots, v_{n+1}\}$. Vertex v_0 is associated with the input u and vertex v_{n+1} with the output y. The remaining vertices $v_1, \ldots, v_n$, correspond to the state variables $x_1, \ldots, x_n$. The edges represent the influence relations among the variables and are constructed as follows.

- $(v_0, v_j) \in E$ if and only if $b_j \neq 0$, $j = 1, \ldots, n$;
- for $i, j = 1, \ldots, n$, $i \neq j$, $(v_i, v_j) \in E$ if and only if $a_{ji} \neq 0$;
- $(v_i, v_{n+1}) \in E$ if and only if $c_i \neq 0$, $i = 1, \ldots, n$.

No other edges are present in the graph.

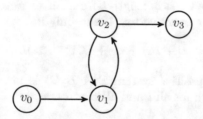

Figure 8.2: The influence graph of the system in Example 8.3.

The corresponding graph matrices are constructed as follows: $\hat{A} = (\hat{a}_{ij})$, where $\hat{a}_{ij} = 1$ if and only if $i \neq j$ and $a_{ji} \neq 0$, otherwise $\hat{a}_{ij} = 0$. The row- and column- matrix $\hat{B}$ and $\hat{C}$, respectively, are constructed in a similar manner. The following $(n+2) \times (n+2)$ matrix

$$\mathbb{A} := \begin{pmatrix} 0 & \hat{B} & 0 \\ 0 & \hat{A} & \hat{C} \\ 0 & 0 & 0 \end{pmatrix}. \tag{8.5}$$

is thus the $0 - 1$ adjacency matrix of the unweighted influence graph.

Many properties of a positive linear system can be described in terms of its influence graph. Observe, for example, that by Proposition 6.1 the following holds.

Corollary 8.7. *A system is irreducible if and only if the subgraph of its influence graph, consisting only of vertices $v_1, \ldots, v_n$ and edges between them, is strongly connected.*

We can also express excitability of the system in terms of the above graph matrices.

Proposition 8.8. *The positive linear system $\Sigma(A, B, C)$ in (8.1) is excitable if and only if there exists at least one walk from the input vertex v_0 to each vertex v_i, $i = 1, \ldots, n$, in the influence graph G, or, equivalently, if and only if*

$$\hat{B} + \hat{B}\hat{A} + \cdots + \hat{B}\hat{A}^{n-1} \gg 0.$$

Proof. Excitability means that each state variable x_i can be influenced by the input u. This implies that there has to be at least one walk from the vertex v_0 to the vertex v_i in the influence graph G.

Note that the powers of the adjacency matrix $\mathbb{A}$ in (8.5) have the same block form:

$$\mathbb{A}^k = \begin{pmatrix} 0 & \hat{B}\hat{A}^{k-1} & \hat{B}\hat{A}^{k-2}\hat{C} \\ 0 & \hat{A}^k & \hat{A}^{k-1}\hat{C} \\ 0 & 0 & 0 \end{pmatrix}, \quad k \in \mathbb{N}.$$

Recall from Proposition 1.1 that the ith component of the row vector $\hat{B}\hat{A}^{k-1}$ represents the number of walks of length k from vertex v_0 to vertex v_i, $i = 1, \ldots, n$. Hence, there is a walk to every vertex, if and only if

$$\hat{B} + \hat{B}\hat{A} + \cdots + \hat{B}\hat{A}^{n-1} \gg 0.$$

Assume that the positive system $\Sigma(A, B, C)$ is not excitable. Then, there exists $i \in \{1, \ldots, n\}$ such for all t and all controls $u \geq 0$,

$$x_i(t) = \left(\int_0^t e^{(t-s)A} Bu(s) \ ds \right)_i = 0.$$

By taking $u(t) = 1$, we obtain

$$b_i = \lim_{t \to 0} \left(\frac{1}{t} \int_0^t e^{sA} \ ds B \right)_i = 0.$$

On the other hand,

$$\dot{x}_i(t) = \left(Bu(t) + A \int_0^t e^{(t-s)A} Bu(s) \ ds \right)_i = 0.$$

This implies that

$$\left(A \int_0^t e^{(t-s)A} Bu(s) \ ds \right)_i = 0.$$

As above, by taking $u(t) = 1$, we obtain $(AB)_i = \sum_{j=1}^n a_{ij} b_j = 0$. Using the positivity, we deduce that $a_{ij} b_j = 0$ for all j (note that, as we have seen above, $b_i = 0$, so also $a_{ii} b_i = 0$). Hence for the graph matrices we have $(\hat{B}\hat{A})_i = 0$. By repeating the same arguments we obtain $(\hat{B}\hat{A}^k)_i = 0$ for all $k = 0, \ldots, n-1$, and this ends the proof of the proposition. $\qquad\qquad\square$

We consider now rather special constant inputs, $u(t) = \bar{u} > 0$.

Theorem 8.9. *An excitable and asymptotically stable positive linear system has a strictly positive equilibrium state.*

Proof. Since by asymptotic stability all the eigenvalues of A have negative real part, A is invertible and $\bar{x} := -A^{-1}B\bar{u}$ is the unique equilibrium of the system, which is asymptotically stable. We only have to show that it is strictly positive.

Suppose that there are indices i such that $\bar{x}_i = 0$, and collect these indices in the set $I := \{i \in \{1, \ldots, n\} : \bar{x}_i = 0\}$. Then, since

$$A\bar{x} + B\bar{u} = 0,$$

we see that

$$\sum_{j \notin I} a_{ij}\bar{x}_j + b_i\bar{u} = 0 \quad \text{for } i \in I.$$

This implies that $b_i = 0$ and that $a_{ij} = 0$ for $i \in I$ and $j \notin I$, since the system is positive. Hence, there is no walk from the input vertex 0 to vertices $i \in I$ and the system is not excitable. □

8.2 Controllability

For simplicity, we consider here systems without observation, i.e., where $Y = X$ and $C = I$. We denote by $\Sigma(A, B)$ the system (8.1) simplified this way.

Definition 8.10. The system $\Sigma(A, B)$ is called *controllable in time* τ if for every initial value $x_0 \in X$ and every state $x_1 \in X$ there is a control u such that for the solution x we have $x(\tau; x_0) = x_1$.

We will briefly call a system *controllable*, if there exists a $\tau > 0$ such that it is controllable in time τ.

Lemma 8.11. *The system $\Sigma(A, B)$ is controllable in time τ if and only if every state $x_1 \in X$ can be reached from $x_0 = 0$ in time τ.*

Proof. We only have to prove the converse. Let us take $x_1 \in X$ and set $x_2 = x_1 - e^{\tau A}x_0$. Then, by assumption, there is a control u such that $x_2 = x(\tau; 0)$. Then

$$x(\tau; 0) = \int_0^\tau e^{(\tau-s)A}Bu(s)\, ds = x_2,$$

hence

$$x(\tau; x_0) = e^{\tau A}x_0 + \int_0^\tau e^{(\tau-s)A}Bu(s)\, ds = e^{\tau A}x_0 + x_2 = x_1,$$

and we are done. □

To investigate the possible reachable states, we take a functional analytic point of view and introduce an operator which maps control functions to states which are reached from the origin by using this control.

Definition 8.12. Fix $\tau > 0$. The *controllability operator* $\mathcal{B}_\tau : L^1([0,\tau], U) \to X$ is defined by

$$\mathcal{B}_\tau(u) := \int_0^\tau e^{(\tau-s)A} Bu(s) \, ds.$$

Hence, the system is controllable in time τ if and only if $\mathcal{B}_\tau$ is surjective.

Let us begin with the following simple properties. The proof is left to the reader.

Lemma 8.13. *The operator $\mathcal{B}_\tau$ has the following properties.*

a) *The operator $\mathcal{B}_\tau$ is linear.*
b) *The operator $\mathcal{B}_\tau : L^1([0,\tau], U) \to X$ is bounded, i.e.,*

$$\sup_{\|u\| \leq 1} \|\mathcal{B}_\tau(u)\| < \infty.$$

Fortunately, there is an important characterization of the range of $\mathcal{B}_\tau$.

Theorem 8.14. *For every $\tau > 0$ we have*

$$\operatorname{im}(\mathcal{B}_\tau) = \operatorname{span}\left\{ x, Ax, A^2 x, \ldots, A^{n-1} x \, : \, x \in \operatorname{im}(B) \right\}.$$

Proof. Let us introduce first some shorthand notation for this proof and introduce

$$X_1 := \operatorname{span}\left\{ x, Ax, A^2 x, \ldots, A^{n-1} x \, : \, x \in \operatorname{im}(B) \right\},$$

as well as two further spaces,

$$X_2^\tau := \operatorname{span}\left\{ e^{tA} y \, : \, 0 \leq t \leq \tau, y \in \operatorname{im}(B) \right\},$$

$$X_3^\tau := \operatorname{span}\left\{ \int_0^t e^{sA} y \, ds \, : \, 0 \leq t \leq \tau, y \in \operatorname{im}(B) \right\}.$$

Since step functions are dense in L^1, and since in X every subspace is automatically closed, we conclude by the continuity of $\mathcal{B}_\tau$ that

$$\operatorname{im}(\mathcal{B}_\tau) = X_3^\tau.$$

Observe also, using Corollary 2.14 and Theorem 2.12, that e^{tA} is a polynomial in A of degree at most $n - 1$, and thus, $X_2^\tau = X_1$.

Now let us take $y \in \operatorname{im}(B)$. Then $x(t) = \int_0^t e^{sA} y \, ds \in X_3^\tau$ for $t \leq \tau$. Clearly, all the derivatives of x lie in X_3^τ, hence

$$x(0) = 0 \in X_3^\tau,$$
$$\dot{x}(0) = y \in X_3^\tau,$$
$$\ddot{x}(0) = Ay \in X_3^\tau,$$

etc.

Thus, $X_1 \subset X_3^\tau$. On the other hand, if $y \in \text{im}(B)$, then $e^{sA}y \in X_1$, implying that

$$\int_0^t e^{sA}y \, ds \in X_1,$$

i.e., $X_3^\tau \subset X_1$.   □

Corollary 8.15 (Kàlmàn criterion). *For a control system $\Sigma(A,B)$ the following are equivalent.*

(i) *The system is controllable in time τ for all $\tau > 0$.*
(ii) *The controllability operator $\mathcal{B}_\tau$ is surjective for every $\tau > 0$.*
(iii) *The rank condition $\text{rank}(B, AB, A^2B, \dots, A^{n-1}B) = n$ is satisfied.*
(iv) *The system is controllable.*

In many applications it is natural to consider only positive initial values, positive controls, and expect the states of the system to remain positive for all times. Hence, we restrict our investigations here to this case.

By $X_+ := \{x \in X : x \geq 0\}$ we denote the *positive cone* of X. The *reachability set* $X_{\tau,+}$ of a positive system $\Sigma(A,B)$ is defined as the set of points that can be reached from the origin in time τ by applying a positive control. In other words,

$$X_{\tau,+} := \left\{ \int_0^\tau e^{(\tau-s)A} Bu(s) \, ds \; : \; u \geq 0 \right\}.$$

By linearity and positivity of the operators, the set $X_{\tau,+} \subseteq X_+$ is a *convex cone* (i.e., for every $x, y \in X_{\tau,+}$ and $\alpha, \beta \geq 0$, $\alpha x + \beta y \in X_{\tau,+}$). Actually, much more can be said.

Theorem 8.16. *The set $X_{\tau,+}$ is a convex cone which is non-degenerate (i.e., it contains an open ball) if and only if the positive system $\Sigma(A,B)$ is controllable, i.e., the Kàlmàn rank condition is satisfied.*

Proof. It can be shown directly that $X_{\tau,+}$ is a convex cone. Assume now that the Kàlmàn rank is less than n. This means that there exists a nonzero $y \in X$ such that $y^\top A^i B = 0$ for $i = 0, \dots, n$. Hence, $y^\top e^{tA}B = 0$ for all $t \geq 0$, since the degree of the minimal polynomial of A is less than n. So, for all $x \in X_{\tau,+}$ we have

$$(y \,|\, x) = \left(y \,\Big|\, \int_0^\tau e^{A(\tau-s)} Bu(s) \right) ds$$

$$= \int_0^\tau \left((e^{A(\tau-s)}B)^\top y \,\Big|\, u(s) \right) ds$$

$$= 0.$$

Therefore, $X_{\tau,+}$ lies in an $(n-1)$-dimensional subspace of X, which means that $X_{\tau,+}$ is degenerate.

Conversely, if $X_{\tau,+}$ is degenerate, it lies in an $(n-1)$-dimensional subspace of X, since $X_{\tau,+}$ is convex. Thus, there exists $y \in X$ such that $(y|x) = 0$ for all $x \in X_{\tau,+}$ and so

$$\int_0^\tau \left((e^{A(\tau-s)}B)^\top y \,\middle|\, u(s) \right) \, ds = 0$$

for all $u \geq 0$ (and hence for all $u \in L^1_{loc}(\mathbb{R}_+, U)$). Taking now a constant function $u(s) = v \in U$ and differentiate the above equation with respect to τ, one obtains

$$(B^\top y \,|\, v) = 0,$$

$$\int_0^\tau \left((Ae^{A(\tau-s)}B)^\top y \,\middle|\, v \right) \, ds = 0$$

for all $v \in U$. Thus, $y^\top B = 0$. Differentiating again the above equation with respect to τ one has

$$((AB)^\top y \,|\, v) = 0,$$

$$\int_0^\tau \left((A^2 e^{A(\tau-s)}B)^\top y \,\middle|\, v \right) \, ds = 0$$

for all $v \in U$. Repeating the above process $(n-1)$-times one gets $y^\top B A^i = 0$ for $i = 0, 1, \ldots, n-1$, which implies that the Kàlmàn rank is less than n.   □

An important case is when $X_{\tau,+}$ is actually the whole positive cone X_+, i.e., when each positive state can be reached by applying a positive control from the origin.

Definition 8.17. A positive system $\Sigma(A, B)$ is called

(i) *(exactly) positive controllable in time τ*, if

$$X_{\tau,+} = X_+$$

(ii) *(exactly) positive controllable*, if

$$\bigcup_{\tau \geq 0} X_{\tau,+} = X_+,$$

(iii) *approximately positive controllable in time τ*, if

$$\overline{X_{\tau,+}} = X_+.$$

(iv) *approximately positive controllable*, if

$$\overline{\bigcup_{\tau \geq 0} X_{\tau,+}} = X_+.$$

Positive controllability is a much more delicate question then usual controllability and the different notions in the definition above do not coincide (as is the case with the usual controllability). In the proof of Theorem 8.14, the range of the controllability operator, that is, the reachability set for the usual case, was characterized via three linear subspaces. Unfortunately $X_{\tau,+}$ is in general not a closed linear subspace and we can only show the following characterization.

By co M we denote he smallest convex set containing M and by cocone M the smallest convex cone containing M and 0.

Proposition 8.18. *Let $u_1, \ldots, u_p$ be the standard basis vectors in $U = \mathbb{C}^p$. For a positive system $\Sigma(A, B)$,*

$$\overline{X_{\tau,+}} = \overline{\mathrm{co}} \left\{ e^{tA} Bu : 0 \le t \le \tau, u \in U_+ \right\}$$
$$= \overline{\mathrm{cocone}} \left\{ e^{tA} Bu_j : 0 \le t \le \tau, 1 \le j \le p \right\}.$$

Proof. By the definition of the integral,

$$\overline{X_{\tau,+}} \subseteq \overline{\mathrm{co}} \left\{ e^{tA} Bu : 0 \le t \le \tau, u \in U_+ \right\}.$$

Now choose any $u \in U_+$. Since the equations are autonomous, it is enough to show that $e^{tA} Bu \in \overline{X_{t,+}}$ for all $0 \le t \le \tau$. To this aim take

$$u_m(s) := \begin{cases} mu, & \text{for } 0 \le s \le \frac{1}{m}, \\ 0, & \text{for } \frac{1}{m} < s \le \tau, \end{cases}$$

and compute

$$\left\| \int_0^t e^{(t-s)A} Bu_m(s) \, ds - e^{tA} Bu \right\| \le m \int_0^{1/m} \left\| e^{(t-s)A} Bu - e^{tA} Bu \right\| \, ds$$

which converges to 0 as $m \to \infty$. Hence, the first equality is proved. The second one now also follows, since $BU_+ = \mathrm{cocone}\{Bu_1, \ldots, Bu_p\}$. $\qquad\square$

8.3 Stabilization

We restrict ourselves again to systems without observation and where the control is given by a suitable feedback K.

Definition 8.19. A system $\Sigma(A, B)$ is called *stabilizable* if there is a feedback K such that the state converges to zero for every initial value, i.e.,

$$\lim_{t \to \infty} x(t) = \lim_{t \to \infty} e^{t(A+BK)} x_0 = 0 \tag{8.6}$$

for every $x_0 \in X$.

Note that by Theorem 4.12 we have the following characterization of stabilizable systems.

Corollary 8.20. *System $\Sigma(A, B)$ is stabilizable if and only if there is a feedback K such that $\mathrm{s}(A + BK) < 0$.*

A positive system $\Sigma(A, B)$ is called *positively stabilizable*, if there is a positive feedback operator K such that (8.6) holds for every $x_0 \geq 0$.

Proposition 8.21. *A positive system is positively stabilizable if and only if it is stabilizable with a positive feedback.*

Proof. Note that for every element in X its real part and imaginary part can be represented as the difference of two positive elements in the positive cone of X. Since

$$\lim_{t\to\infty} \mathrm{e}^{t(A+BK)} x_0 = 0$$

for every x_0 is equivalent to

$$\lim_{t\to\infty} \mathrm{e}^{t(A+BK)}(x_1 - x_2) = 0$$

for every $x_1, x_2 \geq 0$, the statement follows. $\square$

8.4   Notes and Remarks

There are many excellent introductions to systems and control theory. We based our presentation on the monograph by Jacob and Zwart [69], on the work by Mehrmann [93], and on the monograph by Zabczyk [158].

Positivity aspects of control problems are discussed by Schanbacher in [127] and in the monograph by Farina and Rinaldi [45]. Many further interesting topics could be studied here, and in case we succeeded to make you curious, you can look them up in the above-mentioned sources.

8.5   Exercises

1. Prove the basic properties of the controllability operator $\mathcal{B}_\tau$ as stated in Lemma 8.13.

2. Show that a system $\Sigma(A, B)$ is controllable if and only if for every eigenvector v of $A^\top$ we have $vB \neq 0$.

3. Show that a system $\Sigma(A, B)$ is controllable if and only if $\mathrm{rank}(\lambda - A, B) = n$ for all $\lambda \in \mathbb{C}$.

4. Let $U = \mathbb{C}$ and $X = \mathbb{C}^2$, and consider

$$A = \begin{pmatrix} -1 & 0 \\ 1 & -a \end{pmatrix} \quad \text{and} \quad B = \begin{pmatrix} 1 \\ 0 \end{pmatrix}$$

with $a > 0$. What can you say about the reachability set $X_{\tau,+}$ of this positive linear system? In other words, which states can be reached from the origin by applying a positive control u in time τ?

5. Let $U = \mathbb{C}$ and $X = \mathbb{C}^2$, and consider

$$A = \begin{pmatrix} 1 & 0 \\ 0 & 2 \end{pmatrix} \quad \text{and} \quad B = \begin{pmatrix} 1 \\ 1 \end{pmatrix}.$$

Show that $\Sigma(A, B)$ is controllable, but not approximately positive controllable.

6. Let $U = \mathbb{C}$ and $X = \mathbb{C}^2$, and consider

$$A = \begin{pmatrix} 0 & 1 \\ 0 & 0 \end{pmatrix} \quad \text{and} \quad B = \begin{pmatrix} 0 \\ a \end{pmatrix},$$

with $a > 0$. Is the system $\Sigma(A, B)$ stabilizable? Is it positively stabilizable?

Part II

Infinite Dimensions

Chapter 9

A Crash Course on Operator Semigroups

After studying matrix exponential functions, it is natural to ask whether similar properties can be proved in infinite-dimensional spaces. Indeed, we will see shortly that if we have a semigroup which is continuous (in the usual operator norm), then it is the exponential function of a bounded linear operator.

It turns out, however, that in many applications we encounter semigroups which are not continuous. Nevertheless, these examples will still posses a weaker continuity property, what is called in a potentially confusing terminology "strong continuity". Our aim is to motivate this concept with examples and analyze basic properties of such semigroups. One important point will be to show that a semigroup can be considered as the exponential function of an operator, which, however, is no longer bounded.

We analyze in some detail a few fundamental exceptional examples, where the semigroup can be given explicitly: the shift semigroup, multiplication semigroups, and the heat semigroup. We proceed in the canonical way: we collect important properties a semigroup generator must have and provide necessary conditions on an operator to be a semigroup generator.

In contrast to previous chapters, we have to assume some basic knowledge of functional analysis. Nevertheless, we try to keep the prerequisites at a minimum.

9.1 Exponential Functions

In this section we review some basic facts about exponential functions of bounded linear operators. Many results are analogous to the matrix case, and it is actually possible to prove them by using the same functional calculus argument. However, since building up the functional calculus would need some more effort, we leave that line aside.

Definition 9.1. For a Banach space X and $A \in \mathcal{L}(X)$ we define

$$e^{tA} := \sum_{k=0}^{\infty} \frac{t^k A^k}{k!}$$

for each $t \geq 0$.

Observe that we have

$$\|e^{tA}\| \leq \sum_{k=0}^{\infty} \frac{t^k \|A\|^k}{k!} = e^{t\|A\|} \tag{9.1}$$

for all $t \geq 0$. Therefore, the series $\sum_{k=0}^{\infty} \frac{t^k A^k}{k!}$ is absolutely convergent, and since we are in a Banach space, it is convergent. Operator e^{tA} is thus well defined and bounded. For yet another definition of e^{tA}, see Exercise 2.

Proposition 9.2. *For $A \in \mathcal{L}(X)$, the following properties hold for its exponential function $T(t) := e^{tA}$.*

a) *The functional equation*

$$T(0) = I, \qquad T(t+s) = T(t)T(s) \tag{9.2}$$

   is valid for all $t, s \geq 0$.
b) *The function $\mathbb{R}_+ \ni t \mapsto T(t)$ is continuous.*
c) *The function $\mathbb{R}_+ \ni t \mapsto T(t)$ is differentiable and satisfies the differential equation*

$$\dot{T}(t) = AT(t),$$
$$T(0) = I.$$

Proof. Essentially, all the statements follow as in the scalar case, once we justify the appropriate operations for the operators.

a) By the Cauchy formula for the product of infinite series we have

$$\sum_{k=0}^{\infty} \frac{t^k A^k}{k!} \cdot \sum_{k=0}^{\infty} \frac{s^k A^k}{k!} = \sum_{m=0}^{\infty} \sum_{k=0}^{m} \frac{t^{m-k} A^{m-k}}{(m-k)!} \cdot \frac{s^k A^k}{k!}$$

$$= \sum_{m=0}^{\infty} \frac{A^m}{m!} \sum_{k=0}^{m} \frac{m!}{(m-k)!k!} t^{m-k} s^k = \sum_{m=0}^{\infty} \frac{(t+s)^m A^m}{m!}.$$

b) Since equation (9.2) implies

$$e^{(t+h)A} - e^{tA} = e^{tA}(e^{hA} - I),$$

it suffices to prove continuity at 0, which follows from

$$\left\| e^{hA} - I \right\| = \left\| \sum_{k=1}^{\infty} \frac{h^k A^k}{k!} \right\| \leq \sum_{k=1}^{\infty} \frac{|h|^k \cdot \|A\|^k}{k!} = e^{|h| \cdot \|A\|} - 1.$$

c) By a similar argument as above it suffices to see that

$$\left\| \frac{e^{hA} - I}{h} - A \right\| = \left\| \sum_{k=2}^{\infty} \frac{h^{k-1} A^k}{k!} \right\| \leq \sum_{k=2}^{\infty} \frac{|h|^{k-1} \cdot \|A\|^k}{k!}$$

$$= \frac{e^{|h| \cdot \|A\|} - 1}{|h|} - \|A\| \longrightarrow 0$$

as $h \to 0$. $\qquad\qquad\qquad\qquad\qquad\qquad\qquad\qquad\qquad\qquad\qquad\qquad\qquad\square$

The functional equation (9.2) plays a crucial role, hence we give it a name.

Definition 9.3. A map $\mathbb{R}_+ \ni t \mapsto T(t) \in \mathcal{L}(X)$ is called a *one-parameter operator semigroup* (or an *operator semigroup*, or just a *semigroup* for short), if

$$T(0) = I \quad \text{and} \quad T(t+s) = T(t)T(s) \quad \text{for all } t, s \geq 0.$$

The most important property of *continuous* semigroups is that they are nothing but exponential functions.

Proposition 9.4. *Let* $(T(t))_{t \geq 0}$ *be a semigroup which is continuous. Then there is an operator* $A \in \mathcal{L}(X)$ *such that* $T(t) = e^{tA}$.

Proof. Since the function $t \mapsto T(t)$ is continuous and $T(0) = I$, we see that

$$\left\| I - \frac{1}{t_0} \int_0^{t_0} T(s) \, \mathrm{d}s \right\| < 1$$

for sufficiently small[8] $t_0 > 0$. So, by the Neumann series (see formula (A.4)), the operator $\frac{1}{t_0} \int_0^{t_0} T(s) \, \mathrm{d}s$ is invertible, and hence

$$V(t_0) := \int_0^{t_0} T(s) \, \mathrm{d}s$$

is invertible, too. It follows that

$$T(t) = V(t_0)^{-1} V(t_0) T(t) = V(t_0)^{-1} \int_0^{t_0} T(t+s) \, \mathrm{d}s = V(t_0)^{-1} \int_t^{t+t_0} T(s) \, \mathrm{d}s$$

$$= V(t_0)^{-1} (V(t+t_0) - V(t))$$

holds for all $t \geq 0$.

---

[8] For properties of vector-valued Riemann integrals of continuous functions see the Appendix A.7, especially Proposition A.26.

Since V is the integral of a continuous function, it is differentiable and so is the function $t \mapsto T(t)$. Let us introduce the notation $\dot{T}(0) =: A$. Then $A \in \mathcal{L}(X)$ and the functional equation implies that

$$\dot{T}(t) = \lim_{h \to 0} \frac{T(t+h) - T(t)}{h} = \lim_{h \to 0} \frac{T(h) - I}{h} T(t) = AT(t)$$

for all $t \geq 0$. Hence, T satisfies a linear differential equation of the form

$$\dot{T}(t) = AT(t)$$

with $T(0) = I$. But $S(t) = e^{tA}$ also satisfies the same differential equation. Fix $t > 0$ and consider the function $[0, t] \ni s \mapsto T(s)S(t-s) =: u(s)$. Then u is differentiable and its derivative is given by the product rule, see Appendix, Theorem A.18,

$$\frac{\mathrm{d}}{\mathrm{d}s} u(s) = \left(\frac{\mathrm{d}}{\mathrm{d}s} T(t-s) \right) S(s) + T(t-s) \frac{\mathrm{d}}{\mathrm{d}s} (S(s))$$
$$= -AT(t-s)S(s) + T(t-s)AS(s) = 0,$$

and so we obtain the equality $T(t) = u(t) = u(0) = S(t)$. $\square$

9.2   Motivation for Generalizations

First we recall from the previous section that if $(T(t))_{t \geq 0}$ is a continuous semigroup, then there is an operator $A \in \mathcal{L}(X)$ such that $T(t) = e^{tA}$, and $u(t) = T(t)f$ solves the differential equation

$$\begin{cases} \dot{u}(t) = Au(t), & t \geq 0, \\ u(0) = f. \end{cases} \tag{9.3}$$

for all $f \in X$.

It turns out, however, that there are important semigroups which do not satisfy this continuity property, but a weaker one. We will also see that there are important differential equations of the form (9.3), where the operator A is no longer bounded. As a motivation, we work out an important example, the shift (semi)group.

Example 9.5. Let us consider

$$X = \mathrm{C}_{\mathrm{ub}}(\mathbb{R}) := \{ f : \mathbb{R} \to \mathbb{R} : f \text{ is bounded and uniformly continuous} \},$$

which becomes a Banach space with the supremum norm

$$\|f\|_\infty := \sup_{s \in \mathbb{R}} |f(s)|.$$

The additive semigroup structure of $\mathbb{R}$ naturally induces a semigroup on this Banach space by defining

$$(T(t)f)(s) := f(t+s), \quad \text{for } f \in X,\, s \in \mathbb{R},\, t \geq 0.$$

It follows directly from the definition that $T(t)$ is a bounded linear operator on X, actually a linear isometry. The semigroup property also follows from the additive

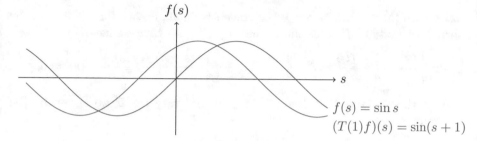

$$f(s) = \sin s$$
$$(T(1)f)(s) = \sin(s+1)$$

Figure 9.1: The shift semigroup.

semigroup structure of $\mathbb{R}$. The uniform continuity of $f \in X$ implies that the mapping

$$t \longmapsto T(t)f$$

is continuous on $X = C_{\mathrm{ub}}(\mathbb{R})$ for every $f \in X$. We will say that $(T(t))_{t \geq 0}$ is a "strongly continuous semigroup" called the *left-shift semigroup*. Note that $T(\cdot)$ is not continuous in the operator norm, because if

$$\|T(t) - I\| = \sup_{\|f\| \leq 1} \|T(t)f - f\| = \sup_{\|f\| \leq 1} \sup_{s \in \mathbb{R}} |f(s+t) - f(s)|$$

would converge to zero as $t \to 0$, then the unit ball of $C_{\mathrm{ub}}(\mathbb{R})$ would be uniformly equicontinuous, which is impossible (and, as a side remark, which would imply by the Arzelà–Ascoli Theorem that the unit ball of $C_{\mathrm{ub}}(\mathbb{R})$ would be compact).

Let us investigate whether this semigroup $(T(t))_{t \geq 0}$ solves an initial value problem such as (9.3). The heuristics of matrix exponential functions helps us here. Given e^{tA} for a matrix $A \in \mathcal{L}(\mathbb{R}^n)$, we can calculate A by differentiating this exponential function at 0:

$$A = \frac{\mathrm{d}}{\mathrm{d}t} e^{tA} \Big|_{t=0}.$$

Let us see if we can use this idea in the case of the shift semigroup $(T(t))_{t \geq 0}$. Note that the semigroup $(T(t))_{t \geq 0}$ is not even continuous for the operator norm, hence we cannot use this idea directly. So let us look at differentiability of the *orbit maps* $t \mapsto T(t)f$ for some $f \in X$, which is referred to as *strong differentiability* of the mapping $T(\cdot)$. The limit

$$\lim_{h \to 0} \frac{1}{h}(T(h)f - f) = \lim_{h \to 0} \frac{f(h + \cdot) - f(\cdot)}{h}$$

must exist in the supremum norm of X. Since the limit must exist pointwise on $\mathbb{R}$, it cannot be anything else but f'. Hence, the function f must be at least differentiable so that the limit exists. Let f be differentiable with f' being uniformly continuous. Then

$$\sup_{s \in \mathbb{R}} \left| \frac{f(s+h) - f(s)}{h} - f'(s) \right| = \sup_{s \in \mathbb{R}} \left| \frac{1}{h} \int_s^{s+h} (f'(r) - f'(s)) \mathrm{d}r \right| \leq \varepsilon,$$

for all h with $|h| \leq \delta$, where $\delta > 0$ is sufficiently small, chosen for $\varepsilon > 0$ from the uniform continuity of f'. This shows that if $f, f' \in X$, then we have

$$\lim_{h \to 0} \left\| \frac{f(\cdot + h) - f(\cdot)}{h} - f'(\cdot) \right\|_\infty = \lim_{h \to 0} \sup_{s \in \mathbb{R}} \left| \frac{f(s+h) - f(s)}{h} - f'(s) \right| = 0.$$

Note that for the derivative of $T(t)f$ at arbitrary $t \in \mathbb{R}$ we obtain by the same argument

$$\frac{\mathrm{d}}{\mathrm{d}t}(T(t)f) = T(t)f'.$$

This means that for $f, f' \in X$ the orbit function $u(t) = T(t)f$ solves the differential equation

$$\begin{cases} \dot{u}(t) = Au(t), & t \in \mathbb{R}, \\ u(0) = f \in D(A), \end{cases}$$

where $(Af)(s) = f'(s)$ is defined with domain

$$D(A) := \{ f : f, f' \in C_{\mathrm{ub}}(\mathbb{R}) \}.$$

Clearly, A is not everywhere defined and cannot be extended to a bounded operator.

Our example motivates the following definition.

Definition 9.6. Let $T : [0, \infty) \to \mathcal{L}(X)$ be a function.

a) We say that $(T(t))_{t \geq 0}$ has the *semigroup property* if for all $t, s \in [0, \infty)$

$$T(t + s) = T(t)T(s) \quad \text{and}$$
$$T(0) = I \quad \text{(i.e., the identity operator on } X\text{)}.$$

b) Suppose $Y \subseteq X$ is a linear subspace and for all $f \in Y$ the mapping

$$t \longmapsto T(t)f \in X$$

is continuous. Then $(T(t))_{t \geq 0}$ is said to be *strongly continuous on* Y. If $Y = X$, we just say *strongly continuous*[9].

---

[9] Here the word *strong* refers to the so-called strong operator topology on $\mathcal{L}(X)$, which characterizes the pointwise convergence of operators (in contrast to the norm, which is responsible for uniform convergence), see Appendix A.5.

c) If a strongly continuous mapping T possesses the semigroup property, then we call $(T(t))_{t\geq 0}$ a *strongly continuous one-parameter semigroup* of bounded linear operators on the Banach space X. Often we shall abbreviate this terminology to C_0-*semigroup*[10].

9.3 Basic Properties

Let us record some elementary consequences of the semigroup property and the strong continuity. The first result we mention here reflects again the exponential function: Semigroups can grow at most exponentially.

Proposition 9.7. *Let* $(T(t))_{t\geq 0}$ *be a* C_0-*semigroup. Then there are* $M \geq 1$ *and* $\omega \in \mathbb{R}$ *such that*

$$\|T(t)\| \leq Me^{\omega t} \quad \text{for all } t \geq 0.$$

We call the semigroup $(T(t))_{t\geq 0}$ of *type* (M,ω) if it satisfies the exponential estimate above with particular constants M and ω. Note already here that the type of a semigroup may change if we pass to an equivalent norm on X.

Proof. For a fixed $f \in X$, the function $s \mapsto T(s)f$ is continuous on $[0,\infty)$, hence bounded on compact intervals $[0,t]$, i.e.,

$$\sup_{s\in[0,t]} \|T(s)f\| < \infty.$$

The uniform boundedness principle, see Appendix, Theorem A.15, implies that

$$M := \sup_{s\in[0,1]} \|T(s)\| < \infty.$$

Taking an arbitrary $t \geq 0$ we write $t = n + r$ with $n \in \mathbb{N}$ and $r \in [0,1)$. From this representation and the semigroup property we infer that

$$\begin{aligned}
\|T(t)\| = \|T(r)T(1)^n\| &\leq M\|T(1)^n\| \\
&\leq M\|T(1)\|^n \leq M(\|T(1)\| + 1)^n \\
&\leq M(\|T(1)\| + 1)^t = Me^{\omega t}
\end{aligned}$$

with $\omega = \log(\|T(1)\| + 1)$. $\qquad\square$

Hence, orbits of C_0-semigroups are *exponentially bounded*. The *greatest lower bound* of these exponential bounds plays a special role in the theory and will be our guide in later chapters.

---

[10]The symbol C_0 or $(C,0)$ stands for Cesàro summable of order zero, which means the continuity property $\lim_{t\to 0} T(t)f = f$ for all $f \in X$.

Definition 9.8. For a C_0-semigroup $(T(t))_{t\geq 0}$ its *growth bound* is defined by

$$\omega_0(T) := \inf\{\omega \in \mathbb{R} : \text{ there is } M = M_\omega \geq 1 \text{ with } \|T(t)\| \leq Me^{\omega t} \text{ for all } t \geq 0\}.$$

Remark 9.9.

a) For $X = \mathbb{C}^n$ and a matrix $A \in \mathcal{L}(X)$, we define $T(t) = e^{tA}$. This semigroup $(T(t))_{t\geq 0}$ is of type $(1, \|A\|)$, as the direct norm estimate

$$\|e^{tA}\| \leq e^{t\|A\|}$$

shows. In contrast to this, in the infinite-dimensional situation it can happen that a semigroup is not of type $(1, \omega)$ for any ω, even though $\omega_0(T) = -\infty$. This is an extremely important fact, which causes major difficulties in many arguments. See Example 9.12 below for a simple demonstration.

b) A C_0-semigroup $(T(t))_{t\geq 0}$ is of type (M, ω) for all $\omega > \omega_0(T)$ and for some $M = M_\omega$. In general, however, it is *not* of type $(M, \omega_0(T))$ for any M. A simple example is the following. Let $X = \mathbb{C}^2$ and let the matrix semigroup given by

$$T(t) = \begin{pmatrix} 1 & t \\ 0 & 1 \end{pmatrix}.$$

Here $\omega_0(T) = 0$, but clearly T is not bounded, i.e., not of type $(M, 0)$ for any M.

The definition of a C_0-semigroup combines the analytic property of strong continuity and the algebraic semigroup property. We show next that these two properties combine well, and we provide some powerful tools for verifying strong continuity.

Proposition 9.10.

a) *Let $T : [0, \infty) \to \mathcal{L}(X)$ be a locally bounded mapping with the semigroup property, and let $f \in X$. If the mapping $T(\cdot)f$ is right continuous at 0, i.e., $T(h)f \to f$ for $h \downarrow 0$, then it is continuous everywhere.*

b) *A mapping T with the semigroup property is strongly continuous on X if and only if it is locally bounded and there is a dense subset $D \subseteq X$ on which T is strongly continuous.*

Proof. a) Fix $f \in X$ and $t > 0$, and set $M_t := \sup_{[0,t]} \|T(s)\|$. Then

$$T(t+h)f - T(t)f = T(t)(T(h)f - f) \qquad \text{if } 0 < h,$$
$$T(t+h)f - T(t)f = T(t+h)(f - T(-h)f) \quad \text{if } -t < h < 0.$$

Summarizing, for $|h| \leq t$ we obtain

$$\|T(t+h)f - T(t)f\| \leq M_t\|f - T(|h|)f\|,$$

which converges to 0 for $|h| \to 0$ by assumption.

b) In view of Proposition 9.7 one implication is straightforward. So we turn to the other, and suppose T to be locally bounded and strongly continuous on a dense subspace D. Take an arbitrary $f \in X$ and $f_k \in D$ such that $f_k \to f$. Then, by local boundedness, for fixed $t_0 > 0$ we obtain that the functions $T(\cdot)f_k$ converge uniformly to the function $T(\cdot)f$ on $[0, t_0]$. Since uniform limits of continuous functions are continuous, the statement follows. $\qquad\square$

Example 9.11. For $f \in L^p(\mathbb{R})$ we define

$$(T(t)f)(s) := f(t+s) \quad \text{for a.e. } s \in \mathbb{R},\ t \geq 0.$$

Then $T(t)$ is a linear isometry on $L^p(\mathbb{R})$. Moreover, the function T has the semi-group property. We call $(T(t))_{t \geq 0}$ the *left-shift semigroup on* $L^p(\mathbb{R})$. We show that for $p \in [1, \infty)$ the left-shift semigroup $(T(t))_{t \geq 0}$ is strongly continuous on $L^p(\mathbb{R})$.

Recall first that the shift semigroup is strongly continuous on the space of bounded uniformly continuous functions, and that the set of continuous functions with compact support is dense in $L^p(\mathbb{R})$. Taking $f \in C_c(\mathbb{R})$ and $\alpha, \beta \in \mathbb{R}$ such that supp $f \subset [\alpha, \beta]$, we see that

$$\|f(\cdot) - f(\cdot + h)\|_p^p = \int_{\mathbb{R}} |f(s) - f(s+h)|^p \, ds \leq (\beta - \alpha) \sup_{s \in [\alpha, \beta]} |f(s) - f(s+h)|^p,$$

which goes to zero as $h \to 0$ by the uniform continuity of f, where the pth norm of $f \in L^p(\mathbb{R})$ is $\|f\|_p := \left(\int_{\mathbb{R}} |f(s)|^p \, ds \right)^{1/p}$. Since $\|T(t)\| \leq 1$, the statement follows by Proposition 9.10.

We have seen in Proposition 9.7 that a semigroup $(T(t))_{t \geq 0}$ is always exponentially bounded, meaning that an estimate of the type

$$\|T(t)\| \leq M e^{\omega t}$$

holds. Let us give here an example to show that in the infinite-dimensional case it is quite possible to have $M > 1$. The example will be a slight modification of the shift semigroup.

Example 9.12. [A bounded semigroup which is not a contraction]

Let us consider the Hilbert space $L^2((0, 1), \mu)$, where μ denotes the measure defined by

$$\mu(\Omega) := 2\lambda(\Omega \cap (0, \tfrac{1}{2})) + \lambda(\Omega \cap (\tfrac{1}{2}, 1))$$

for all Lebesgue measurable sets $\Omega \subset (0, 1)$, where λ is the Lebesgue measure. Furthermore, let $(T(t))_{t \geq 0}$ be the nilpotent left-shift semigroup, defined by

$$(T(t)f)(s) := \begin{cases} f(s+t) & \text{for } s+t \leq 1, \\ 0 & \text{for } s+t > 1. \end{cases}$$

Obviously, $(T(t))_{t \geq 0}$ satisfies the semigroup property, and since the norm $\|\cdot\|_\mu$ is equivalent to the norm $\|\cdot\|_\lambda$, the semigroup $(T(t))_{t \geq 0}$ is strongly continuous by similar arguments as in the previous example. Clearly, $\|T(t)\| \leq 2$. In addition we see that $T(t) = 0$ for all $t > 1$.

Finally, consider the function

$$f_t = \left(\frac{1}{\sqrt{t}}\right) \chi_{(\frac{1}{2}, \frac{1}{2}+t)}$$

for $t \in (0, \frac{1}{2})$, which satisfies $\|f_t\|_\mu = 1$ and

$$\|T(t)f_t\|_\mu = 2.$$

Hence, $\|T(t)\| = 2$ for $t \in (0, \frac{1}{2})$. This means that an estimate of the type

$$\|T(t)\| \leq Me^{\omega t}$$

cannot hold for $M < 2$ independently of the value of $\omega \in \mathbb{R}$.

9.4   The Infinitesimal Generator

One message what we would like to deliver is that if we have a *semigroup*, then there is a *differential equation* so that the semigroup provides its solutions. Looking for the equation, we now consider the differentiability of orbit maps as in Example 9.5.

Lemma 9.13. *Take a semigroup $(T(t))_{t \geq 0}$ and an element $f \in X$. For the orbit map $u : t \mapsto T(t)f$, the following properties are equivalent:*

(i) *u is differentiable on $[0, \infty)$;*

(ii) *u is right differentiable at 0.*

If u is differentiable, then

$$\dot{u}(t) = T(t)\dot{u}(0).$$

Proof. Since (ii) is clearly a special case of (i), we only have to show that (ii) implies (i). We proceed analogously to the proof of Proposition 9.10. First, we have

$$\lim_{h \downarrow 0} \frac{1}{h}\left(u(t+h) - u(t)\right) = \lim_{h \downarrow 0} \frac{1}{h}\left(T(t+h)f - T(t)f\right) = T(t)\lim_{h \downarrow 0} \frac{1}{h}\left(T(h)f - f\right)$$

$$= T(t)\lim_{h \downarrow 0} \frac{1}{h}\left(u(h) - u(0)\right) = T(t)\dot{u}(0),$$

by the continuity of $T(t)$. Hence, u is right differentiable on $[0, \infty)$.

On the other hand, for $-t \leq h < 0$, we write

$$\frac{1}{h}\left(T(t+h)f - T(t)f\right) - T(t)\dot{u}(0)$$

$$= T(t+h)\left(\frac{1}{h}\left(f - T(-h)f\right) - \dot{u}(0)\right) + T(t+h)\dot{u}(0) - T(t)\dot{u}(0).$$

As $h \uparrow 0$, the first term on the right-hand side converges to zero by the first part and by the boundedness of $\|T(t+h)\|$ for $h \in [-t, t]$. The other term converges to zero by the strong continuity of T. Hence, u is also left differentiable at $t > 0$, and its derivative is

$$\dot{u}(t) = T(t)\,\dot{u}(0)$$

for all $t \geq 0$.   □

We thus see that the derivative $\dot{u}(0)$ of the orbit map $u(t) = T(t)f$ at $t = 0$ determines the derivative at each point $t \in [0, \infty)$. For later reference, we give a name to the operator which maps f into the derivative of the map $t \mapsto T(t)f$ at $t = 0$.

Definition 9.14. The *infinitesimal generator*, or simply the *generator*, A of a strongly continuous semigroup $(T(t))_{t \geq 0}$ is defined as follows. Its *domain* is given by

$$D(A) := \{f \in X : \text{ the function } t \mapsto T(t)f \text{ is differentiable on } [0, \infty)\}$$

and for $f \in D(A)$ we set

$$Af := \frac{\mathrm{d}}{\mathrm{d}t} T(t)f \Big|_{t=0} = \lim_{t \downarrow 0} \tfrac{1}{t}\left(T(t)f - f\right).$$

It is probably not a great surprise now that a semigroup yields solutions to some linear initial value problem in a Banach space X.

Proposition 9.15. *The generator A of a C_0-semigroup $(T(t))_{t \geq 0}$ has the following properties.*

a) $A : D(A) \subseteq X \to X$ *is linear.*

b) *If $f \in D(A)$, then $T(t)f \in D(A)$ for $t > 0$, and $\dot{T}(t)f = T(t)Af = AT(t)f$ for all $t \geq 0$.*

c) *For a given $f \in D(A)$, the semigroup $(T(t))_{t \geq 0}$ provides a solution to the initial value problem*

$$\begin{cases} \dot{u}(t) = Au(t), & t \geq 0, \\ u(0) = f, \end{cases}$$

via $u(t) := T(t)f$.

Proof. a) Linearity follows directly from the definition, because we take a limit of linear objects as $h \downarrow 0$.

b) Take $f \in D(A)$ and $t \geq 0$. We have to show that $T(\cdot)T(t)f$ is right differentiable at 0, with derivative $T(t)Af$. From the strong continuity of $T(t)$ we obtain

$$T(t)Af = T(t) \lim_{h \downarrow 0} \frac{T(h)f - f}{h} = \lim_{h \downarrow 0} \frac{T(h)T(t)f - T(t)f}{h}.$$

By the definition of A, this further equals $AT(t)f$.

c) This is just a reformulation of b) in the language of differential equations.

□

We investigate now infinitesimal generators further and establish a generalization of the Newton–Leibniz formula.

Proposition 9.16. *The generator A of a C_0-semigroup $(T(t))_{t \geq 0}$ has the following properties.*

a) *For all $t \geq 0$ and $f \in X$, one has*

$$\int_0^t T(s)f\,\mathrm{d}s \in D(A),$$

where the integral has to be understood as the Riemann integral of the continuous function $s \mapsto T(s)f$, see Appendix A.7.

b) *For all $t \geq 0$, one has*

$$T(t)f - f = A \int_0^t T(s)f\,\mathrm{d}s \quad \text{if } f \in X,$$

$$= \int_0^t T(s)Af\,\mathrm{d}s \quad \text{if } f \in D(A).$$

Proof. a) For $g := \int_0^t T(s)f\,\mathrm{d}s$ we calculate the difference quotient:

$$\frac{T(h)g - g}{h} = \frac{1}{h}\left(T(h)\int_0^t T(s)f\,\mathrm{d}s - \int_0^t T(s)f\,\mathrm{d}s\right)$$

$$= \frac{1}{h}\left(\int_0^t T(h+s)f\,\mathrm{d}s - \int_0^t T(s)f\,\mathrm{d}s\right)$$

$$= \frac{1}{h}\left(\int_h^{t+h} T(s)f\,\mathrm{d}s - \int_0^t T(s)f\,\mathrm{d}s\right)$$

$$= \frac{1}{h}\left(\int_t^{t+h} T(s)f\,\mathrm{d}s - \int_0^h T(s)f\,\mathrm{d}s\right)$$

$$= \frac{1}{h}\int_t^{t+h} T(s)f\,\mathrm{d}s - \frac{1}{h}\int_0^h T(s)f\,\mathrm{d}s.$$

Since the integrands here are continuous, we can take limits as $h \downarrow 0$ and obtain

$$\lim_{h \downarrow 0} \frac{T(h)g - g}{h} = T(t)f - f.$$

This yields $g \in D(A)$ and $Ag = T(t)f - f$.

b) Taking $f \in D(A)$, by Proposition 9.15.b) we see that the identity $AT(t)f = T(t)Af$ holds, hence $v(t) := AT(t)f$ defines a continuous function. For $h > 0$ define the continuous functions $v_h(t) := \frac{1}{h}(T(t+h)f - T(t)f)$. Then we have the estimate

$$\|v_h(t) - v(t)\| \leq \|T(t)\|\left\|\frac{1}{h}(T(h)f - f) - Af\right\|.$$

From this and the definition of A we conclude (by using the exponential bound-edness of T) that v_h converges to v uniformly on every compact interval. This yields

$$\int_0^t v_h(s)\mathrm{d}s \longrightarrow \int_0^t v(s)\mathrm{d}s \quad \text{as } h \downarrow 0.$$

In part a) of this proof we have calculated the limit of the left-hand side and seen that it equals $T(t)f - f$. Hence,

$$T(t)f - f = \int_0^t AT(s)f\mathrm{d}s,$$

which completes the proof. $\qquad\qquad\qquad\qquad\qquad\qquad\qquad\qquad\qquad\square$

Before turning our attention to the most fundamental result of this section, let us introduce a new notation and define what a *closed operator* is. For a linear operator A defined on a linear subspace $D(A)$ of a Banach space X, we define the *graph norm* of A by

$$\|f\|_A := \|f\| + \|Af\| \quad \text{for } f \in D(A).$$

Then, indeed, $\|\cdot\|_A$ is a norm[11] on $D(A)$. The operator A is called *closed* if $D(A)$ is complete with respect to this graph norm, i.e., if $D(A)$ is a Banach space with this graph norm $\|\cdot\|_A$.

The following lemma yields simple, yet useful reformulations of the closedness of a linear operator.

Lemma 9.17. *Let A be a linear operator with domain $D(A)$ in X. The following assertions are equivalent.*

(i) *A is a closed operator.*

(ii) *For every sequence $(f_k) \subseteq D(A)$ with $f_k \to f$ and $Af_k \to g$ in X for some $f, g \in X$, one has $f \in D(A)$ and $Af = g$.*

If A is injective, the properties above are further equivalent to the following:

(iii) *A^{-1} (defined on the range of A) is a closed operator.*

Proof. (i) $\Longrightarrow$ (ii): Assume A is a closed operator. Let $f_k \to f$ and $Af_k \to g$ in X. Then

$$\|f_k - f_\ell\|_A = \|f_k - f_\ell\| + \|A(f_k - f_\ell)\|$$
$$\leq \|f_k - f_\ell\| + \|Af_k - g\| + \|g - Af_\ell\|$$

---

[11] If X is a Hilbert space, it is customary to define the graph norm as $\|f\|_A^2 := \|f\|^2 + \|Af\|^2$, which makes $D(A)$ a pre-Hilbert space. Clearly, the two definitions yield equivalent norms.

shows that (f_k) is a Cauchy sequence in $D(A)$, and thus for some $h \in D(A)$ we have $f_k \to h$ in $(D(A), \| \cdot \|_A)$, since this space is complete. Note that

$$\|f_k - h\| \leq \|f_k - h\| + \|A(f_k - h)\| = \|f_k - h\|_A.$$

Thus $f_k \to h$ in $(X, \| \cdot \|)$ as well, which means, by uniqueness of the limit, that $h = f$. Then $f_k \to f$ in $(D(A), \| \cdot \|)$, and so

$$\|Af_k - Af\| \leq \|Af_k - Af\| + \|f_k - f\| = \|f_k - f\|_A \longrightarrow 0.$$

Therefore $Af_k \to Af$ in X, and so $Af = g$, by uniqueness of the limit.

(ii) $\implies$ (i): We wish to show that A is a closed operator, i.e., $D(A)$ is complete with respect to $\| \cdot \|_A$. Let (f_k) be a Cauchy sequence in $D(A)$. Then

$$\|f_k - f_\ell\| \leq \|f_k - f_\ell\| + \|Af_k - Af_\ell\| = \|f_k - f_\ell\|_A$$

and

$$\|Af_k - Af_\ell\| \leq \|f_k - f_\ell\| + \|Af_k - Af_\ell\| = \|f_k - f_\ell\|_A$$

and so (f_k) and (Af_k) are Cauchy sequences in X. Since X is complete, we see that $f_k \to f$ and $Af_k \to g$ in X for some $f, g \in X$. Consequently, using the assumptions in (ii), $f_k \to f$ with $f \in D(A)$ and $Af = g$. Thus $D(A)$ is complete with respect to $\| \cdot \|_A$.

(ii) $\implies$ (iii): Note that the linearity of A implies the linearity of A^{-1}. Let A be a closed operator. We need to show $(D(A^{-1}), \| \cdot \|_{A^{-1}})$ is complete. Therefore, take an arbitrary Cauchy sequence (g_k) in $D(A^{-1})$ and show its limit g exists. Using the fact that $D(A^{-1}) = \operatorname{im} A$ and hence there exist f_k, f_ℓ, such that $Af_k = g_k$, $Af_\ell = g_\ell$, we obtain the relations

$$\|g_k - g_\ell\|_{A^{-1}} = \|g_k - g_\ell\| + \|A^{-1}g_k - A^{-1}g_\ell\|$$
$$= \|Af_k - Af_\ell\| + \|f_k - f_\ell\|.$$

Thus (Af_k) and (f_k) are Cauchy sequences in X, which is complete, and so $f_k \to f$ and $Af_k \to g$. Then $g = Af$ by (ii) and so

$$\|g_k - g\|_{A^{-1}} = \|Af_k - Af\| + \|f_k - f\| \longrightarrow 0.$$

Thus, g_k converges to g in $D(A^{-1})$.

(iii) $\implies$ (i): The above argument shows that $A = (A^{-1})^{-1}$ is closed whenever A^{-1} is closed. $\square$

The main result of this section summarizes the basic properties of the generator.

Theorem 9.18. *The generator of a C_0-semigroup is a closed and densely defined linear operator that determines the semigroup uniquely.*

Proof. To show the closedness of A, let $(f_k) \subset D(A)$ be a sequence and $f, g \in X$ such that

$$f_k \longrightarrow f \quad \text{and} \quad Af_k \longrightarrow g.$$

We have to show that $f \in D(A)$ and $Af = g$.

For $t > 0$ we have

$$T(t)f_k - f_k = \int_0^t T(s)Af_k \mathrm{d}s,$$

thanks to Proposition 9.16. If we set $u_k(s) := T(s)Af_k$ and $u(s) := T(s)g$, then $u_k \to u$ uniformly on $[0, t]$ because the semigroup $(T(t))_{t \geq 0}$ is locally bounded. So we can pass to the limit in the identity above, and obtain

$$T(t)f - f = \int_0^t T(s)g \mathrm{d}s.$$

From this we deduce that the function $t \mapsto T(t)f$ is differentiable at 0 with derivative $u(0) = g$. This means precisely that $f \in D(A)$ and $Af = g$, which implies that A is a closed operator.

We now show that $D(A)$ is dense in X. Let $f \in X$ be arbitrary and define

$$v(t) := \frac{1}{t} \int_0^t T(s)f \mathrm{d}s, \quad t > 0.$$

By Proposition 9.16, we obtain that $v(t) \in D(A)$. Since the function $s \mapsto T(s)f$ is continuous, we have $v(t) \to T(0)f = f$ for $t \downarrow 0$.

Suppose $(S(t))_{t \geq 0}$ is a C_0-semigroup with the same generator A as $(T(t))_{t \geq 0}$. Let $f \in D(A)$ and $t > 0$ be fixed, and consider the function $u : [0, t] \to X$ given by $u(s) := T(t - s)S(s)f$. Then u is differentiable and its derivative is given by the product rule, see Appendix, Theorem A.18:

$$\frac{\mathrm{d}}{\mathrm{d}s}u(s) = \left(\frac{\mathrm{d}}{\mathrm{d}s}T(t - s)\right) S(s)f + T(t - s)\frac{\mathrm{d}}{\mathrm{d}s}(S(s)f)$$

$$= -AT(t - s)S(s)f + T(t - s)AS(s)f.$$

Recalling that the semigroup and its generator commute on $D(A)$, see Proposition 9.15.b), we obtain that the right-hand term is 0, so u must be constant. This implies that

$$S(t)f = u(t) = u(0) = T(t)f,$$

i.e., the bounded linear operators $S(t)$ and $T(t)$ coincide on the dense subspace $D(A)$, hence they must be equal everywhere. $\qquad \square$

We conclude the section with the most important property of the generator from the point of view of applications. The *abstract Cauchy problem*

$$\begin{cases} \dot{u}(t) = Au(t), & t \geq 0, \\ u(0) = f, \end{cases} \tag{9.4}$$

is *well posed*, if the domain $D(A)$ is dense and for each $f \in D(A)$ there exists a unique *classical solution* $u = u(\cdot, f)$ depending continuously on the initial value f. More precisely, $u(\cdot, f) : \mathbb{R}_+ \to X$ is continuously differentiable, $u(t, f) \in D(A)$ for all $t \geq 0$, it satisfies equation (9.4), and for every sequence $(f_k) \subset D(A)$ converging to 0 one has $\lim_{k \to \infty} u(t, f_k) = 0$ uniformly for t on compact intervals of $\mathbb{R}_+$.

Theorem 9.19. *A closed linear operator A on a Banach space X generates a C_0-semigroup $(T(t))_{t \geq 0}$ if and only if the abstract Cauchy problem (9.4) is well posed.*

Proof. If A is a generator, then the well-posedness of the abstract Cauchy problem follows from Theorem 9.18 and Proposition 9.15.

Conversely, let the abstract Cauchy problem (9.4) be well posed. Using the unique classical solution $u = u(\cdot, f)$, where $f \in D(A)$, for every $t \geq 0$ we can define
$$T(t)f := u(t, f).$$

Observe that all $T(t)$ are bounded operators on X. Their linearity is a consequence of the linearity of A and the uniqueness of u. Since $T(0)f = u(0, f) = f$ and by the uniqueness of the solutions, we also have
$$\begin{aligned} T(t + s)f &= u(t + s, f) \\ &= u(t, u(s, f)) = T(t)T(s)f \end{aligned}$$

for every $f \in D(A)$ and $t, s \geq 0$. Thus the semigroup property holds.

Furthermore, $\|T(t)\|$ is uniformly bounded on every compact interval $[0, \tau]$. Otherwise, there would exist a sequence $(t_k) \subset [0, \tau]$ with $\|T(t_k)\| \to \infty$ as $k \to \infty$ and we could choose $(f_k) \in D(A)$ converging to 0 such that
$$\|u(t_k, f_k)\| = \|T(t_k)f_k\| \geq 1,$$

which contradicts the assumptions on u.

Since the mapping $t \mapsto T(t)f$ is continuous for every f in the dense set $D(A)$, Proposition 9.10.b) now implies that $(T(t))_{t \geq 0}$ is a C_0-semigroup on X. Denote its generator by B. The operators A and B are both closed, have dense domains, and coincide on the $T(t)$-invariant dense set $D(A)$, hence they are equal. $\square$

9.5 Multiplication Semigroups

As an example, we consider now semigroups generated by multiplication operators in $X = \ell^p$ for $p \in [1, \infty)$. Let the sequence $(a_n) \subset \mathbb{C}$ be given and define the multiplication operator by (a_n) as

$$M_{(a_n)}(x_n) := A(x_n) := (a_n x_n), \tag{9.5}$$

with $D(A) := \{(x_n) \in \ell^p : (a_n x_n) \in \ell^p\}$.

Proposition 9.20. *The operator $A = M_{(a_n)}$ defined in (9.5) has the following properties:*

a) $A \in \mathcal{L}(X)$ *if and only if* $(a_n) \in \ell^\infty$.
b) $D(A)$ *is dense.*
c) A *is closed.*

Proof. a) Note first that A is clearly linear. If $(a_n) \in \ell^\infty$, then

$$\|A(x_n)\|_p \leq \|(a_n)\|_\infty \|(x_n)\|_p,$$

therefore A is bounded.

Suppose now that $(a_n) \notin \ell^\infty$. Then $\|A\| \geq \|Au_k\| = |a_k|$, where u_k is the standard basis vector with 1 in the kth component and 0's everywhere else. Hence, in this case A is not bounded.

b) Take an arbitrary element of $(y_n) \in X = \ell^p$. Then the sequence

$$\begin{pmatrix} y_1 \\ 0 \\ 0 \\ 0 \\ \vdots \end{pmatrix}, \begin{pmatrix} y_1 \\ y_2 \\ 0 \\ 0 \\ \vdots \end{pmatrix}, \begin{pmatrix} y_1 \\ y_2 \\ y_3 \\ 0 \\ \vdots \end{pmatrix}, \dots$$

converges to (y_n) and belongs to $D(A)$.

c) Let $\varphi_k = \left(f_n^{(k)} \right) \in D(A)$ such that $\varphi_k \to \varphi = (f_n)$ and $A\varphi_k \to \psi = (g_n)$ in X. Then coordinatewise, for each $n \in \mathbb{N}$, we have $f_n^{(k)} \to f_n$ as $k \to \infty$, and $a_n f_n^{(k)} \to g_n$. Thus, $a_n f_n = g_n$. Since $\psi \in \ell^p$, this gives $f \in D(A)$ and $Af = g$. $\qquad\square$

From the previous proof we immediately obtain the identity

$$\|A\| = \|(a_n)\|_\infty$$

for $(a_n) \in \ell^\infty$. Let us characterize now when a multiplication operator is a semigroup generator.

Proposition 9.21. *Suppose that there is $\omega \in \mathbb{R}$ such that $\operatorname{Re} a_n \leq \omega$ for all $n \in \mathbb{N}$. Then $A = M_{(a_n)}$ defined by rule (9.5) generates a C_0-semigroup $(T(t))_{t \geq 0}$ given by*

$$T(t)(x_n) = \left(e^{t a_n} x_n \right).$$

Proof. Assume $\operatorname{Re} a_n \leq \omega$ and let us check the desired properties. Clearly,

$$\|T(t)\| = \left\| (e^{t a_n}) \right\|_\infty = \sup_{n \in \mathbb{N}} e^{t \operatorname{Re} a_n} \leq e^{t\omega}$$

holds and hence $T(t) \in \mathcal{L}(X)$. Showing the semigroup property is straightforward. By Proposition 9.10, it suffices to show strong continuity at zero and on a dense subset, say c_{00}, which contains all sequences that have only finitely many nonzero terms. Also note that for the standard basis vectors u_n we have

$$\|T(t)u_n - u_n\| = \|((e^{ta_n} - 1)u_n)\| = |e^{ta_n} - 1| \longrightarrow 0$$

as $t \to 0$. Then by linearity,

$$\left\| (T(t) - I)\left(\sum_{n=1}^{N} x_n u_n \right) \right\| \leq \sum_{n=1}^{N} \|x_n\|\, \|(T(t) - I)u_n\| \longrightarrow 0$$

as $t \to 0$. Thus we have a strong convergence on c_{00} and therefore on ℓ^p.

Let B be the generator of $(T(t))_{t \geq 0}$. We have to show that $B = A$. If $x = (x_n) \in D(B)$, then

$$\lim_{t \to 0} \frac{T(t)x - x}{t} = \lim_{t \to 0}\left(\frac{e^{ta_n} - 1}{t} x_n \right) = \left(\lim_{t \to 0} \frac{e^{ta_n} - 1}{t} x_n \right)$$

exists in ℓ^p. Then by elementary calculus, for each n it holds that

$$\left(\frac{e^{ta_n} - 1}{t} \right) x_n \longrightarrow a_n x_n \quad \text{for } t \downarrow 0,$$

which implies that $x \in D(A)$ and $Bx = Ax$. Therefore, $B \subseteq A$.

Let us take now $x = (x_n) \in D(A)$. Then for each $n \in \mathbb{N}$, we define

$$y_n := \lim_{t \to 0}\left(\frac{e^{ta_n} - 1}{t} \right) x_n = a_n x_n,$$

and by assumption we have $y = (y_n) \in \ell^p$. Moreover, for each $n \in \mathbb{N}$ we see that

$$\left| \frac{e^{ta_n} - 1}{ta_n} \right| = \left| \frac{1}{t} \int_0^t e^{sa_n} ds \right| \leq e^{t \max\{\omega, 0\}}$$

since $\operatorname{Re} a_n \leq \omega$ holds. This implies that

$$\left| \left(\frac{T(t)x - x}{t} \right)_n \right| \leq |a_n x_n| e^{t \max\{\omega, 0\}} = |y_n| e^{t \max\{\omega, 0\}}.$$

Recall that $(y_n) \in \ell^p$. Hence, $\frac{1}{t}(T(t)x - x)$ converges in ℓ^p as $t \to 0$, which means that $x \in D(B)$. $\qquad\square$

9.6 Gaussian Semigroup

In this section we analyse another fundamental example where one can determine the semigroup explicitly.

Let us consider the heat equation on the entire line $\mathbb{R}$:

$$\begin{cases} \partial_t w(t, x) = \partial_{xx} w(t, x), & t \geq 0,\ x \in \mathbb{R}, \\ w(0, x) = w_0(x), & x \in \mathbb{R}, \end{cases} \tag{9.6}$$

where $w(t, x)$ can be interpreted as the heat density and w_0 is a function on $\mathbb{R}$ providing the initial heat profile. We look for the solution to this problem as an orbit map of some C_0-semigroup. To find a candidate for this semigroup, we first make some calculations by using the *Fourier transform*, which is given for $f \in L^1(\mathbb{R})$ by the Fourier integral

$$\hat{f}(\xi) := \mathcal{F}(f)(\xi) := \frac{1}{\sqrt{2\pi}} \int_{-\infty}^{\infty} e^{-i\xi x} f(x) \mathrm{d}x. \tag{9.7}$$

Let us recall here the important fact that the operator $\mathcal{F}$ maps differentiation to multiplication by the Fourier variable $i\xi$, that is,

$$\mathcal{F}(f')(\xi) = i\xi \mathcal{F}(f)(\xi).$$

If we take the Fourier transform of equation (9.6) with respect to x and interchange the actions of $\mathcal{F}$ and ∂_t, we obtain

$$\begin{cases} \partial_t \hat{w}(t, \xi) = -\xi^2 \hat{w}(t, \xi), & t \geq 0,\ \xi \in \mathbb{R}, \\ \hat{w}(0, \xi) = \hat{w}_0(\xi), & \xi \in \mathbb{R}. \end{cases}$$

This is an ordinary differential equation for $\hat{w}$ in each point ξ, which can be solved directly:

$$\hat{w}(t, \xi) = e^{-t|\xi|^2} \hat{w}_0(\xi).$$

To get back our unknown function w, we take the inverse Fourier transform of this solution:

$$w(t, \cdot) = \mathcal{F}^{-1}(\hat{w}(t, \cdot)) = \frac{1}{\sqrt{2\pi}} \mathcal{F}^{-1}\left(e^{-t|\cdot|^2}\right) * \mathcal{F}^{-1}(\hat{w}_0),$$

where we used that $\mathcal{F}^{-1}$ maps products to convolutions. Let us recall that

$$\mathcal{F}^{-1}\left(e^{-t|\cdot|^2}\right)(x) = \frac{1}{\sqrt{2t}} e^{-\frac{|x|^2}{4t}}.$$

Setting

$$g_t(x) := \frac{1}{\sqrt{4\pi t}} e^{-\frac{|x|^2}{4t}}, \quad t > 0,$$

we see that the solution w to equation (9.6) is of the form

$$w(t) = g_t * w_0 \quad \text{for } t > 0.$$

Let us pause here and collect some fundamental properties of the function g_t.

Remark 9.22.

a) The *standard Gaussian function*

$$g(x) := \frac{1}{\sqrt{4\pi}} e^{-\frac{x^2}{4}}$$

satisfies $g \geq 0$, $\|g\|_1 = 1$ and the function g belongs to $L^p(\mathbb{R})$ for all $p \in [1, \infty]$.

b) We have $g_t(x) = \frac{1}{\sqrt{t}} g\left(\frac{x}{\sqrt{t}}\right)$, hence $g_t \geq 0$, $\|g_t\|_1 = 1$ and

$$\lim_{t \downarrow 0} \int_{|x| > r} g_t(s) \, ds = 0 \quad \text{for all } r > 0 \text{ fixed.}$$

The function

$$G(t, x, y) := g_t(x - y), \quad t > 0, \ x \in \mathbb{R}, \ y \in \mathbb{R},$$

is called the *Gaussian kernel* on $\mathbb{R}$ and gives a rise to a semigroup, called the *Gaussian semigroup*.

Proposition 9.23. *Let $p \in [1, \infty)$. For $f \in L^p(\mathbb{R})$ and $t > 0$ define*

$$(T(t)f)(x) := (g_t * f)(x) = \frac{1}{\sqrt{4\pi t}} \int_{\mathbb{R}} f(y) e^{-\frac{(x-y)^2}{4t}} \, dy = \int_{\mathbb{R}} f(y) G(t, x, y) dy$$

and set $T(0)f := f$. Then $T(t)$ is a linear operator on $L^p(\mathbb{R})$, satisfies the norm estimate $\|T(t)\| \leq 1$, and $(T(t))_{t \geq 0}$ is a C_0-semigroup. Its generator A is given by

$$D(A) := \{f \in L^p(\mathbb{R}) : f'' \in L^p(\mathbb{R})\},$$
$$Af := f'',$$

where f'' is the second derivative of $f \in L^p(\mathbb{R})$ in the sense of distributions (see A.11).

Proof. Let $f \in L^p(\mathbb{R})$. By Young's inequality (see Lemma A.14) and since $g_t \in L^1(\mathbb{R})$, we obtain that the convolution $g_t * f$ exists and

$$\|g_t * f\|_p \leq \|g_t\|_1 \cdot \|f\|_p = \|f\|_p.$$

In particular, $g_t * f$ belongs to $L^p(\mathbb{R})$. Since linearity of $f \mapsto g_t * f$ follows immediately from the definition, we obtain that $T(t)$ is a linear contraction.

To prove the semigroup property, fix $f \in L^1(\mathbb{R}) \cap L^p(\mathbb{R})$. Taking the Fourier transform of $g_t * (g_s * f)$ we obtain

$$\mathcal{F}(g_t * (g_s * f)) = \sqrt{2\pi}\,\mathcal{F}(g_t) \cdot \mathcal{F}(g_s * f) = (2\pi)\mathcal{F}(g_t) \cdot \mathcal{F}(g_s) \cdot \mathcal{F}(f).$$

Recall that

$$\mathcal{F}(g_t)(\xi) = \frac{1}{\sqrt{2\pi}} e^{-t\xi^2},$$

which implies

$$\mathcal{F}(g_t)(\xi) \cdot \mathcal{F}(g_s)(\xi) = \frac{1}{2\pi} e^{-(t+s)\xi^2} = \frac{1}{\sqrt{2\pi}} \mathcal{F}(g_{t+s})(\xi).$$

This yields

$$\mathcal{F}(g_t * (g_s * f)) = \sqrt{2\pi}\mathcal{F}(g_{t+s}) \cdot \mathcal{F}(f) = \mathcal{F}(g_{t+s} * f),$$

hence $g_t * (g_s * f) = g_{t+s} * f$ by the injectivity of the Fourier transform. Therefore, the equality $T(t)T(s)f = T(t+s)f$ holds for $f \in L^1(\mathbb{R}) \cap L^p(\mathbb{R})$. By the uniform boundedness of the semigroup operators and by the denseness of this subspace in $L^p(\mathbb{R})$, we obtain the equality everywhere.

From the properties of the function g_t listed in Remark 9.22.b), it is possible to prove that $g_t * f \to f$ in $L^p(\mathbb{R})$ if $t \downarrow 0$ for every step function f. Since step functions are dense in $L^p(\mathbb{R})$, and $\|T(t)\| \leq 1$, we see that the semigroup $(T(t))_{t\geq0}$ is strongly continuous.

Now we identify the generator of $(T(t))_{t\geq0}$. Denote by B the operator defined by

$$D(B) := \{f \in L^p(\mathbb{R}) : f'' \in L^p(\mathbb{R})\},$$
$$Bf := f''.$$

It is easy to see that B is a closed linear operator on $L^p(\mathbb{R})$. On the other hand, by Proposition A.47, we deduce that $D(B) = W^{2,p}(\mathbb{R})$. Let now $f \in D(B)$. By the denseness of $C_c^\infty(\mathbb{R})$ in $W^{2,p}(\mathbb{R})$, there is $(f_k) \subset C_c^\infty(\mathbb{R})$ such that $f_k \to f$ and $Bf_k \to Bf$ as $k \to \infty$. If we denote by $a(x) := -x^2$, then

$$\mathcal{F}\left(\frac{\mathrm{d}^2}{\mathrm{d}x^2} T(t)f_k\right) = a\mathcal{F}(T(t)f_k) = \mathcal{F}(g_t)a\mathcal{F}(f_k)$$

$$= \mathcal{F}(g_t)\mathcal{F}(f_k'') = \mathcal{F}(T(t)f_k'').$$

So, $BT(t)f_k = T(t)Bf_k$, and by the closedness of B, we have $T(t)f \in D(B)$ and

$$BT(t)f = T(t)Bf, \quad t \geq 0. \tag{9.8}$$

Take now the *Schwartz space* $\mathcal{S}(\mathbb{R})$ of rapidly decreasing functions, defined in (A.7), and recall that for all $\varphi \in \mathcal{S}(\mathbb{R})^*$ and $f \in \mathcal{S}(\mathbb{R})$ the identity

$$\mathcal{F}(f * \varphi) = \sqrt{2\pi}\,\mathcal{F}(f) \cdot \mathcal{F}(\varphi)$$

holds. For any $\varphi \in \mathcal{S}(\mathbb{R})$, we thus have

$$\mathcal{F}\left(\int_0^t T(s)\varphi''\mathrm{d}s\right)(x) = \int_0^t \mathcal{F}(T(s)\varphi'')(x)\mathrm{d}s$$

$$= \int_0^t \mathrm{e}^{-sx^2} a(x)\mathcal{F}(\varphi)(x)\mathrm{d}s$$

$$= (\mathrm{e}^{-tx^2} - 1)\mathcal{F}(\varphi)(x)$$

$$= \mathcal{F}(T(t)\varphi - \varphi)(x).$$

Hence,

$$\int_0^t T(s)\varphi''\mathrm{d}s = T(t)\varphi - \varphi, \quad \varphi \in \mathcal{S}(\mathbb{R}), \ t \geq 0. \tag{9.9}$$

So, using Fubini's theorem (see Theorem A.24) and relation (9.9) we deduce that

$$\left\langle \varphi, B \int_0^t T(s)f\mathrm{d}s \right\rangle = \left\langle \varphi, \partial_{xx} \int_0^t T(s)f\mathrm{d}s \right\rangle = \left\langle \varphi'', \int_0^t T(s)f\mathrm{d}s \right\rangle$$

$$= \int_0^t \langle \varphi'', T(s)f \rangle \mathrm{d}s = \int_0^t \langle T(s)\varphi'', f \rangle \mathrm{d}s$$

$$= \langle T(t)\varphi - \varphi, f \rangle = \langle \varphi, T(t)f - f \rangle$$

for any $\varphi \in \mathcal{S}(\mathbb{R})$, $f \in \mathrm{L}^p(\mathbb{R})$, and $t \geq 0$. This proves that $\int_0^t T(s)f\mathrm{d}s \in D(B)$ for any $f \in \mathrm{L}^p(\mathbb{R})$ and

$$B \int_0^t T(s)f\mathrm{d}s = T(t)f - f, \quad f \in \mathrm{L}^p(\mathbb{R}), \ t \geq 0.$$

This together with (9.8) proves that $A = B$. $\qquad\square$

Remark 9.24. More generally, one can see, using the same arguments, that the rule

$$(T(t)f)(x) := (4\pi t)^{-N/2} \int_{\mathbb{R}^N} f(y)\mathrm{e}^{-\frac{|x-y|^2}{4t}}\,\mathrm{d}y =: \int_{\mathbb{R}^N} f(y)G(t,x,y)\mathrm{d}y, \quad t > 0,$$

$$T(0)f := f,$$

defines a C_0-semigroup on $\mathrm{L}^p(\mathbb{R}^N)$, $1 \leq p < \infty$. Its generator can be identified with the Laplacian Δ with the maximal domain

$$D(\Delta) = \{f \in \mathrm{L}^p(\mathbb{R}^N) : \Delta f \in \mathrm{L}^p(\mathbb{R}^N)\}.$$

We remark that, using deep results from harmonic analysis, it is possible to show for $1 < p < \infty$ that

$$D(\Delta) = \mathrm{W}^{2,p}(\mathbb{R}^N).$$

9.7 Resolvent of a Generator

We have seen in Part I that spectral analysis of matrices, more precisely, the determination of eigenvalues and eigenvectors, led to a construction of the semigroup generated by them. We investigate now some basic spectral properties of semigroup generators. Let us begin with the following fundamental spectral theoretic notions.

Definition 9.25. Let A be a linear operator defined on a linear subspace $D(A)$ of a Banach space X.

a) The *spectrum* of A is the set

$$\sigma(A) := \{\lambda \in \mathbb{C} : \lambda - A : D(A) \to X \text{ is not bijective}$$
$$\text{or its inverse is not continuous}\}.$$

Its subset

$$\sigma_{\mathrm{p}}(A) := \{\lambda \in \mathbb{C} : \lambda - A \text{ is not injective }\}$$

is called the *point spectrum* of A and consists of *eigenvalues*.

b) The *resolvent set* of A is $\rho(A) := \mathbb{C} \setminus \sigma(A)$, i.e.,

$$\rho(A) := \{\lambda \in \mathbb{C} : \lambda - A : D(A) \to X \text{ is bijective with a continuous inverse}\}.$$

c) If $\lambda \in \rho(A)$, then $\lambda - A$ is bijective, hence has an algebraic inverse $(\lambda - A)^{-1}$. We call this operator the *resolvent of A at the point* λ and denote it by

$$(\lambda - A)^{-1} =: R(\lambda, A) \in \mathcal{L}(X).$$

It is important to note that if A is closed and if λ is such that $\lambda - A : D(A) \to X$ is bijective, then its algebraic inverse

$$(\lambda - A)^{-1} : X \longrightarrow D(A)$$

is defined on the entire X. Since A is closed, so are $\lambda - A$ and its inverse. As a consequence of the closed graph theorem, see Theorem A.21, we immediately obtain that the operator $(\lambda - A)^{-1}$ is bounded. Then the following holds.

Proposition 9.26. *For a closed linear operator A one has*

$$\rho(A) := \{\lambda \in \mathbb{C} : \lambda - A : D(A) \to X \text{ is bijective}\}.$$

Remark 9.27. A linear operator A on a Banach space X with a nonempty resolvent set is always closed. To see this, let $\lambda \in \rho(A)$ and consider a sequence $(f_k) \subset D(A)$ and $f, g \in X$ such that $\lim_{k \to \infty} f_k = f$ and $\lim_{k \to \infty} Af_k = g$. Then,

$$\lim_{k \to \infty} R(\lambda, A)Af_k = \lim_{k \to \infty} \lambda R(\lambda, A)f_k - f_k$$
$$= \lambda R(\lambda, A)f - f$$
$$= R(\lambda, A)g.$$

Hence, $f \in D(A)$ and $Af = g$.

Here we summarize some fundamental properties of spectrum and resolvent.

Proposition 9.28. *Let X be a Banach space and let A be a linear operator with domain $D(A) \subseteq X$. Then the following assertions are true.*

a) *The resolvent set $\rho(A)$ is open, hence its complement, the spectrum $\sigma(A)$, is closed.*

b) *The mapping*
$$\rho(A) \ni \lambda \longmapsto R(\lambda, A) \in \mathcal{L}(X)$$
is complex differentiable. Moreover, for $k \in \mathbb{N}$ we have
$$\frac{d^k}{d\lambda^k} R(\lambda, A) = (-1)^k k!\, R(\lambda, A)^{k+1}.$$

c) *If $A \in \mathcal{L}(X)$, then for every $\lambda \in \mathbb{C}$ with $|\lambda| > r(A)$ we have $\lambda \in \rho(A)$ and the Neumann series representation of the resolvent holds:*
$$R(\lambda, A) = \sum_{k=0}^{\infty} \frac{A^k}{\lambda^{k+1}}.$$

d) *Let $\lambda_k \in \rho(A)$ with $\lim_{k \to \infty} \lambda_k = \lambda_0$. Then $\lambda_0 \in \sigma(A)$ if and only if*
$$\lim_{k \to \infty} \|R(\lambda_k, A)\| = \infty.$$

Proof. a) follows from the Neumann series representation of the resolvent: For $\mu \in \rho(A)$ we have
$$\lambda - A = (I - (\mu - \lambda)R(\mu, A))(\mu - A).$$
Hence, for $|\lambda - \mu| < \frac{1}{\|R(\mu, A)\|}$ we obtain, by (A.4),
$$R(\lambda, A) = \sum_{k=0}^{\infty} (\lambda - \mu)^k R(\mu, A)^{k+1}. \tag{9.10}$$

b) follows from the power series representation in a) and from the fact that a power series is always a Taylor series.

c) follows by similar Neumann series arguments as in the proof of a) since
$$(\lambda - A) = \lambda \left(I - \frac{A}{\lambda} \right)$$
formally yields the series, which converges if $|\lambda| > \limsup \|A^k\|^{1/k} =: r(A)$.

d) Suppose that $\lambda_0 \in \rho(A)$. The resolvent map is continuous and remains bounded on the compact set $\{\lambda_k : k \geq 0\}$, which contradicts the assertion on the limit of $\|R(\lambda_k, A)\|$, hence $\lambda_0 \in \sigma(A)$. For the converse implication observe that our considerations at the beginning of the proof yield $\|R(\mu, A)\| \geq \frac{1}{\text{dist}(\mu, \sigma(A))}$ for all $\mu \in \rho(A)$ (see also Corollary 9.30). $\qquad\square$

The following result is also known as *the spectral mapping theorem for the resolvent*.

Proposition 9.29. *Let A be a linear operator with $\rho(A) \neq \emptyset$. Then for any $\lambda \in \rho(A)$ the identity*

$$\sigma(R(\lambda, A)) \setminus \{0\} = \left\{ \frac{1}{\lambda - \mu} : \mu \in \sigma(A) \right\}$$

holds.

Proof. For $0 \neq \alpha \in \mathbb{C}$ and $\lambda \in \rho(A)$, we have

$$
\begin{aligned}
(\alpha - R(\lambda, A))f &= \alpha \left[(\lambda - \tfrac{1}{\alpha}) - A \right] R(\lambda, A)f \quad \text{for all } f \in X, \\
&= \alpha R(\lambda, A) \left[(\lambda - \tfrac{1}{\alpha}) - A \right] f \quad \text{for all } f \in D(A).
\end{aligned}
$$

Thus, $\alpha \in \sigma(R(\lambda, A))$ if and only if $\lambda - \tfrac{1}{\alpha} \in \sigma(A)$. $\qquad\square$

As a corollary one determines the spectral radius of $R(\lambda, A)$.

Corollary 9.30. *For $\lambda \in \rho(A)$ one has*

$$\mathrm{dist}(\lambda, \sigma(A)) = \frac{1}{\mathrm{r}(R(\lambda, A))} \geq \frac{1}{\|R(\lambda, A)\|}.$$

Proof. Let $\lambda \in \rho(A)$. Then, by Proposition 9.29,

$$
\begin{aligned}
\mathrm{dist}(\lambda, \sigma(A)) &= \inf\{|\lambda - \mu| : \mu \in \sigma(A)\} \\
&= \left(\sup \left\{ \left| \frac{1}{\lambda - \mu} \right| : \mu \in \sigma(A) \right\} \right)^{-1} \\
&= (\max\{|\alpha| : \alpha \in \sigma(R(\lambda, A))\})^{-1} \\
&= \frac{1}{\mathrm{r}(R(\lambda, A))} \geq \frac{1}{\|R(\lambda, A)\|}
\end{aligned}
$$

which proves the assertion. $\qquad\square$

Having revealed some properties of the resolvent of a general closed linear operator, we now focus on generators of strongly continuous semigroups. Our aim is to prove that the resolvent set of a generator A is non-empty and to relate the resolvent of A to the semigroup $(T(t))_{t \geq 0}$. The first step is provided by the next lemma.

Lemma 9.31. *Let $(T(t))_{t \geq 0}$ be a C_0-semigroup with generator A. Then for all $\lambda \in \mathbb{C}$ and $t > 0$ the identities*

$$
\mathrm{e}^{-\lambda t} T(t)f - f = (A - \lambda) \int_0^t \mathrm{e}^{-\lambda s} T(s)f \, \mathrm{d}s \qquad \text{if } f \in X,
$$

$$
= \int_0^t \mathrm{e}^{-\lambda s} T(s)(A - \lambda)f \, \mathrm{d}s \qquad \text{if } f \in D(A)
$$

hold.

Proof. Observe that $S(t) = e^{-\lambda t}T(t)$ is also a C_0-semigroup with generator $B = A - \lambda$, see Exercise 4. Applying Proposition 9.16.b) to $S(t)$, the result follows. $\square$

As a consequence of the second identity, we obtain a very useful spectral inclusion property for the point spectra.

Corollary 9.32. *Let A be the generator of a C_0-semigroup $(T(t))_{t\geq 0}$. If λ is an eigenvalue of A with a corresponding eigenvector $f \in D(A)$, then for all $t \geq 0$ $e^{\lambda t}$ is an eigenvalue of $T(t)$ with corresponding eigenvector f.*

With the help of Lemma 9.31 we also obtain the following important relations between the resolvent of the generator and the semigroup.

Proposition 9.33. *Let $(T(t))_{t\geq 0}$ be a C_0-semigroup of type (M, ω) with generator A. Then the following assertions are true.*

a) *For all $f \in X$ and $\lambda \in \mathbb{C}$ with $\mathrm{Re}\,\lambda > \omega$,*

$$R(\lambda, A)f = \int_0^\infty e^{-\lambda s}T(s)f\,ds = \lim_{N\to\infty}\int_0^N e^{-\lambda s}T(s)f\,ds. \qquad (9.11)$$

b) *For all $f \in X$, $\lambda \in \mathbb{C}$ with $\mathrm{Re}\,\lambda > \omega$ and $k \in \mathbb{N}$,*

$$R(\lambda, A)^k f = \frac{1}{(k-1)!}\int_0^\infty s^{k-1}e^{-\lambda s}T(s)f\,ds.$$

c) *For all $\lambda \in \mathbb{C}$ with $\mathrm{Re}\,\lambda > \omega$ and $k \in \mathbb{N}$,*

$$\|R(\lambda, A)^k\| \leq \frac{M}{(\mathrm{Re}\,\lambda - \omega)^k}. \qquad (9.12)$$

Proof. a) By Lemma 9.31, the closedness of A (see Lemma 9.17), and by taking the limit as $t \to \infty$, we conclude that for $\mathrm{Re}\,\lambda > \omega$ we have

$$-f = (A - \lambda)\int_0^\infty e^{-\lambda s}T(s)f\,ds \qquad \text{if } f \in X,$$

$$= \int_0^\infty e^{-\lambda s}T(s)(A - \lambda)f\,ds \qquad \text{if } f \in D(A).$$

Since this expression gives a bounded operator, a) is proved.

b) Notice that

$$R(\lambda, A)^k f = \frac{(-1)^{k-1}}{(k-1)!}\frac{d^{k-1}}{d\lambda^{k-1}}R(\lambda, A)f = \frac{1}{(k-1)!}\int_0^\infty s^{k-1}e^{-\lambda s}T(s)f\,ds.$$

c) By a simple norm estimate we obtain

$$\|R(\lambda, A)^k f\| \leq \frac{1}{(k-1)!}\int_0^\infty s^{k-1}e^{-\mathrm{Re}\,\lambda s}Me^{\omega s}\|f\|ds$$

$$\leq \frac{M\|f\|}{(k-1)!}\int_0^\infty s^{k-1}e^{(\omega - \mathrm{Re}\,\lambda)s}ds = \frac{M}{(\mathrm{Re}\,\lambda - \omega)^k}\|f\|,$$

which finishes the proof. $\square$

Let us summarize the above as follows.

If A is the generator of an operator semigroup $(T(t))_{t\geq 0}$, then it is closed, densely defined, and a suitable right half-plane belongs to its resolvent set, where the estimate (9.12) holds. Using (9.11), we see that the resolvent operators are given by the *Laplace transform* of the semigroup, also called the *integral representation of the resolvent*.

9.8 Adjoint Semigroups

There are many ways to construct a new C_0-semigroup from a given one. Some examples can be found in Exercise 4. Here we briefly discuss another construction we shall need later. Recall the definitions of a dual space and of an adjoint operator in Appendix, Section A.8.

Definition 9.34. Let $(T(t))_{t\geq 0}$ be a C_0-semigroup on a Banach space X. The *adjoint semigroup* $(T(t)^*)_{t\geq 0}$ consists of all adjoint operators $T(t)^*$ on the dual space X^*.

Since
$$\langle f, T(t)^* f^* \rangle = \langle T(t)f, f^* \rangle$$
holds for every $f \in X$, $f^* \in X^*$, and $t \geq 0$, the adjoint of a C_0-semigroup is always *weak*-continuous*. However, it is in general not strongly continuous, as the following example shows.

Example 9.35. Take the left-shift semigroup $(T(t))_{t\geq 0}$ from Example 9.11 on the space $L^1(\mathbb{R})$. Then for $g \in L^\infty(\mathbb{R})$ one has that
$$\langle T(t)f, g \rangle = \int_{\mathbb{R}} f(t+s)g(s)\, \mathrm{d}s = \int_{\mathbb{R}} f(s)g(s-t)\, \mathrm{d}s.$$

Hence, the adjoint semigroup of the left shift is the *right-shift semigroup* on $L^\infty(\mathbb{R})$. It follows from the discussion in Example 9.5 that $t \mapsto T(t)f$ is continuous in the supremum norm only if f is uniformly continuous. Hence, the right-shift semigroup is not strongly continuous on $L^\infty(\mathbb{R})$.

To overcome this problem, we restrict the adjoint semigroup to an appropriate subspace of the dual space.

Definition 9.36. The *semigroup dual space* or *sun dual* of a C_0-semigroup $(T(t))_{t\geq 0}$ on a Banach space X is defined as
$$X^\odot := \left\{ f^* \in X^* \ : \ \lim_{t\downarrow 0} \|T(t)^* f^* - f^*\| = 0 \right\}.$$

This is a closed and $T(t)^*$-invariant subspace of X^*, hence the *sun semigroup* is the semigroup of the adjoint operators restricted to this space,
$$T(t)^\odot := T(t)^*|_{X^\odot}.$$

Since $\|T(t)^*\| = \|T(t)\|$, by Proposition 9.10 the sun dual $X^\odot \subset X^*$ is a closed subspace and $(T(t)^\odot)_{t\geq 0}$ a C_0-semigroup on it. We show that the space $X^\odot$ is big enough.

Lemma 9.37. *We have $D(A^*) \subset X^\odot$.*

Proof. Without loss of generality we may assume that $\|T(t)\| \leq M$ for all $t \geq 0$. For any $g^* \in D(A^*)$ we have by Proposition 9.16 that

$$|\langle f, T(t)^*g^* - g^*\rangle| = |\langle T(t)f - f, g^*\rangle|$$
$$= \left|\left\langle A\int_0^t T(s)f \, ds, g^*\right\rangle\right|$$
$$\leq tM\|f\|\|A^*g^*\|$$

for all $f \in X$. This converges uniformly to zero as $t \to 0$ for $\|f\| \leq 1$, hence

$$\lim_{t\downarrow 0}\|T(t)^*g^* - g^*\| = 0$$

and $g^* \in X^\odot$. $\square$

Denote by $A^\odot$ the generator of the sun semigroup $(T(t)^\odot)_{t\geq 0}$. There is a nice relation between this operator and the adjoint A^* of the generator of $(T(t))_{t\geq 0}$.

Proposition 9.38. *For a C_0-semigroup $(T(t))_{t\geq 0}$ on X with generator A the following holds.*

a) *The generator $A^\odot$ is the part of A^* in $X^\odot$, i.e.,*

$$A^\odot f^* = A^* f^* \quad \text{for } f^* \in D(A^\odot) = \{f^* \in D(A^*) \ : \ A^* f^* \in X^\odot\}.$$

b) *$X^\odot = \overline{D(A^*)}$.*

Proof. a) Let $f^* \in D(A^\odot)$. Then

$$\left\langle f, \frac{1}{t}(T(t)^\odot f^* - f^*)\right\rangle = \left\langle f, \frac{1}{t}(T(t)^* f^* - f^*)\right\rangle = \left\langle \frac{1}{t}(T(t)f - f), f^*\right\rangle.$$

So, letting $t \to 0^+$, we obtain

$$\langle Af, f^*\rangle = \langle f, A^\odot f^*\rangle \quad \text{for all } f \in D(A).$$

Hence, from the definition of A^*, see Appendix A.8, we have $f^* \in D(A^*)$ and $A^\odot f^* = A^* f^*$.

b) Since the domain of the generator $D(A^\odot) \subset D(A^*)$ is dense in $X^\odot$, this follows by Lemma 9.37. $\square$

To obtain the sun semigroup it thus suffices to restrict the adjoint semigroup to the closure of the domain of the adjoint A^* of the original generator A.

Example 9.39. Let us continue Example 9.35 and take again the left-shift semi-group on $X = L^1(\mathbb{R})$. Since the largest subspace of $L^\infty(\mathbb{R})$ on which the right-shift operators form a C_0-semigroup is the space $C_{ub}(\mathbb{R})$ of bounded uniformly continuous functions, we have $X^\odot = C_{ub}(\mathbb{R})$. For the generator $Af = f'$ we have

$$D(A) = \{f \in L^1(\mathbb{R}) \quad : \; f \text{ absolutely continuous and } f' \in L^1(\mathbb{R})\},$$
$$D(A^*) = \{f \in L^\infty(\mathbb{R}) \quad : \; f \text{ absolutely continuous and } f' \in L^\infty(\mathbb{R})\},$$
$$D(A^\odot) = \{f \in C_{ub}(\mathbb{R}) \; : \; f \in C^1(\mathbb{R}) \text{ and } f' \in C_{ub}(\mathbb{R})\}.$$

Also the spectra of the operators A, A^*, and $A^\odot$ coincide, which is a very useful property.

Proposition 9.40. *For a C_0-semigroup $(T(t))_{t\geq 0}$ on X with generator A we have the following equalities.*

a) $\sigma(A) = \sigma(A^*) = \sigma(A^\odot)$.

b) $s(A) = s(A^*) = s(A^\odot)$.

Proof. a) Since the first equality holds by Proposition A.32, it suffices to show that $\rho(A^*) = \rho(A^\odot)$. First notice that

$$D(A^\odot) \subset D(A^*) \subset X^\odot \subset X^*.$$

For any $\lambda \in \rho(A^*)$ we have

$$R(\lambda, A^*)X^\odot \subset R(\lambda, A^*)X^* = D(A^*) \subset X^\odot$$

and the part of $R(\lambda, A^*)$ in $X^\odot$ equals $R(\lambda, A^\odot)$. Hence, we have $\rho(A^*) \subset \rho(A^\odot)$.

For the converse inclusion, observe that the part A_1 of A^* in $D(A^*)$ coincides with the part of $A^\odot$ in $D(A^*)$. By the same argument as above, one obtains $\rho(A^\odot) \subset \rho(A_1)$, and since $\rho(A_1) = \rho(A^*)$, we have the desired equality.

b) This is a consequence of a). $\qquad\square$

9.9 Notes and Remarks

Operator semigroups have been widely studied during the last decades and there are many monographs dealing with them. We mention here the excellent graduate texts by Engel and Nagel [43, 44], which motivated many parts of this manuscript. The first milestone in the theory was the opus of Hille and Phillips [66]. An important later reference is the book of Pazy [110], which was written from a PDE perspective, and Goldstein [54], which contains lots of other applications as well.

For the Fourier transform used in Section 9.6 we suggest the monograph by Stein and Weiss [132]. Distributions and the Schwartz space, used in the proof of Proposition 9.23, are briefly recalled in Section A.11 of the Appendix. For the domain of the Laplacian in higher dimensions as mentioned at the end of Section 9.6 we refer to Stein [131, Theorem VI.4.4].

9.10   Exercises

1. Let $A \in \mathcal{L}(X)$. Show that the Cauchy problem

$$\dot{F}(t) = AF(t),$$
$$F(0) = B \in \mathcal{L}(X),$$

   has a unique solution $F : \mathbb{R}_+ \to \mathcal{L}(X)$.

2. For $A \in \mathcal{L}(X)$ prove Euler's formulas

$$\lim_{k \to \infty} \left(I + \frac{t}{k} A \right)^k = e^{tA} \quad \text{and} \quad \lim_{k \to \infty} \left(I - \frac{t}{k} A \right)^{-k} = e^{tA}$$

   for $t \geq 0$.

3. Let $X = C_0(\mathbb{R})$ and $q \in C(\mathbb{R})$. Consider the operator $(Af)(s) := q(s)f(s)$ with $D(A) := \{f \in X : qf \in X\}$ and make analogous statements as in Section 9.5. Prove these statements.

4. Consider a C_0-semigroup $(T(t))_{t \geq 0}$ of type (M, ω) with generator A on a Banach space X. For each $(S(t))_{t \geq 0}$ defined below, prove that it is a C_0-semigroup, and determine its type and its generator.

   a) $S(t) := R^{-1}T(t)R$ for a boundedly invertible transformation $R \in \mathcal{L}(X)$.

   b) $S(t) := e^{tz}T(t)$ for some $z \in \mathbb{C}$.

   c) $S(t) := T(\alpha t)$ for some $\alpha \geq 0$.

5. Consider the closed subspace

$$C_0([0,1)) := \{f \in C([0,1]) : f(1) = 0\}$$

   of the Banach space $C([0,1])$ of continuous functions on $[0,1]$. Define the nilpotent left-shift semigroup on it and determine its generator.

6. Let $F_b(\mathbb{R})$ denote the linear space of all bounded functions $\mathbb{R} \to \mathbb{R}$. Define

$$(T(t)f)(s) := f(t+s) \quad \text{for } f \in F_b(\mathbb{R}),\ s \in \mathbb{R},\ t \geq 0.$$

   Prove that each of the following spaces is a Banach space with the supremum norm $\|\cdot\|_\infty$ and invariant under $T(t)$ for all $t \geq 0$. Is $(T(t))_{t \geq 0}$ a C_0-semigroup on these spaces?

   a) $F_b(\mathbb{R})$.

   b) $C_b(\mathbb{R}) = $ the space bounded and continuous functions.

   c) $C_0(\mathbb{R}) = $ the space bounded and continuous functions vanishing at infinity.

7. Determine the set of those $f \in C_{ub}(\mathbb{R})$ for which the orbit of the left-shift semigroup is differentiable.

8. Determine whether the following operators are closed or not.
   a) $X := C[0, 1]$, $Af(s) := \frac{1}{s(1-s)} f(s)$, $D(A) := \{f \in X : Af \in X\}$
   b) $X := C[0, 1]$, $Bf(s) := f'(s)$, $D(B) := \{f \in C^1[0, 1] : f'(1) = 0\}$
   c) $X := C[0, 1]$, $Cf(s) := f'(s)$, $D(C) := \{f \in C^1[0, 1] : f(0) = f(1)\}$
   d) $X := C[0, 1]$, $Df(s) := f''(s)$, $D(D) := C^2[0, 1]$
   e) $X := C[0, 1]$, $Ef(s) := f''(s)$, $D(E) := \{f \in C^2[0, 1] : f(0) = f(1) = 0\}$
   f) $X := C[0, 1]$, $Ff(s) := f''(s)$, $D(F) := \{f \in C^2[0, 1] : f''(0) = 0\}$

9. Calculate the spectrum and the point spectrum of the following operators on the Banach space $X := C[0, 1]$.
   a) $Af(s) := f'(s)$, $D(A) := \{f \in C^1[0, 1] : f'(0) = f'(1)\}$.
   b) $Bf(s) := f''(s)$, $D(B) := C^2[0, 1]$.
   c) $Cf(s) := f''(s)$, $D(C) := \{f \in C^2[0, 1] : f(0) = f(1) = 0\}$.

10. Let $X = C_0(\mathbb{R})$, $q \in C_b(\mathbb{R})$, and

$$T(t)f(s) := e^{\int_{s-t}^{t} q(\tau) d\tau} \cdot f(s - t).$$

Show that $(T(t))_{t \geq 0}$ is a C_0-semigroup and the identity operator its generator.

11. Let $(T(t))_{t \geq 0}$ be a semigroup on the Banach space X with generator A. Prove that for all $f \in D(A^2)$ we have the Taylor formula

$$T(t)f = f + tAf + \int_0^t (t - s)T(s)A^2 f ds.$$

Find a general Taylor formula for $f \in D(A^n)$.

12. Let $(T(t))_{t \geq 0}$ be a contraction semigroup on the Banach space X with generator A. Prove that
$$\|Af\|^2 \leq 4\|A^2 f\| \cdot \|f\|$$
holds for all $f \in D(A^2)$.

13. Let $X := L^p[1, \infty)$, $1 \leq p < \infty$ and $(T(t)f)(s) := f(se^t)$. Show that $(T(t))_{t \geq 0}$ is a C_0-semigroup and that $\omega_0(T) = -\frac{1}{p}$. Can you identify its generator?

14. Consider some examples of semigroups appearing in this chapter and write down the corresponding abstract initial value problems. Can you associate partial differential equations to these initial value problems?

Chapter 10

Banach Lattices and Positive Operators

In the remaining chapters we shall try to extend the theory of positive matrices to infinite-dimensional spaces. One of the first questions is how to generalize concepts like positivity of vectors, or positivity, irreducibility, and imprimitivity of matrices. We have tried to have an abstract look at the finite-dimensional case, to motivate infinite-dimensional concepts. Still, the transition from finite to infinite dimensions is not easy. This is why we decided to focus in this chapter only on the order relation and explore basic properties of infinite-dimensional ordered vector spaces, more precisely, Banach lattices.

We also continue the investigation of positive operators and positive exponential functions on Banach lattices. We shall be guided by the finite-dimensional situation and there will be many results and proofs which will be essentially reappearances from previous chapters.

10.1 Ordered Function Spaces

Let us first summarize the order structure of $\mathbb{R}^n$. Note that vectors in $\mathbb{R}^n$ can be identified with functions:

$$\mathbb{R}^n \equiv \{f : \{1, \ldots, n\} \to \mathbb{R}\}.$$

Positivity of a vector is thus nothing but pointwise positivity of the representing function:

$$f \geq 0 \text{ if and only if } f(k) \geq 0 \text{ for all } k = 1, \ldots, n.$$

Hence, if we have a vector space of real-valued functions, it is natural to introduce an order relation by pointwise ordering. Let us illustrate this with the most important example.

For a compact Hausdorff space K we take the space of continuous functions

$$X := C(K, \mathbb{R}) := \{f : K \to \mathbb{R} : f \text{ is continuous}\},$$

which is a Banach space with the norm

$$\|f\| = \|f\|_\infty = \max_{x \in K} |f(x)|.$$

The pointwise ordering in this case is

$$f \geq g \iff f(x) \geq g(x) \text{ for all } x \in K.$$

This clearly generalizes the finite-dimensional case with $K = \{1, \ldots, n\} \subset \mathbb{R}$ and the usual maximum norm.

It is straightforward from the definition that the ordering is compatible with the vector space operations in the sense that

$$f \leq g \text{ implies } f + h \leq g + h \text{ for all } h \in C(K, \mathbb{R})$$

and

$$0 \leq f \text{ implies } 0 \leq tf \text{ for all } t \geq 0.$$

We can also define the supremum and infimum of two functions as

$$(f \vee g)(x) := \max\{f(x), g(x)\} \quad \text{and} \quad (f \wedge g)(x) := \min\{f(x), g(x)\}$$

for all $x \in K$. The *positive part*, *negative part*, and *absolute value* of a function can be then given as

$$f^+ := f \vee 0, \quad f^- := (-f) \vee 0, \quad |f| := f \vee (-f).$$

An important property of the positive and negative part of a function is that they live separate lives: if $f^+(x) \neq 0$, then $f^-(x) = 0$ and vice versa. This property is sometimes called *orthogonality* or *disjointness*.

Note that the following properties also follow from the fact that we defined the order relation pointwise and that the order behaves well on the real numbers:

$$
\begin{aligned}
f &= f^+ - f^-, \\
|f| &= f^+ + f^-, \\
f \leq g &\iff f^+ \leq g^+ \text{ and } g^- \leq f^-, \\
|f - g| &= (f \vee g) - (f \wedge g), \\
|f| \leq |g| &\implies \|f\| \leq \|g\|.
\end{aligned}
\tag{10.1}
$$

Recall that for reducibility in Chapter 5 (see Definition 5.8) we needed the invariance of a subspace of the form

$$J_M := \left\{ (\xi_1, \ldots, \xi_n)^\top : \xi_i = 0 \text{ for } i \in M \right\} \subset \mathbb{R}^n$$

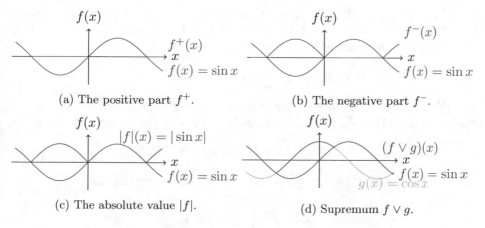

(a) The positive part f^+.

(b) The negative part f^-.

(c) The absolute value $|f|$.

(d) Supremum $f \vee g$.

Figure 10.1: Examples of f^+, f^-, $|f|$, and $f \vee g$.

for some $\emptyset \neq M \subsetneq \{1, \ldots, n\}$. In analogy, we define the following. Suppose that $F \subset K$ is a closed set and set

$$J_F := \{f \in C(K, \mathbb{R}) \ : \ f(x) = 0 \text{ for all } x \in F\}. \tag{10.2}$$

Subspaces of the above form are also called *ideals*. It is important that such ideals can be characterized by order theoretic concepts.

Proposition 10.1. *For a closed subspace $I \subset C(K, \mathbb{R})$ the following assertions are equivalent.*

(i) $$f \in I \text{ implies } |f| \in I,$$

and

$$0 \leq g \leq f \in I \text{ implies } g \in I.$$

(ii) *There is a closed set $F \subset K$ such that $I = J_F$.*

Proof. Since the case $I = \{0\}$ (where 0 stands here for the constant zero function) is obvious, we may assume that $I \neq \{0\}$.

Clearly, if $I = J_F$ for a closed subset F, then the properties listed in (i) hold. For the other direction, define

$$F := \{x \in K \ : \ f(x) = 0 \text{ for all } f \in I\}$$

and for $\alpha \in \mathbb{R}$ and $f \in C(K, \mathbb{R})$ denote

$$[f \geq \alpha] := \{x \in K \ : \ f(x) \geq \alpha\}.$$

Obviously, $I \subset J_F$. Take now a positive nonzero function $0 \neq f \in J_F$. Our aim is to show that $f \in I$.

For $\varepsilon > 0$ let $B_\varepsilon^f := [f \geq \varepsilon]$. Observe that B_ε^f is a closed set satisfying $B_\varepsilon^f \cap F = \emptyset$. Thus, for every $x \in B_\varepsilon^f$ there is $0 \leq g_x \in I$ such that $g_x(x) > 0$. Since B_ε^f is compact, there are finitely many $x_1, \ldots, x_r \in B_\varepsilon^f$ such that

$$B_\varepsilon^f \subset [g_{x_1} > 0] \cup [g_{x_2} > 0] \cup \cdots \cup [g_{x_r} > 0].$$

We construct now an approximation of f in the set I. First observe that (i) and (10.1) show that $f_1, f_2 \in I$ implies $f_1 \vee f_2 \in I$ and $f_1 \wedge f_2 \in I$. We define

$$g := g_{x_1} \vee g_{x_2} \vee \cdots \vee g_{x_r} \in I,$$

and take $\delta > 0$ such that $g(x) \geq \delta$ for all $x \in B_\varepsilon^f$. Then the function

$$h := f \wedge \left(\frac{\|f\|}{\delta} g \right) \in I$$

satisfies $0 \leq h \leq f$ and $h(x) = f(x)$ for all $x \in B_\varepsilon^f$. By the definition of the set B_ε^f, we see that $\|f - h\| \leq \varepsilon$. Hence, for every $f \in J_F$ and every $\varepsilon > 0$ we found $h \in I$ such that h approximates f with an error less than ε. By the closedness of I we obtain the desired conclusion. $\square$

Thus, a closed subspace I of $C(K, \mathbb{R})$ is an ideal if any of the equivalent conditions in Proposition 10.1 is satisfied. Let us only remark that this is also equivalent to saying that I is an *algebraic ideal* of the Banach algebra $C(K, \mathbb{R})$.

An operator T on $C(K, \mathbb{R})$ is called *reducible* if there exists a nontrivial ideal which is invariant under T. An operator which is not reducible, is called *irreducible*.

Another important observation concerning ideals is the following. Taking $f \geq 0$, we build the smallest ideal containing f, and denote it by E_f. It is then straightforward to check using Proposition 10.1 that

$$E_f = \bigcup_{k \in \mathbb{N}} [-kf, kf]$$

holds, where $[f_1, f_2] := \{g : f_1 \leq g \leq f_2\}$ denotes the *order interval* determined by f_1 and f_2, see Figure 10.2. We call E_f *the ideal generated by* f.

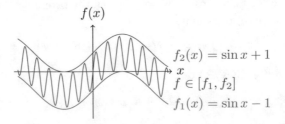

Figure 10.2: The order interval.

In some proofs (like in Corollary 7.4) strictly positive vectors (or the vector 1) played an important role. A natural observation is that the ideal generated by a strictly positive function is the whole space $C(K, \mathbb{R})$. A function with this property is sometimes also called an *order unit*. Unfortunately, as we shall see in the next section, not all function spaces possess order units. We will be able to introduce a weaker notion that will be almost as satisfactory for our proofs, see Example 10.16 and the considerations before that.

Finally, let us note that for statements in spectral theory we need complex vector spaces. Observe that we can make the identification

$$C(K, \mathbb{C}) \cong C(K, \mathbb{R}) \oplus i \cdot C(K, \mathbb{R}),$$

meaning that for a complex-valued continuous function its real and imaginary parts are real-valued continuous functions.

10.2   Vector Lattices

Now we take an abstract point of view and try to axiomatize what we have seen in the previous section. Our main examples, besides the finite-dimensional vector spaces, are $C(K)$ spaces, $L^p(\Omega, \mu)$ spaces, and $C_0(\Omega)$ spaces (see Example 10.6 later on). If you are uncomfortable with abstract terminology, you should pick one of these spaces and keep it in mind for the rest of this chapter.

We start by ordering. A non empty set M with a relation $\leq$ is said to be an *ordered set* if the following conditions are satisfied:

a) $f \leq f$ for every $f \in M$,

b) $f \leq g$ and $g \leq f$ imply $f = g$, and

c) $f \leq g$ and $g \leq h$ imply $f \leq h$.

First examples of ordered sets are the number sets $\mathbb{N}$, $\mathbb{Z}$, $\mathbb{Q}$, and $\mathbb{R}$.

Having an ordering at hand, we can consider order boundedness. Let F be a subset of an ordered set M. The element $f \in M$ (resp. $h \in M$) is called *an upper bound* (resp. *lower bound*) of F if $g \leq f$ for all $g \in F$ (resp. $h \leq g$ for all $g \in F$). Moreover, if there is an upper bound (resp. lower bound) of F, then F is said to be *bounded from above* (resp. *bounded from below*). If F is bounded from above and from below, then it is called an *order bounded set*.

We can introduce the concept of an order interval analogous to the intervals on the real line. Let $f, h \in M$ such that $f \leq h$. We denote by

$$[f, h] := \{g \in M : f \leq g \leq h\}$$

the *order interval* between f and g. We infer that a subset F is order bounded if and only if it is contained in some order interval.

Definition 10.2. A real vector space E which is ordered by some order relation $\leq$ is called a *vector lattice* if any two elements $f, g \in E$ have a least upper bound, denoted by $f \vee g = \sup(f, g) \in E$, and a greatest lower bound, denoted by $f \wedge g = \inf(f, g) \in E$, and the following properties are satisfied:

a) if $f \leq g$, then $f + h \leq g + h$ for all $f, g, h \in E$,

b) if $0 \leq f$, then $0 \leq tf$ for all $f \in E$ and $0 \leq t \in \mathbb{R}$.

Let E be a vector lattice. We denote by $E_+ := \{f \in E : 0 \leq f\}$ the *positive cone* of E. For $f \in E$, we define

$$f^+ := f \vee 0, \quad f^- := (-f) \vee 0, \quad \text{and} \quad |f| := f \vee (-f)$$

the *positive part*, the *negative part*, and the *absolute value* of f, respectively. Two elements $f, g \in E$ are called *orthogonal* (or *lattice disjoint*) (denoted by $f \perp g$) if $|f| \wedge |g| = 0$.

For a vector lattice E we have the following properties, which we will use frequently.

Proposition 10.3. *For all $f, g, h \in E$ the following assertions hold true.*

a) $f + g = (f \vee g) + (f \wedge g)$.

b) $f \vee g = -(-f) \wedge (-g)$.

c) $(f \vee g) + h = (f + h) \vee (g + h)$ *and* $(f \wedge g) + h = (f + h) \wedge (g + h)$.

d) $(f \vee g) \wedge h = (f \wedge h) \vee (g \wedge h)$ *and* $(f \wedge g) \vee h = (f \vee h) \wedge (g \vee h)$.

e) *For all* $f, g, h \in E_+$ *we have* $(f + g) \wedge h \leq (f \wedge h) + (g \wedge h)$.

Proof. We shall only prove a). The proof of the other properties is left to the reader (see Exercise 1). We have $f \wedge g \leq g \Longrightarrow f \leq f + g - f \wedge g$. In a similar way we have $g \leq f + g - f \wedge g$. Hence, $f \vee g \leq f + g - f \wedge g$, which gives

$$f \vee g + f \wedge g \leq f + g.$$

For the reverse inequality we note that $g \leq f \vee g \Longrightarrow f + g - f \vee g \leq f$, and similarly $f + g - f \vee g \leq g$. Thus,

$$f + g - f \vee g \leq f \wedge g. \qquad \square$$

For the positive part, negative part, and absolute value of $f \in E$ we have the following properties (compare with Properties (10.1) of functions in $C(K, \mathbb{R})$).

Proposition 10.4. *If $f, g \in E$, then*

a) $f = f^+ - f^-$.

b) $|f| = f^+ + f^-$.

c) $f^+ \perp f^-$ *and the decomposition of f into the difference of two orthogonal positive elements is unique.*

d) $f \leq g$ *is equivalent to* $f^+ \leq g^+$ *and* $g^- \leq f^-$.

e) $|f - g| = (f \vee g) - (f \wedge g)$.

Proof. a) Using Proposition 10.3 a) and b), we obtain

$$f = f + 0 = f \vee 0 + f \wedge 0$$
$$= f \vee 0 - (-f) \vee 0 = f^+ - f^-.$$

b) Applying Proposition 10.3.c) and a) proved above, we have

$$|f| = f \vee (-f) = (2f \vee 0) - f = 2(f \vee 0) - f$$
$$= 2f^+ - f^+ + f^- = f^+ + f^-.$$

c) Let us prove first that $f^+ \wedge f^- = 0$. To this purpose we apply Proposition 10.3.c) and deduce

$$f^+ \wedge f^- = (f^+ - f^-) \wedge 0 + f^- = (f \wedge 0) + f^-$$
$$= -[(-f) \vee 0] + f^- = 0.$$

Let now $f = g - h$ with $g \wedge h = 0$. By c) and a) of Proposition 10.3, we have

$$f^+ = (g - h) \vee 0 = g \vee h - h = (g + h - (g \wedge h)) - h = g.$$

In a similar way we obtain $f^- = h$.

d) Using a), this is straightforward.

e) This can be proved using the identities

$$f \vee g = \frac{1}{2}(f + g + |f - g|) \quad \text{and} \quad f \wedge g = \frac{1}{2}(f + g - |f - g|)$$

(see Exercise 2). $\qquad \square$

10.3   Banach Lattices

We finally arrived at the main objects of this chapter and consider Banach spaces which are ordered and whose norm is compatible with this ordering. First, let us explain what we mean by compatible.

A norm on a vector lattice E is called a *lattice norm* if

$$|f| \leq |g| \text{ implies } \|f\| \leq \|g\| \quad \text{for } f, g \in E.$$

Definition 10.5. A *Banach lattice* is a real Banach space E endowed with an ordering $\leq$ such that $(E, \leq)$ is a vector lattice and the norm on E is a lattice norm.

We will see that this combination of properties of a complete normed vector space and a compatible ordering leads to many fruitful results.

As already mentioned, apart from finite dimensional vector spaces (such as $\mathbb{R}$ or $\mathbb{R}^n$), there are many interesting infinite-dimensional examples of Banach lattices.

Examples 10.6. The following Banach spaces are Banach lattices for the pointwise (almost everywhere) ordering.

a) Let (Ω, μ) be a measure space and take $L^p(\Omega, \mu; \mathbb{R})$, $1 \leq p \leq \infty$, endowed with the norm

$$\|f\|_p = \left(\int_\Omega |f(x)|^p \, d\mu \right)^{1/p} \quad \text{if } 1 \leq p < \infty,$$

$$\|f\|_\infty = \inf\{M : |f(x)| \leq M \text{ for } \mu\text{-a.e. } x \in \Omega\} \quad \text{if } p = \infty,$$

and with the order

$$f \geq 0 \iff f(x) \geq 0 \text{ for } \mu\text{-a.e. } x \in \Omega.$$

We furthermore define

$$(f \vee g)(x) := \max\{f(x), g(x)\} \quad \text{and} \quad (f \wedge g)(x) := \min\{f(x), g(x)\} \quad (10.3)$$

for μ-a.e. $x \in \Omega$, which are both measurable functions. Note that for the absolute value of f this imposes

$$|f|(x) = (f \vee (-f))(x) = \max\{f(x), -f(x)\} = |f(x)| \text{ for } \mu\text{-a.e. } x \in \Omega.$$

Since

$$|(f \vee g)(x)| \leq |f(x)| + |g(x)| \quad \text{and} \quad |(f \wedge g)(x)| \leq |f(x)| + |g(x)| \quad (10.4)$$

for μ-a.e. $x \in \Omega$, we see, that

$$\|f \vee g\|_p \leq \|f\|_p + \|g\|_p \quad \text{and} \quad \|f \wedge g\|_p \leq \|f\|_p + \|g\|_p,$$

hence $f \vee g, f \wedge g \in L^p(\Omega, \mu)$ for every $1 \leq p \leq \infty$. Clearly, the properties in Definition 10.2 are fulfilled and the p-norm is a lattice norm.

b) For a locally compact noncompact Hausdorff topological space Ω we take $C_0(\Omega)$, the space of all real-valued continuous functions vanishing at infinity, endowed with the supremum norm

$$\|f\|_\infty = \sup_{x \in \Omega} |f(x)|,$$

and with the natural order

$$f \geq 0 \iff f(x) \geq 0 \text{ for all } x \in \Omega.$$

We define $f \vee g$, $f \wedge g$ as in (10.3), but now for every $x \in \Omega$. We obtain continuous functions and using inequalities (10.4) we see that $f \vee g, f \wedge g \in C_0(\Omega)$. Again, the properties in Definition 10.2 are fulfilled and the supremum norm is a lattice norm.

c) The space of real-valued continuous functions $C(K)$ on a compact Hausdorff space K, endowed with the supremum norm and with the order defined above was already investigated in Section 10.1.

Note that there are many ordered function spaces which are not Banach lattices. Let us give the following simple example.

Examples 10.7.

a) Consider the Banach space $C^1([0,1])$ of continuously differentiable functions on $[0,1]$ with the norm

$$\|f\| = \max_{s\in[0,1]} |f(s)| + \max_{s\in[0,1]} |f'(s)|$$

and the natural order $f \geq 0$ if $f(s) \geq 0$ for all $s \in [0,1]$. Since $\sup\{t, 1-t\} \notin C^1([0,1])$, the space $C^1([0,1])$ is not a vector lattice. Moreover the above norm is not compatible with the order. In fact, let $f \equiv 1$ and $g(s) = \sin(2s)$, $s \in [0,1]$. Then, $0 \leq g \leq f$ and $\|g\| \geq |g'(0)| = 2 > 1 = \|f\|$.

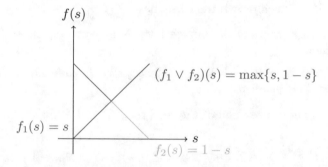

Figure 10.3: $C^1[0,1]$ with the norm in Example 10.7.a) is not a lattice.

b) Consider the Sobolev space $H^1(0,1)$. Using similar arguments as in the previous example, we see that the norm is again not compatible with the order. As a difference, however, note that $H^1(0,1)$ is a vector lattice, see Exercise 4.

Now we list some further properties of Banach lattices.

Proposition 10.8. *For a Banach lattice E the following hold.*

a) *The lattice operations are continuous.*
b) *The positive cone E_+ is closed.*
c) *The order intervals are closed and bounded.*

Proof. a) Consider $(f_k), (g_k) \subset E$ and $f, g \in E$ such that $\lim_{k\to\infty} f_k = f$ and $\lim_{k\to\infty} g_k = g$. Applying Birkhoff's inequality, see Exercise 2.f), we have

$$|f_k \wedge g_k - f \wedge g| \leq |f_k \wedge g_k - f_k \wedge g| + |f_k \wedge g - f \wedge g|$$
$$\leq |g_k - g| + |f_k - f|.$$

Thus,

$$\|f_k \wedge g_k - f \wedge g\| \leq \|g_k - g\| + \|f_k - f\|.$$

This yields the continuity of $\wedge$. Analogously, one obtains the continuity of $\vee$.

b) Take $(f_k) \subset E_+$ such that $\lim_{k\to\infty} f_k = f \in E$. Then, by a),

$$\lim_{k\to\infty} f_k = \lim_{k\to\infty} (f_k \vee 0) = f \vee 0.$$

Hence, $f = f \vee 0 \in E_+$.

c) Let $f, g, h \in E$ with $h \in [f, g]$. Then, $0 \le h - f \le g - f$. So, using the triangle inequality from Exercise 2.d), one has

$$\|h\| - \|f\| \le \|h - f\| \le \|g - f\|, \quad \text{whence } \|h\| \le \|f\| + \|g - f\|.$$

Therefore, order intervals are bounded. We prove now that order intervals are closed. Take $(h_k) \subset E$ and $f, g \in E$ with $f \le h_k \le g$ for all $k \in \mathbb{N}$. Since E_+ is closed, by b), $\lim_{k\to\infty}(h_k - f) = h - f \ge 0$ and $\lim_{k\to\infty}(g - h_k) = g - h \ge 0$. Hence, $f \le h \le g$, which proves the closedness of $[f, g]$. $\qquad \square$

The following property of Banach lattices is a consequence of the Hahn–Banach theorem.

Proposition 10.9. *In a Banach lattice E every weakly convergent increasing sequence (f_k) is norm convergent.*

Proof. Consider the convex hull of the set $\{f_k\}$,

$$F := \left\{ \sum_{i=1}^{m} a_i f_i : m \in \mathbb{N}, \ a_i \ge 0, \ a_1 + \cdots + a_m = 1 \right\}.$$

By the Hahn–Banach theorem, Theorem A.27, the norm closure of F coincides with the weak closure. This implies that $f \in \overline{F}$, where $f := \text{weak-}\lim_{k\to\infty} f_k$. Thus, for $\varepsilon > 0$, there exist $g \in F$, i.e.,

$$g = a_1 f_1 + \cdots + a_m f_m, \quad \text{with } a_1, \ldots, a_m \ge 0 \text{ and } a_1 + \cdots + a_m = 1,$$

such that $\|g - f\| < \varepsilon$. Since $g \le f_k \le f$, we infer that $\|f - f_k\| \le \|f - g\| < \varepsilon$ for all $k \ge m$. $\qquad \square$

Here we state a result that we shall need later. A Banach lattice E is *totally ordered* if for every $f \in E$ one has either $0 \le f$ or $f \le 0$.

Lemma 10.10. *A totally ordered real Banach lattice E is at most one-dimensional.*

Proof. Take $e \in E_+$ with $e \ne 0$, and $f \in E$. Consider the closed subsets of $\mathbb{R}$

$$C_+ := \{\alpha \in \mathbb{R} : \alpha e \ge f\} \quad \text{and} \quad C_- := \{\alpha \in \mathbb{R} : \alpha e \le f\}.$$

It is obvious that C_+ and C_- are non-empty and $C_+ \cup C_- = \mathbb{R}$. Since $\mathbb{R}$ is connected, it follows that $C_+ \cap C_- \ne \varnothing$. Hence, there is $\alpha \in \mathbb{R}$ such that $f = \alpha e$. $\qquad \square$

10.4 Sublattices and Ideals

We want to equip a vector subspace of a vector or Banach lattice with some order structure. Therefore we define two kinds of subspaces (compare with Proposition 10.1).

Definition 10.11. A vector subspace F of a vector lattice E is a *vector sublattice* if for all $f \in F$ we have $|f| \in F$. A subspace I of a Banach lattice E is called an *ideal* if

$$f \in I \text{ implies } |f| \in I \text{ and } 0 \le g \le f \in I \text{ implies } g \in I.$$

Consequently, a vector sublattice F is an ideal in E if $f \in F$ and $0 \le g \le f$ implies $g \in F$. Note also that if F is a vector sublattice, then $f^+ \in F$ and $f^- \in F$ for all $f \in F$.

Since the notions of sublattice and ideal are invariant under the formation of arbitrary intersections, there exists, for any subset M of E, a unique smallest sublattice (resp. ideal) of E containing M. This will be called *the sublattice (resp. the ideal) generated by M*.

We summarize all properties of sublattices and ideals which we will need in the sequel.

Proposition 10.12. *If E is a Banach lattice, then the following properties hold.*

a) *The closure of every sublattice of E is a sublattice.*
b) *The closure of every ideal of E is an ideal.*
c) *For every $f \in E_+$, the ideal generated by $\{f\}$ is*

$$E_f = \bigcup_{k \in \mathbb{N}} k[-f, f].$$

Proof. The first two assertions follow from the continuity of the lattice operations, see Proposition 10.8. For the last assertion one can see easily that $I = \bigcup_{k \in \mathbb{N}} k[-f, f]$ is an ideal while any ideal included in I and containing f equals I. This means that $I = E_f$. □

For examples of closed ideals we again pay a visit to our function spaces and start by restating Proposition 10.1 in this context.

Proposition 10.13. *If $E = C(K)$, where K is a compact Hausdorff space, then a subspace J of E is a closed ideal if and only if there is a closed subset $F \subset K$ such that*

$$J = \{\varphi \in E : \varphi(x) = 0 \text{ for all } x \in F\}.$$

The arguments of the proof of Proposition 10.1 can be modified accordingly to obtain the following characterization.

Proposition 10.14. *If $E = C_0(\Omega)$, where Ω is a locally compact Hausdorff topological space, then a subspace J of E is a closed ideal if and only if there is a closed subset F of Ω such that*

$$J = \{\varphi \in E : \varphi(x) = 0 \ \text{for all} \ x \in F\}.$$

Finally, we close this set of examples by characterizing closed ideals of L^p-spaces.

Proposition 10.15. *If $E = \mathrm{L}^p(\Omega, \mu)$, $1 \le p < \infty$, where (Ω, μ) is a σ-finite measure space, then a subspace I of E is a closed ideal if and only if there exists a measurable subset Y of Ω such that*

$$I = \{\psi \in E : \psi(x) = 0 \ \ a.e. \ \ x \in Y\}.$$

Proof. First we show that for a measurable set $Y \subset \Omega$, the set

$$I_Y := \{\psi \in E : \psi(x) = 0 \ \text{a.e.} \ x \in Y\}$$

is a closed ideal. Clearly, I_Y is a linear subspace and if $f \in I_Y$, then $|f| \in I_Y$. The definition implies directly that if $f \in I_Y$ and $0 \le g \le f$, then $g \in I_Y$. Hence it only remains to show the closedness.

Let $(f_k) \subset I_Y$ be a sequence such that $f_k \to f \in E$. Then there is a subsequence (f_{k_m}) such that $f_{k_m}(x) \to f(x)$ for a.e. $x \in \Omega$. In particular, $f_{k_m}(x) \to f(x)$ for a.e. $x \in Y$, hence, $f \in I_Y$.

Conversely, suppose that $\{0\} \ne I \subset E$ is an ideal. We have to show the existence of a measurable set $Y \subset \Omega$ such that $I = I_Y$. Since (Ω, μ) is a σ-finite measure space, there is an increasing sequence (Ω_k) of sets of finite measure with $\bigcup_{k \in \mathbb{N}} \Omega_k = \Omega$. For each $k \in \mathbb{N}$ we define

$$\mathcal{B}_k := \{M \subset \Omega_k : \chi_M \in I\}.$$

Since I is a non-trivial ideal, we infer that there is $k \in \mathbb{N}$ such that $\mathcal{B}_k \ne \emptyset$. Observe that if $\mathcal{M} \subset \mathcal{B}_k$ is a finite set, then

$$\sup_{M \in \mathcal{M}} \chi_M \in I.$$

We also have that

$$s_k := \sup_{\mathcal{M} \subset \mathcal{B}_k, \, \mathcal{M} \ \text{finite}} \left\| \sup_{M \in \mathcal{M}} \chi_M \right\| \le \mu(\Omega_k)^{1/p} < \infty.$$

Take a sequence $\mathcal{M}_m \subset \mathcal{B}_m$, where $\mathcal{M}_m$ is finite and

$$\left\| \sup_{M \in \mathcal{M}_m} \chi_M \right\| \ge s_k - \frac{1}{m}.$$

holds for every $m \in \mathbb{N}$. Observe that for $m_1 \leq m_2$ one has $\mathcal{M}_{m_2} \subseteq \mathcal{M}_{m_1}$. Now we define

$$C_k := \bigcup_{m \in \mathbb{N},\, M \in \mathcal{M}_m} M \in \mathcal{B}_k, \quad C := \bigcup_{k \in \mathbb{N}} C_k, \text{ and } Y := \Omega \setminus C.$$

Clearly, the sets C_k, C, and Y are measurable. Moreover, the sequence

$$\left(\sup_{M \in \mathcal{M}_m} \chi_M \right) \subset I$$

is bounded and monotone, and since I is closed, the Dominated Convergence Theorem (see Theorem A.23) implies that its limit $\chi_{C_k} \in I$ for all $k \in \mathbb{N}$.

Take now $f \in I$ and show that $f \in I_Y$. Since I and I_Y are both ideals it suffices to consider positive f only. Assume on the contrary that there is $M \subset Y$ such that $\mu(M) > 0$ and $f(x) > 0$ for a.e. $x \in M$. Fix k_0 such that $\mu(M \cap \Omega_{k_0}) > 0$. Since f is strictly positive on $M \cap \Omega_{k_0}$, there exists $j \in \mathbb{N}$ such that

$$\mu(M \cap \Omega_{k_0} \cap \{f \geq 1/j\}) > 0.$$

For such j we introduce the function $g_j := \chi_{M \cap \Omega_{k_0} \cap \{f \geq 1/j\}}$. Then, $0 \leq g_j \leq jf$ and hence, $g_j \in I$ and $\tilde{\mathcal{B}} := M \cap \Omega_{k_0} \cap \{f \geq 1/j\} \in \mathcal{B}_{k_0}$. Since $\tilde{\mathcal{B}} \cap C_{k_0} = \emptyset$ for sufficiently large $m \in \mathbb{N}$, we must have

$$\left\| \chi_{\tilde{\mathcal{B}}} + \sup_{M \in \mathcal{M}_m} \chi_M \right\| > s_{k_0},$$

which is a contradiction. Hence, $I \subset I_Y$.

To show that $I_Y \subset I$, take $0 \leq f \in I_Y$ and fix $\varepsilon > 0$. The sequences $(f - \chi_{\Omega_k} f)$ and $(f - \chi_{\{f \leq k\}} f)$ of positive functions converge to zero almost everywhere. The Dominated Convergence Theorem (see Theorem A.23) implies that there is $k_0 \in \mathbb{N}$ such that

$$\| f - \chi_{\Omega_{k_0}} f \| \leq \frac{\varepsilon}{2} \quad \text{and} \quad \| f - \chi_{\{f \leq k_0\}} f \| \leq \frac{\varepsilon}{2}.$$

Since $k_0 \chi_{C_{k_0}} \in I$, we infer that

$$h := \chi_{\Omega_{k_0} \cap \{f \leq k_0\}} f \wedge k_0 \chi_{C_{k_0}} \in I.$$

By the choice of k_0 and the fact that $f = 0$ almost everywhere on Y, we have that $\| f - h \| \leq \varepsilon$. The closedness of I now implies that $f \in I$. $\qquad\square$

Sometimes a Banach lattice E is generated by a single positive element. If $E_e = E$ holds for some $e \in E_+$ then e is called an *order unit*. If $\overline{E}_e = E$, then $e \in E_+$ is called a *quasi-interior point* of E_+.

It follows that e is an order unit of E if and only if e is an interior point of E_+. Quasi-interior points of the positive cone exist, for example, in every separable Banach lattice.

Examples 10.16.

a) If $E = C(K)$, where K is a compact Hausdorff space, then the constant function $\mathbb{1}$, $\mathbb{1}(x) \equiv 1$, is an order unit. In fact, for every $f \in E$, there is $k \in \mathbb{N}$ such that $\|f\|_\infty \leq k$. Hence, $|f(s)| \leq k\mathbb{1}(s)$ for all $s \in K$. This implies $f \in k[-\mathbb{1}, \mathbb{1}]$.

b) Let $E = L^p(\Omega, \mu)$ with a σ-finite measure μ such that $\mu(\{x\}) = 0$ for every $x \in \Omega$ and $1 \leq p < \infty$. Then the quasi-interior points of E_+ coincide with the μ-a.e. strictly positive functions, while E_+ does not contain any interior point.

10.5 Complexification of Real Banach Lattices

It is often necessary to consider complex vector spaces (for instance in spectral theory). Therefore, we introduce the concept of a *complex Banach lattice*.

The complexification of a real Banach lattice E is the complex Banach space $E_{\mathbb{C}}$ whose elements are pairs $(f, g) \in E \times E$, with addition and scalar multiplication defined by $(f_0, g_0) + (f_1, g_1) := (f_0 + f_1, g_0 + g_1)$ and $(a + ib)(f, g) := (af - bg, ag + bf)$, and norm

$$\|(f, g)\| := \| |(f, g)| \|,$$

where

$$|(f, g)| := \sup_{0 \leq \theta \leq 2\pi} (f \sin \theta + g \cos \theta)$$

is the natural extension of the modulus $|\cdot|$ in E. Note that the existence of the above supremum in E is in this generality a nontrivial fact, but we accept it here. However, in the standard function spaces, which are our main examples, this is a straightforward fact.

By identifying $(f, 0) \in E_{\mathbb{C}}$ with $f \in E$, the space E is isometrically isomorphic to a real linear subspace $E_{\mathbb{R}}$ of $E_{\mathbb{C}}$. We write $0 \leq f \in E_{\mathbb{C}}$ if and only if $f \in E_+$.

A complex Banach lattice is an ordered complex Banach space $(E_{\mathbb{C}}, \leq)$ that arises as the complexification of a real Banach lattice E. The underlying real Banach lattice E is called the real part of $E_{\mathbb{C}}$ and is uniquely determined as the closed linear span of all $f \in (E_{\mathbb{C}})_+$.

Instead of the notation (f, g) for elements of $E_{\mathbb{C}}$, we usually write $f + ig$. The complex conjugate of an element $h = f + ig \in E_{\mathbb{C}}$ is the element $\overline{h} = f - ig$. We use also the notation $\mathrm{Re}(h) := f$ for $h = f + ig \in E_{\mathbb{C}}$. All concepts introduced for real Banach lattices have a natural extension to complex Banach lattices.

10.6 Positive Operators

This section is concerned with positive operators on Banach lattices, that is, operators that preserve positive cones.

Definition 10.17. Let E and F be two complex Banach lattices. A linear operator $T : E \to F$ is called *positive* if $TE_+ \subset F_+$. Notation: $T \geq 0$.

Let us immediately give an alternative characterization of a positive operator (compare with the matrix case given in Lemma 5.3).

Lemma 10.18. *The following assertions for a linear operator $T : E \to F$ between the Banach lattices E and F are equivalent.*

(i) T *is positive.*
(ii) *For all $f \in E_{\mathbb{R}}$, we have $(Tf)^+ \leq Tf^+$ and $(Tf)^- \leq Tf^-$.*
(iii) $|Tf| \leq T|f|$ *for all $f \in E$.*

Proof. (i) $\implies$ (ii): For $f \in E_{\mathbb{R}}$ we have $Tf = Tf^+ - Tf^- \leq Tf^+$ and $(Tf)^+ = Tf \vee 0$, which imply $(Tf)^+ \leq Tf^+$. The second property now follows since

$$Tf^+ - (Tf)^+ = Tf^- - (Tf)^-.$$

(ii) $\implies$ (iii): Using $f = f^+ + f^-$ for $f \in E_{\mathbb{R}}$, and (ii) we obtain

$$|Tf| = (Tf)^+ + (Tf)^- \leq Tf^+ + Tf^- = T|f|.$$

For general $f \in E$ the assertion follows from the definition of $|f|$.

(iii) $\implies$ (i): Let $f \in E_+$. Then $T|f| = Tf$ and by assumption we have

$$Tf = T|f| \geq |Tf| \geq 0. \qquad \square$$

We shall need a stronger property than the one given in Lemma 10.18.(iii), i.e., preserving the absolute value.

Definition 10.19. Let E and F be two complex Banach lattices. A linear operator $T : E \to F$ is called a *lattice homomorphism* if $|Tf| = T|f|$ for all $f \in E$.

All positive operators are bounded, as the following result shows.

Theorem 10.20. *Every positive linear operator $T : E \to F$ is continuous.*

Proof. Assume by contradiction that T is not bounded. Then there is $(f_k) \subset E$ such that $\|f_k\| = 1$ and $\|Tf_k\| \geq k^\gamma$ for each $k \in \mathbb{N}$ and some $\gamma > 2$. Since $|Tf_k| \leq T|f_k|$, one can assume that $f_k \geq 0$ for all $k \in \mathbb{N}$. From $\sum_{k=1}^{\infty} \frac{\|f_k\|}{k^{\gamma-1}} < \infty$ we infer that $\sum_{k=1}^{\infty} \frac{f_k}{k^{\gamma-1}}$ is norm convergent in E. Set $f = \sum_{k=1}^{\infty} \frac{f_k}{k^{\gamma-1}}$. Then

$$0 \leq \frac{f_k}{k^{\gamma-1}} \leq f \quad \text{for all } k \in \mathbb{N}.$$

So

$$k \le \left\| T\left(\frac{f_k}{k^{\gamma-1}}\right)\right\| \le \|Tf\| < \infty \quad \text{for all } k \in \mathbb{N},$$

which is a contradiction. Thus $T \in \mathcal{L}(E, F)$.   □

As a consequence, we obtain the equivalence of Banach lattice norms (recall also Tikhonov's theorem, Theorem 1.5).

Corollary 10.21. *Let E be a vector lattice and $\| \cdot \|_1$ and $\| \cdot \|_2$ two norms such that $E_1 = (E, \| \cdot \|_1)$ and $E_2 = (E, \| \cdot \|_2)$ are both Banach lattices. Then the norms $\| \cdot \|_1$ and $\| \cdot \|_2$ are equivalent.*

Proof. This follows from the positivity of the identity operators $I : E_1 \to E_2$ and $I : E_2 \to E_1$ and Theorem 10.20.   □

We denote by $\mathcal{L}(E, F)_+$ the set of all positive linear operators from a Banach lattice E into a Banach lattice F. For positive operators one has

Proposition 10.22. *Let $T \in \mathcal{L}(E, F)_+$. Then the following properties hold.*

 a) $\|T\| = \sup\{\|Tf\| : f \in E_+, \|f\| \le 1\}$.
 b) *If $S \in \mathcal{L}(E, F)$ is such that $0 \le S \le T$ (this means that $0 \le Sf \le Tf$ for all $f \in E_+$), then $\|S\| \le \|T\|$.*

Proof. a) holds by Lemma 10.18 (iii).

b) Since $0 \le S \le T$ we have $|Sf| \le S|f| \le T|f|$ for all $f \in E$. The assertion now follows by a).   □

Another property of positive operators is that they have positive resolvent. The converse in not always true, see also Proposition 10.29.

Proposition 10.23. *Let $T \in \mathcal{L}(E)$ be a positive operator with spectral radius $\mathrm{r}(T)$.*

 a) *The resolvent $R(\mu, T)$ is positive whenever $\mu > \mathrm{r}(T)$.*
 b) *If $|\mu| > \mathrm{r}(T)$, then*

$$|R(\mu, T)f| \le R(|\mu|, T)|f|, \quad f \in E.$$

Proof. We use the Neumann series representation

$$R(\mu, T) = \sum_{k=0}^{\infty} \frac{T^k}{\mu^{k+1}}$$

for the resolvent, which is valid for $|\mu| > \mathrm{r}(T)$, see Proposition 9.28.c).

a) If $T \ge 0$, then $T^k \ge 0$ for all k, hence for $\mu > \mathrm{r}(T)$, we have for every $f \in E_+$ that

$$R(\mu, T)f = \lim_{N \to \infty} \sum_{k=0}^{N} \frac{T^k f}{\mu^{k+1}} \ge 0,$$

since the finite sums are positive.

b) We have for $|\mu| > r(T)$ and $f \in E$ that

$$|R(\mu,T)f| = \left| \lim_{N\to\infty} \sum_{k=0}^{N} \frac{T^k f}{\mu^{k+1}} \right| \leq \lim_{N\to\infty} \sum_{k=0}^{N} \left| \frac{T^k f}{\mu^{k+1}} \right|$$

$$\leq \lim_{N\to\infty} \sum_{k=0}^{N} \frac{T^k}{|\mu|^{k+1}} |f| = R(|\mu|,T)|f|. \qquad \square$$

The following is an easy version of Perron's theorem (see Theorem 5.6) for the infinite-dimensional case.

Theorem 10.24. *If $T \in \mathcal{L}(E)$ is positive, then $r(T) \in \sigma(T)$.*

Proof. Assertion b) of Proposition 10.23 implies that

$$\|R(\mu,T)\| \leq \|R(|\mu|,T)\| \quad \text{for } |\mu| > r(T).$$

Let now $\lambda \in \sigma(T)$ such that $|\lambda| = r(T)$. Then, Proposition 9.28 implies that $\|R(\mu,T)\| \to \infty$ whenever μ approaches λ. Putting $\mu = s\lambda$ with $s > 1$ the above estimate yields

$$\|R(sr(T),T)\| \geq \|R(s\lambda,T)\| \longrightarrow \infty \quad \text{as } s \downarrow 1,$$

hence, by Corollary 9.30, $r(T)$ must be in the spectrum of T. $\qquad \square$

Combining Proposition 10.23 and Theorem 10.24 we have the following useful characterization of positivity of the operator $R(1,T) = (I - T)^{-1}$.

Lemma 10.25. *Let T be a positive linear operator on E. Then*

$$r(T) < 1 \iff 1 \in \rho(T) \text{ and } R(1,T) \geq 0.$$

Proof. The implication is a consequence of Proposition 10.23. For the converse, assume that $1 \in \rho(T)$ and $R(1,T) \geq 0$. For any $k \in \mathbb{N}$ we have

$$(I - T) \sum_{j=0}^{k} T^j = I - T^{k+1}.$$

Hence,

$$\sum_{j=0}^{k} T^j = R(1,T)(I - T^{k+1}) \leq R(1,T), \qquad (10.5)$$

since $T \geq 0$. So, in particular $T^k \leq R(1,T)$ for all $k \in \mathbb{N}$. Now Proposition 10.22 implies that

$$\|T^k\| \leq \|R(1,T)\|, \quad k \in \mathbb{N}.$$

Using the above estimate and the definition of $r(T)$ we obtain $r(T) \leq 1$. If $r(T) = 1$, then Theorem 10.24 yields $1 \in \sigma(T)$, which contradicts our assumption. $\qquad \square$

Now we define irreducible operators on a Banach lattice.

Definition 10.26. An operator $T \in \mathcal{L}(E)$ is called *reducible* if there exists a non-trivial ideal which is invariant under T. Operators that are not reducible are called *irreducible*.

As in the finite-dimensional situation, positive irreducible operators enjoy some special spectral properties (see, e.g., Theorem A.38). However, we shall not discuss these properties here. We study them in the case of semigroups of positive irreducible operators in Section 14.3.

We end this section by reconsidering the Banach lattice of continuous functions on a compact Hausdorff space K.

Lemma 10.27. *Suppose that K is a compact Hausdorff topological space and $T : C(K) \to C(K)$ is a linear operator satisfying $T\mathbb{1} = \mathbb{1}$. Then $0 \leq T$ if and only if $\|T\| \leq 1$.*

Proof. If $0 \leq T$, then

$$|Tf| \leq T|f| \leq T(\|f\|_\infty \mathbb{1}) = \|f\|_\infty \mathbb{1}.$$

Hence $\|T\| \leq 1$.

To prove the converse, we first observe that

$$-\mathbb{1} \leq f \leq \mathbb{1} \iff \|f - ir\mathbb{1}\|_\infty \leq \rho_r := \sqrt{1 + r^2} \quad \text{for all } r \in \mathbb{R}. \tag{10.6}$$

Let $0 \leq f \in C(K)$. Then there is $k \in \mathbb{N}$ such that $0 \leq f \leq k\mathbb{1}$. Set $g = \frac{2}{k}f$. Then $0 \leq g \leq 2\mathbb{1}$, and so $-\mathbb{1} \leq g - \mathbb{1} \leq \mathbb{1}$. By (10.6), $\|g - \mathbb{1} - ir\mathbb{1}\|_\infty \leq \rho_r$ for all $r \in \mathbb{R}$. Since $T\mathbb{1} = \mathbb{1}$ and $\|T\| \leq 1$, $\|Tg - \mathbb{1} - ir\mathbb{1}\|_\infty \leq \rho_r$ for all $r \in \mathbb{R}$. So by (10.6) we obtain $-\mathbb{1} \leq Tg - \mathbb{1} \leq \mathbb{1}$. This implies $0 \leq Tg \leq 2\mathbb{1}$ and hence $Tf \geq 0$. $\square$

Indeed, operators satisfying $T\mathbb{1} = \mathbb{1}$ occur quite often and have a special name. Recall that in the finite-dimensional case we have shown this property for the transition matrix P of a Markov chain (see Lemma 6.6).

Definition 10.28. Let K and L be compact Hausdorff spaces. A linear operator $T : C(K) \to C(L)$ is called a *Markov operator* if $T\mathbb{1}_K = \mathbb{1}_L$.

10.7   Positive Exponential Functions

In the following, let E be a Banach lattice and $A \in \mathcal{L}(E)$. We investigate the positivity and asymptotic properties of the exponential function of A, and start with a characterization through the resolvent of A.

Proposition 10.29. *The semigroup $T(t) = e^{tA}$ is positive if and only if*

$$R(\lambda, A) = (\lambda - A)^{-1} \geq 0$$

for all $\lambda > \omega_0(T)$.

Proof. When $T(\cdot)$ is a positive semigroup, then $R(\lambda, A)$ is positive for $\lambda > \omega_0(T)$ by the Laplace transform representation in (9.11).

For the other direction notice that, by Exercise 9.10.2, the Euler formula

$$\lim_{k \to \infty} \left(I - \frac{t}{k} A \right)^{-k} = e^{tA}$$

holds for $t \geq 0$. Since $(I - \frac{t}{k} A)^{-k} = \left(\frac{k}{t} R(\frac{k}{t}, A) \right)^k \geq 0$ for k sufficiently large by assumption, the positivity of the operators $T(t)$ follows. For an alternative proof where the Euler formula is not needed we refer to Remark 11.3 and Corollary 11.4. □

Recall the notation already used in the case of matrices. For $A \in \mathcal{L}(E)$ we define its *spectral bound* as

$$s(A) := \sup\{\operatorname{Re} \lambda \,:\, \lambda \in \sigma(A)\}. \tag{10.7}$$

The following is a fundamental technical result on positive exponential functions. It tells us that the Laplace transform representation in the case of positive bounded generators holds on an even larger set.

Proposition 10.30. *For a positive exponential function $T(t) = e^{tA}$ we have*

$$R(\lambda, A) = \int_0^\infty e^{-\lambda s} T(s) \, \mathrm{d}s$$

for all $\lambda > s(A)$. Hence, for all $\lambda > s(A)$ we have $0 \leq R(\lambda, A)$.

Proof. It is clear from the previous considerations that $R(\lambda, A)$ is positive for $\lambda > \|A\|$. The Neumann series representation from Proposition 9.28.c) implies that $R(\lambda, A) \geq 0$ for $\lambda > s(A)$.

Since we can always consider $A - \mu$ instead of A for $\mu > 0$, for simplicity we restrict our proof to the case where $s(A) < 0$ and $\operatorname{Re} \lambda > 0$. The assumption implies

$$0 \leq V(t) = \int_0^t T(s) \, \mathrm{d}s = R(0, A) - R(0, A)T(t) \leq R(0, A),$$

so $\|V(t)\| \leq M$ for some constant M for all $t \geq 0$. Hence the improper integral

$$\int_0^\infty e^{-\lambda s} V(s) \, \mathrm{d}s$$

exists for all $\operatorname{Re} \lambda > 0$. Integration by parts yields

$$\int_0^t e^{-\lambda s} T(s) \, \mathrm{d}s = e^{-\lambda t} V(t) + \lambda \int_0^t e^{-\lambda s} V(s) \, \mathrm{d}s.$$

This last expression converges as $t \to \infty$, hence we infer that

$$R(\lambda, A) = \int_0^\infty e^{-\lambda s} T(s) \, ds$$

holds for all $\operatorname{Re} \lambda > 0$. $\square$

As a corollary we obtain a version of Perron's theorem for the positive exponential function.

Corollary 10.31. *For a positive exponential function $T(t) = e^{tA}$ we have*

$$s(A) \in \sigma(A).$$

Proof. The positivity of the operators $T(t)$ means that

$$|T(t)f| \le T(t)|f|$$

for all $f \in E$ and $t \ge 0$. Therefore,

$$|R(\lambda, A)f| \le \int_0^\infty e^{-\operatorname{Re} \lambda s} T(s)|f| \, ds$$

for all $\operatorname{Re} \lambda > s(A)$ and $f \in E$. Hence,

$$\|R(\lambda, A)\| \le \|R(\operatorname{Re} \lambda, A)\|.$$

Recall that since $A \in \mathcal{L}(E)$ is a bounded operator, we have that $\sigma(A) \ne \emptyset$. Further, there is $\lambda_k \in \rho(A)$ such that $\operatorname{Re} \lambda_k \to s(A)$, $\operatorname{Re} \lambda_k > s(A)$ and $\|R(\lambda_k, A)\| \to \infty$. This implies $\|R(\operatorname{Re} \lambda_k, A)\| \to \infty$, and hence $s(A) \in \sigma(A)$ (see Corollary 9.30). $\square$

Compare the following with Corollary 7.4 from Chapter 7.

Corollary 10.32. *Let K be a compact Hausdorff topological space and $E = C(K)$. For a positive exponential function $T(t) = e^{tA}$, $A \in \mathcal{L}(E)$, the following assertions are equivalent.*

(i) $s(A) < 0$.

(ii) $-A^{-1}$ *exists and it is positive.*

(iii) *There exists $0 \le f \in E$ such that $Af = -\mathbb{1}$.*

Proof. The equivalence (i) $\iff$ (ii) follows from Proposition 10.30. Since $-A^{-1} = R(0, A)$, (ii) $\implies$ (iii) follows by taking $f := -A^{-1}\mathbb{1}$.

We close the loop by showing (iii) $\implies$ (i). Assume that $Af = -\mathbb{1}$ for some $0 \le f \in E$. Then for $\lambda > \max\{s(A), 0\}$ we have

$$0 \le R(\lambda, A)\mathbb{1} = -AR(\lambda, A)f$$
$$= f - \lambda R(\lambda, A)f \le f.$$

Hence

$$\sup_{\lambda > \max\{s(A),0\}} \|R(\lambda, A)\| \leq \|f\|_\infty.$$

Since by Corollary 10.31, $s(A) \in \sigma(A)$, it follows from Corollary 9.30 that $s(A) < 0$. $\qquad\square$

We close this chapter by a minimum principle characterization of positive exponential functions.

Theorem 10.33. *Let Ω be a locally compact Hausdorff space and let $A \in \mathcal{L}(E)$, where $E = C_0(\Omega)$. Then the following are equivalent.*

(i) *A generates a positive exponential function, i.e., $e^{tA} \geq 0$ for $t \geq 0$.*
(ii) *For $0 \leq f \in E$ and $x \in \Omega$, $f(x) = 0$ implies that $(Af)(x) \geq 0$.*
(iii) *$A + \|A\|I \geq 0$.*

Proof. (i) $\Longrightarrow$ (ii): Take $0 \leq f \in E$ and $x \in \Omega$ with $f(x) = 0$. Then

$$(Af)(x) = \lim_{t \downarrow 0} \frac{e^{tA}f - f}{t}(x) = \lim_{t \downarrow 0} \frac{e^{tA}f(x) - f(x)}{t} = \lim_{t \downarrow 0} \frac{e^{tA}f(x)}{t} \geq 0.$$

(ii) $\Longrightarrow$ (iii): Consider $x \in \Omega$. We have to show that $(Af)(x) + \|A\|f(x) \geq 0$ for all $f \in E$. Define

$$A^*\delta_x = \mu + c\delta_x,$$

where $\mu \in M(\Omega)$ is such that $\mu(\{x\}) = 0$, and $c \in \mathbb{R}$. We claim that $\mu \geq 0$. Take $0 \leq f \in E$ such that $f(x) = 0$. Then

$$\langle f, \mu \rangle = \langle f, A^*\delta_x \rangle = \langle Af, \delta_x \rangle = (Af)(x) \geq 0.$$

It can be shown (see Exercise 9) that this implies that $\langle g, \mu \rangle \geq 0$ for all $0 \leq g \in E$. Hence $\mu \geq 0$.

Moreover,

$$|c| = \|c\delta_x\| \leq \|c\delta_x + \mu\| = \|A^*\delta_x\| \leq \|A\|.$$

Hence, for $0 \leq f \in E$, we have that

$$(Af)(x) + \|A\|f(x) = \langle Af + \|A\|f, \delta_x \rangle = \langle f, A^*\delta_x + \|A\|\delta_x \rangle$$
$$= \langle f, \mu + (c + \|A\|)\delta_x \rangle \geq 0.$$

(iii) $\Longrightarrow$ (i): The same argument as in the proof of Theorem 7.1 applies. We know that if $B := A + \|A\|I \in \mathcal{L}(E)$ is positive, then

$$e^{tB} = \sum_{k=0}^\infty \frac{(tB)^k}{k!} \geq 0.$$

Hence,

$$e^{tA} = e^{-t\|A\|}e^{tB} \geq 0. \qquad\square$$

Remark 10.34. The equivalence of (i) and (iii) in Theorem 10.33 is in complete analogy to Theorem 7.1 in the matrix case. This result is true in general Banach lattices, but the proofs are more involved. Conditions (ii) and (iii) are nothing but generalizations of the "positive off-diagonal" property for matrices.

10.8   Notes and Remarks

The investigation of ordered algebraic structures is a classical subject and of great interest in the literature, we mention here the monograph by Fuchs [50]. Most results of this chapter can be found for example in the monographs by Schaefer [126], Meyer-Nieberg [95] or Aliprantis and Burkinshaw [2]. For Proposition 10.12 see [95, Propositions 1.1.5, 1.2.3, and 1.2.5]. For complexification of real Banach lattices, we refer to Schaefer [126, Section II.11] or Meyer-Nieberg [95, Section 2.2].

For Theorem 10.33 we refer to Nagel (ed.) [101, Theorem B-II.1.3]. For the generalization to arbitrary Banach lattices, see [101, Theorem C-II.1.11].

10.9   Exercises

1. Prove the properties b)–e) in Proposition 10.3 and d) in Proposition 10.4.

2. Let E be a vector lattice and $f, g, h \in E$.

   a) Prove that $f \vee g = \frac{1}{2}(f + g + |f - g|)$ and $f \wedge g = \frac{1}{2}(f + g - |f - g|)$.

   b) Show that $|f| \vee |g| = \frac{1}{2}(|f + g| + |f - g|)$ and deduce that

   $$|f| \wedge |g| = \frac{1}{2} \, \big||f + g| - |f - g|\big| \, .$$

   c) Deduce that $f \perp g$ is equivalent to $|f - g| = |f + g|$.

   d) Prove this variant of the triangle inequality:

   $$\big||f| - |g|\big| \leq |f + g| \leq |f| + |g|.$$

   e) Deduce that $f \perp g$ is equivalent to $|f| \vee |g| = |f| + |g|$, and that in this case
   $$\big||f| - |g|\big| = |f + g| = |f| + |g|.$$

   f) Show Birkhoff's inequalities:

   $$|f \vee h - g \vee h| \leq |f - g| \quad \text{and} \quad |f \wedge h - g \wedge h| \leq |f - g|.$$

3. Prove that a subspace I of a Banach lattice is an ideal if and only if

   $$(f \in I, \ |g| \leq |f|) \Longrightarrow g \in I.$$

4. Prove that $\mathrm{H}^1(0,1)$ endowed with the natural order, $f \geq 0$ if $f(s) \geq 0$ for a.e. $s \in [0,1]$, is a vector lattice.

5. Consider $E := \mathrm{C}^1([0,1])$ equipped with the norm

$$\|f\| = \max_{s \in [0,1]} |f'(s)| + |f(0)|$$

and the order $f \geq 0$ whenever $f(0) \geq 0$ and $f' \geq 0$. Show that E is a Banach lattice.

6. Let E be a Banach lattice. Use the Hahn–Banach theorem to prove that

   a) $0 \leq f$ is equivalent to $\langle f, f^* \rangle \geq 0$ for all $f^* \in E_+^*$;

   b) for each $f \in E$ there exists $f^* \in E_+^*$ such that $\|f^*\| \leq 1$ and $\langle f, f^* \rangle = \|f^+\|$.

7. Consider the Banach lattice $\mathrm{C}^1([0,1])$ as in Exercise 5 and define the operator

$$(Tf)(t) := \int_0^t g(s)f(s)\,\mathrm{d}s$$

with a given $g \in \mathrm{C}([0,1])$. Calculate $\|T\|$. For which g is T positive?

8. Let $T \in \mathcal{L}(E,F)$, where E and F are two Banach lattices. Show that T is a lattice homomorphism if and only if one of the following equivalent properties holds.

   (i) $T(f \vee g) = Tf \vee Tg$ and $T(f \wedge g) = Tf \wedge Tg$ for all $f, g \in E$.

   (ii) $Tf^+ \wedge Tf^- = 0$ for all $f \in E$.

9. Let Ω be a locally compact Hausdorff space, $x \in \Omega$, and μ a regular bounded Borel measure on Ω such that $\mu(\{x\}) = 0$. Show that $\mu \geq 0$ if and only if $\langle f, \mu \rangle \geq 0$ for all $f \in C_0(\Omega)$ satisfying $f \geq 0$ and $f(x) = 0$.

Chapter 11

Generation Properties

In this chapter we continue to study the connection between semigroups and their generators. We characterize generators by the properties of their resolvents. As an important byproduct, we derive a characterization of positive semigroups by the positivity of the resolvent. This is a fundamental result and will be frequently used in the sequel.

As a first application of the characterization theorem we show a simple perturbation result. We give also the explicit form of the perturbed semigroup.

Then we focus on generators of positive contraction semigroups on Banach lattices and characterize them using an important property called dispersivity. We present some examples and conclude with a simple positive minimum principle. These are all technical tools needed to investigate properties of positive semigroups, which will be the topic of the following chapters.

11.1 The Hille–Yosida Generation Theorem

In Chapter 9 some properties of semigroup generators and their resolvents were collected. It turns out that some of these properties actually characterize semigroup generators. The following theorem was proved independently by E. Hille and K. Yosida in 1948.

Theorem 11.1 (Hille–Yosida). *Let A be a linear operator on a Banach space X. Then the following properties are equivalent.*

(i) *A generates a C_0-semigroup $(T(t))_{t\geq 0}$ of type (M,ω).*

(ii) *A is closed, densely defined, and there exist $M \geq 1$ and $\omega \in \mathbb{R}$ such that for every $\lambda \in \mathbb{C}$ with $\operatorname{Re}\lambda > \omega$ one has $\lambda \in \rho(A)$ and*

$$\left\| R(\lambda, A)^k \right\| \leq \frac{M}{(\operatorname{Re}\lambda - \omega)^k} \quad \textit{for all } k \in \mathbb{N}. \tag{11.1}$$

We will not prove this theorem here in its full generality, but only follow through the main idea of its proof. Let us note that, by replacing A by $A - \omega$, one can take $\omega = 0$ in (11.1). As a first step we need the following approximation result.

Lemma 11.2. *Let A be a closed, densely defined operator. Assume that there exists $M \geq 1$ such that for all $\lambda > 0$, we have $\lambda \in \rho(A)$ and $\|\lambda R(\lambda, A)\| \leq M$. Then*

a) $\lambda R(\lambda, A)g \to g$ *for all* $g \in X$ *as* $\lambda \to \infty$, *and*
b) $\lambda A R(\lambda, A)f \to Af$ *for all* $f \in D(A)$ *as* $\lambda \to \infty$.

Proof. Taking $g \in D(A)$, we see that $\lambda R(\lambda, A)g = R(\lambda, A)Ag + g$. By assumption,

$$\|R(\lambda, A)Ag\| \leq \frac{M}{\lambda}\|Ag\|,$$

and hence $\lambda R(\lambda, A)g \to g$ as $\lambda \to \infty$. By the denseness of $D(A)$ and the boundedness of $R(\lambda, A)$, the convergence follows for all $g \in X$. Assertion b) is an immediate consequence of a), inserting $g = Af$ and using the fact that A and $R(\lambda, A)$ commute. $\qquad\square$

The operators $\lambda A R(\lambda, A)$, $\lambda > 0$, are called *Yosida approximants*.

Proof of Theorem 11.1. Since every generator is closed and densely defined, the implication (i) $\Longrightarrow$ (ii) follows by Proposition 9.33.c).

The core of the proof of the implication (ii) $\Longrightarrow$ (i) is the case $M = 1$, $\omega = 0$. As mentioned above, one may assume, without loss of generality, that $\omega = 0$. The general situation can be reduced to this case by considering the equivalent norm

$$\|\!|f|\!\| := \sup_{\mu > 0} \|f\|_\mu, \tag{11.2}$$

where $\|f\|_\mu := \sup_{k \geq 0} \|\mu^k R(\mu, A)^k f\|$, see Exercises 4 and 5.
Let

$$A_k := k A R(nk, A) = k^2 R(k, A) - kI,$$

which are bounded operators for each $k \in \mathbb{N}$ commuting with each other. By Lemma 11.2, the sequence A_k converges to A pointwise on $D(A)$. Consider the uniformly continuous, mutually commuting semigroups given by

$$T_k(t) := \mathrm{e}^{t A_k}, \quad t \geq 0.$$

First of all, each $T_k(\cdot)$ is a contraction semigroup, since

$$\|T_k(t)\| \leq \mathrm{e}^{-kt}\mathrm{e}^{\|k^2 R(k,A)\|t} \leq \mathrm{e}^{-kt}\mathrm{e}^{kt} = 1 \qquad \text{for } t \geq 0.$$

By Theorem A.17 it thus suffices to prove convergence of $(T_k(t)f)$ for $f \in D(A)$ only. Using the vector-valued version of the fundamental theorem of calculus applied to the functions

$$s \longmapsto T_m(t - s)T_n(s)f, \quad s \in [0, t],$$

for some $f \in D(A)$, $m, n \in \mathbb{N}$, we see that

$$T_n(t)f - T_m(t)f = \int_0^t \frac{\mathrm{d}}{\mathrm{d}s}\left(T_m(t-s)T_n(s)f\right)\,\mathrm{d}s$$
$$= \int_0^t T_m(t-s)T_n(s)(A_n f - A_m f)\,\mathrm{d}s.$$

By the contractivity of all $T_k(t)$ it thus follows that

$$\|T_n(t)f - T_m(t)f\| \le t\,\|A_n f - A_m f\|.$$

Since by Lemma 11.2 $(A_k f)$ is a Cauchy sequence for each $f \in D(A)$, the sequence $(T_k(t)f)$ converges uniformly on finite intervals $[0, \tau]$ for every $f \in D(A)$, and therefore converges for every $f \in X$.

By the above, the limit $T(t)f := \lim_{k\to\infty} T_k(t)f$ exists for every $f \in X$, satisfies the semigroup property, and consists of contractions. Furthermore, for each $f \in D(A)$, the corresponding orbit map $t \mapsto T(t)f$, $0 \le t \le \tau$, is continuous, hence by Proposition 9.10, the semigroup $(T(t))_{t\ge0}$ is strongly continuous.

Now denote by B the generator of $(T(t))_{t\ge0}$. We show that $B = A$. Fix $f \in D(A)$ and define

$$\xi_k(t) := T_k(t)f \quad \text{and} \quad \xi(t) := T(t)f.$$

On each compact interval $[0, \tau]$ the functions ξ_k converge uniformly to ξ and $\dot{\xi}_k(t) = T_k(t)A_k f$ converge uniformly to $\eta(t) := T(t)Af$. Hence, the function ξ is differentiable with $\dot{\xi}(t) = \eta(t)$ and

$$Af = \eta(0) = \dot{\xi}(0) = Bf \quad \text{for } f \in D(A).$$

On the other hand, for every $\lambda > 0$, we have $\lambda \in \rho(A)$ by assumption, as well as $\lambda \in \rho(B)$, because B generates a contraction semigroup. By the above, $\lambda - A \subset \lambda - B$, where the first operator is a bijection from $D(A)$ onto X, and the second a bijection from $D(B)$ onto X. This is only possible if $D(A) = D(B)$ and $A = B$. $\qquad\square$

Remark 11.3. For a C_0-semigroup $(T(t))_{t\ge0}$ with generator A on a Banach space X we have

$$T(t)f = \lim_{k\to\infty} e^{tA_k} f$$

in the norm of X for each $f \in X$. To see this, we assume without loss of generality that $\|T(t)\| \le M$ for all $t \ge 0$, and endow X with the norm $\vertiii{\cdot}$, defined in (11.2). Then, $\vertiii{T(t)} \le 1$ (see Exercises 4 and 5). So, from the above proof we have that

$$\lim_{k\to\infty} \vertiii{T(t)f - e^{tA_k} f} = 0,$$

and the claim follows thanks to the equivalence of the norms $\|\cdot\|$ and $\vertiii{\cdot}$.

From this remark we immediately obtain a characterization of the generators of positive semigroups.

Corollary 11.4. *Let E be a Banach lattice. A C_0-semigroup $(T(t))_{t\geq 0}$ with generator A is positive if and only if $R(\lambda, A) \geq 0$ for each λ sufficiently large.*

Proof. If $(T(t))_{t\geq 0}$ is a positive C_0-semigroup, then by the Laplace transform in relation (9.11) also $R(\lambda, A) \geq 0$ for all $\lambda > \omega_0(T)$.

Conversely, if for k sufficiently large, $R(k, A) \geq 0$, then since

$$A_k := kAR(k, A) = k^2 R(k, A) - kI,$$

we see that

$$e^{tA_k} = e^{-kt}e^{tk^2 R(k, A)} \geq 0.$$

The statement then follows from Remark 11.3.   □

11.2   Bounded Perturbations

As an application of the generation theorem, we mention some basic perturbation results. The idea is always the same: We start with a generator A, assume that the operator B is "nice enough", and want to conclude that $A + B$ generates a semigroup.

Let us start with the simplest and most used perturbation result, in which the perturbation is bounded. Some further perturbations will be considered in Chapter 13.

Theorem 11.5. *If A generates a semigroup $(T(t))_{t\geq 0}$ of type (M, ω) and $B \in \mathcal{L}(X)$, then $A + B$ with $D(A + B) = D(A)$ generates a semigroup $(S(t))_{t\geq 0}$ of type $(M, \omega + M\|B\|)$.*

Proof. First we change the operator to $A - \omega$ and then use the renorming procedure (11.2) mentioned in the proof of the Hille–Yosida theorem. So, we only need to consider the case when A generates a semigroup of type $(1, 0)$, i.e., a contraction semigroup.

As a next step, we show that the operator $A + B$ has a non-empty resolvent set. More precisely, if $\lambda > 0$, we can use the identity

$$\lambda - A - B = (I - BR(\lambda, A))(\lambda - A), \tag{11.3}$$

yielding that if $\|BR(\lambda, A)\| < 1$, then $\lambda \in \rho(A + B)$ and

$$R(\lambda, A + B) = R(\lambda, A) \sum_{k=0}^{\infty} (BR(\lambda, A))^k. \tag{11.4}$$

By assumption, A is the generator of a contraction semigroup, so $\lambda\|R(\lambda, A)\| \leq 1$. Thus, if $\lambda > \|B\|$, then $\lambda \in \rho(A + B)$ and relation (11.4) holds.

For the norm of the resolvent we see by estimating the series term-by-term that

$$\|R(\lambda, A+B)\| \leq \frac{1}{\lambda} \sum_{k=0}^{\infty} \left(\frac{\|B\|}{\lambda} \right)^k = \frac{1}{\lambda} \cdot \frac{1}{1-\frac{\|B\|}{\lambda}} = \frac{1}{\lambda - \|B\|}. \tag{11.5}$$

By the Hille–Yosida theorem, this means that the operator $A+B$ generates a semigroup $(S(t))_{t\geq 0}$ of type $(1, \|B\|)$, which proves the statement. □

We know from Theorem 11.5 that $A+B$ with domain $D(A)$ generates a C_0-semigroup $(S(t))_{t\geq 0}$ on a Banach space X whenever A generates a C_0-semigroup $(T(t))_{t\geq 0}$ and $B \in \mathcal{L}(X)$. The perturbed semigroup $(S(t))_{t\geq 0}$ can be given by a series called *Dyson–Phillips expansion*, as the following result shows.

Proposition 11.6. *Let A with domain $D(A)$ be the generator of a C_0-semigroup $(T(t))_{t\geq 0}$ on a Banach space X and let $B \in \mathcal{L}(X)$. Then the semigroup $(S(t))_{t\geq 0}$ generated by $A+B$ with domain $D(A)$ is given by the Dyson–Phillips series*

$$S(t) = \sum_{k=0}^{\infty} U_k(t), \quad t \geq 0, \tag{11.6}$$

where

$$U_0(t) = T(t) \quad and \quad U_{k+1}(t) = \int_0^t U_k(t-s)BT(s)\,ds, \quad t \geq 0, \ k \in \mathbb{N}.$$

Moreover, the following variation of constant formulas hold:

$$S(t)f = T(t)f + \int_0^t S(t-s)BT(s)f\,ds \tag{11.7}$$

$$= T(t)f + \int_0^t T(t-s)BS(s)f\,ds, \quad f \in X, t \geq 0. \tag{11.8}$$

Proof. Since $U_0(t) = T(t)$, we have $\|U_0(t)\| \leq Me^{\omega t}$ for $t \geq 0$ and some constants $M \geq 1$ and $\omega \in \mathbb{R}$. Thus,

$$\|U_1(t)\| \leq Me^{\omega t} \int_0^t e^{-\omega s} \|BT(s)\|\,ds \leq M^2 \|B\| t e^{\omega t} \quad \text{for all } t \geq 0.$$

By induction one can verify that

$$\|U_k(t)\| \leq M \frac{(M\|B\|t)^k}{k!} e^{\omega t} \quad \text{for all } t \geq 0.$$

Therefore the series $\sum_{k=0}^{\infty} U_k(t)$ converges in $\mathcal{L}(X)$ uniformly on compact intervals of $\mathbb{R}_+$. Moreover,

$$\left\| \sum_{k=0}^{\infty} U_k(t) \right\| \leq Me^{(\omega+M\|B\|)t} \quad \text{for all } t \geq 0.$$

Set $U(t) := \sum_{k=0}^{\infty} U_k(t)$ for $t \geq 0$. Since $t \mapsto U_k(t)f$ is continuous and the convergence is uniform on compact subsets of $\mathbb{R}_+$, we deduce that the function $t \mapsto U(t)f$ is continuous for any $f \in X$.

Let us prove now that $(U(t))_{t \geq 0}$ is a C_0-semigroup which coincides with the perturbed semigroup $(S(t))_{t \geq 0}$. For this purpose we let $f \in X$, $t, s \geq 0$, and first verify that

$$\sum_{j=0}^{k} U_j(s) U_{k-j}(t) f = U_k(t+s) f. \tag{11.9}$$

This is obviously true for $k = 0$. Assuming (11.9) holds for some $k \in \mathbb{N}$, we compute

$$\sum_{j=0}^{k+1} U_j(s) U_{k+1-j}(t) f$$

$$= \sum_{j=0}^{k} U_j(s) \int_0^t U_{k-j}(t-r) BT(r) f \ \mathrm{d}r + \int_0^s U_k(s-r) BT(r+t) f \, \mathrm{d}r$$

$$= \int_0^t U_k(s+t-r) BT(r) f \ \mathrm{d}r + \int_t^{t+s} U_k(s+t-r) BT(r) f \, \mathrm{d}r$$

$$= \int_0^{t+s} U_k(s+t-r) BT(r) f \, \mathrm{d}r = U_{k+1}(t+s) f, \quad t, s \geq 0, \ f \in X.$$

Hence,

$$U(t)U(s)f = \sum_{i=0}^{\infty} U_i(t) \sum_{j=0}^{\infty} U_j(s) f = \sum_{m=0}^{\infty} \sum_{k=0}^{m} U_k(t) U_{m-k}(s) f$$

$$= \sum_{m=0}^{\infty} U_m(t+s) f = U(t+s) f, \quad t, s \geq 0, \ f \in X.$$

Therefore, $(U(t))_{t \geq 0}$ is a C_0-semigroup on X. Let us denote its generator by C with domain $D(C)$. Using the definition of $U(t)$ we obtain

$$U(t) = T(t) + \sum_{k=0}^{\infty} U_{k+1}(t)$$

$$= T(t) + \sum_{k=0}^{\infty} \int_0^t U_k(t-s) BT(s) \ \mathrm{d}s$$

$$= T(t) + \int_0^t \left(\sum_{k=0}^{\infty} U_k(t-s) BT(s) \right) \ \mathrm{d}s$$

$$= T(t) + \int_0^t U(t-s) BT(s) \ \mathrm{d}s \quad \text{for all } t \geq 0. \tag{11.10}$$

Observe now that proving

$$\lim_{t \to 0} \frac{1}{t} \int_0^t U(t-s)BT(s)f \, ds = Bf$$

for every f yields that $D(C) = D(A)$ and $Cf = Af + Bf$ for $f \in D(A)$, and thus the semigroups $(U(t))_{t \geq 0}$ and $(S(t))_{t \geq 0}$ coincide.

To finish the proof, let us take $f \in D(A)$ and compute

$$\lim_{t \to 0} \left(\frac{1}{t} \int_0^t U(t-s)BT(s)f \, ds - Bf \right)$$

$$= \lim_{t \to 0} \left(\frac{1}{t} \int_0^t U(t-s)B(T(s)f - f) \, ds + \frac{1}{t} \int_0^t (U(t-s)Bf - Bf) \, ds \right)$$

$$= \lim_{t \to 0} \left(\frac{1}{t} \int_0^t U(t-s)B \int_0^s T(r)Af \, dr \, ds + \frac{1}{t} \int_0^t (U(s)Bf - Bf) \, ds \right) = 0.$$

where we used (11.10) and Proposition 9.16.b).

By (11.10), also the variation of constants formula (11.7) holds. To show the second formula, take $f \in D(A)$ and define

$$\xi(s) := T(t-s)S(s)f, \quad s \in [0,t].$$

Note that the function ξ is continuously differentiable, with

$$\frac{d}{ds}\xi(s) = T(t-s)CS(s)f - T(t-s)AS(s)f = T(t-s)BS(s)f.$$

Thus, by taking the integral, we see that

$$\int_0^t T(t-s)BS(s)f \, ds = \xi(t) - \xi(0) = S(t)f - T(t)f.$$

Since $D(A)$ is dense, this holds for every $f \in X$. □

Using either the above Dyson–Phillips series, or the resolvent series in (11.4), we obtain the following positivity result.

Corollary 11.7. *If A generates a positive C_0-semigroup on a Banach lattice E and $B \in \mathcal{L}(E)$ is a positive operator, then the semigroup generated by $A + B$ is positive.*

11.3 Positive Contraction Semigroups

Let us now focus our attention on the case when our operators act on a real Banach lattice E. We would like to characterize when the operator A generates a positive contraction semigroup.

To motivate what follows, let us first note that, by Theorem 19.5 and Corollary 11.4, A generates a positive contraction semigroup if and only if $R(\lambda, A) \geq 0$ and

$$\|\lambda R(\lambda, A)\| \leq 1$$

holds for all $\lambda > 0$.

Lemma 11.8. *A bounded linear operator T on a real Banach lattice E is a positive contraction if and only if*

$$\|(Tf)^+\| \leq \|f^+\| \tag{11.11}$$

for all $f \in E$.

Proof. Assume first that T is a positive contraction. Using Lemma 10.18 we see that

$$\|(Tf)^+\| \leq \|Tf^+\| \leq \|f^+\|.$$

Conversely, let (11.11) hold for a bounded operator T. Then, $\|(T(-f))^+\| \leq \|(-f)^+\|$ for all $f \in E$. Since $(T(-f))^+ = (-Tf)^+ = (Tf)^-$ and $(-f)^+ = f^-$, we obtain

$$\|(Tf)^-\| \leq \|f^-\|, \quad f \in E.$$

Take now $f \in E_+$. Since $f^- = 0$, from the above estimate we infer that $(Tf)^- = 0$ and thus $Tf \in E_+$. So, T is a positive operator.

Now, take any $f \in E$. By Lemma 10.18, the positivity of T, and inequality (11.11), respectively, we have that

$$\|Tf\| \leq \|T|f|\| = \|(T|f|)^+\| \leq \||f|^+\| = \|f\|. \qquad \square$$

This motivates introducing the following property.

Definition 11.9. A linear operator A on E is called *dispersive* if for every $f \in D(A)$ and $\lambda > 0$, we have that

$$\|(\lambda f - Af)^+\| \geq \lambda \|f^+\|.$$

The main result in this section was first proved by Ralph Phillips in 1962.

Theorem 11.10. *Let A be a densely defined linear operator on a real Banach lattice E. Then the following assertions are equivalent.*

  (i) *A generates a positive contraction C_0-semigroup.*
  (ii) *A is dispersive and $\operatorname{im}(\lambda - A) = E$ for some (and then for all) $\lambda > 0$.*

Proof. (i) $\Longrightarrow$ (ii) It suffices to prove that A is dispersive. Since $\lambda R(\lambda, A)$ is a positive contraction, Lemma 11.8 implies that

$$\|(\lambda R(\lambda, A)g)^+\| \leq \|g^+\|$$

for all $\lambda > 0$ and $g \in E$. Hence, Taking $f \in D(A)$ and defining $g := (\lambda - A)f$, we obtain that

$$\lambda \|f^+\| \leq \|(\lambda f - Af)^+\|$$

for all $f \in D(A)$ and $\lambda > 0$.

(ii) $\Longrightarrow$ (i) Let $\lambda_0 > 0$ such that $\operatorname{im}(\lambda_0 - A) = E$. Since A is dispersive,

$$\|(\lambda_0 f - Af)^+\| \geq \lambda_0 \|f^+\| \quad \text{and} \quad \|(Af - \lambda_0 f)^+\| \geq \lambda_0 \|(-f)^+\|$$

for all $f \in D(A)$. This implies that $\lambda_0 - A$ is injective, and hence

$$\|(\lambda_0 R(\lambda_0, A)f)^+\| \leq \|f^+\|$$

for all $f \in E$. By Lemma 11.8, we obtain $\lambda_0 \in \rho(A)$ and $\|\lambda_0 R(\lambda_0, A)\| \leq 1$. We will show that $(0, \infty) \subset \rho(A)$ and $\|R(\lambda, A)\| \leq \frac{1}{\lambda}$ for all $\lambda > 0$.

Let $\lambda \in (0, 2\lambda_0)$. Then,

$$(\lambda - A) = (\lambda_0 - A) - (\lambda_0 - \lambda)$$
$$= (I - (\lambda_0 - \lambda)R(\lambda_0, A))(\lambda_0 - A)$$

and $\|(\lambda_0 - \lambda)R(\lambda_0, A)\| \leq \frac{|\lambda_0 - \lambda|}{\lambda_0} < 1$ since $0 < \lambda < 2\lambda_0$. Thus,

$$R(\lambda, A) = R(\lambda_0, A)(I - (\lambda_0 - \lambda)R(\lambda_0, A))^{-1} \in \mathcal{L}(X)$$

and from the dispersivity of A one obtains that

$$\|\lambda R(\lambda, A)\| \leq 1 \quad \text{and} \quad R(\lambda, A) \geq 0$$

for all $\lambda \in (0, 2\lambda_0)$. Taking now $\mu \in [\lambda_0, 2\lambda_0)$ and arguing as before, we see that

$$(0, 2\mu) \subset \rho(A) \quad \text{and} \quad \|\lambda R(\lambda, A)\| \leq 1, \quad \text{and} \quad R(\lambda, A) \geq 0$$

for all $\lambda \in (0, 2\mu)$.

Repeating the same process we obtain

$$(0, \infty) \subset \rho(A) \quad \text{and} \quad \|\lambda R(\lambda, A)\| \leq 1 \quad \text{for all } \lambda > 0.$$

Further, $R(\lambda, A) \geq 0$. So, the claim follows by the Hille–Yosida theorem (see Theorem 11.1) and Corollary 11.4. $\qquad \square$

As it turns out, dispersivity can be characterized in a very elegant way which is useful for many applications. For this, note that the Hahn–Banach theorem implies that the set

$$\mathcal{I}^+(f) := \{f^* \in E_+^* : \|f^*\| \leq 1, \langle f, f^* \rangle = \|f^+\|\}$$

is non-empty (see Exercise 10.9.6).

Let us first list some examples for $\mathcal{I}^+(f)$.

Example 11.11.

a) Let $E := C_0(\Omega)$, with Ω a locally compact Hausdorff space. Given $0 \neq f \in E_+$, there exists $x_0 \in \Omega$ such that $f(x_0) = \|f^+\|_\infty$. One can see that $\delta_{x_0} \in \mathcal{I}^+(f)$. If $f \in E_-$, then we choose $f^* = 0$.

b) Let $E := L^p(\Omega, \mu)$, $1 < p < \infty$, where (Ω, Σ, μ) is a σ-finite measure space, and let $f \in E$ with $f^+ \neq 0$. Set $\varphi := \|f^+\|^{-p/q}(f^+)^{p-1}$, where $\frac{1}{p} + \frac{1}{q} = 1$. Then it can be seen that $\varphi \in \mathcal{I}^+(f)$. We remark that it is known that in L^p-spaces with $1 < p < \infty$, the sets $\mathcal{I}^+(f)$ are singletons. Hence,

$$\mathcal{I}^+(f) = \{\varphi\}.$$

c) In the case $E := L^1(\Omega, \mu)$, where (Ω, Σ, μ) is a σ-finite measure space, the function $\varphi := \chi_{[f>0]}$ belongs to $\mathcal{I}^+(f)$ for any $f \in E$.

Proposition 11.12. *An operator A on the real Banach lattice E is dispersive if and only if for every $f \in D(A)$, there is $f^* \in \mathcal{I}^+(f)$ such that*

$$\langle Af, f^* \rangle \leq 0. \tag{11.12}$$

Proof. We start by proving that (11.12) implies dispersivity. If (11.12) holds, then for any $\lambda > 0$ and $f \in D(A)$, we have

$$\begin{aligned}
\lambda \|f^+\| = \langle \lambda f, f^* \rangle &= \langle \lambda f - Af + Af, f^* \rangle \\
&\leq \langle \lambda f - Af, f^* \rangle \\
&\leq \langle (\lambda f - Af)^+, f^* \rangle \\
&\leq \|(\lambda f - Af)^+\|,
\end{aligned}$$

showing that A is dispersive.

For the converse, let $f \in D(A)$, $\lambda > 0$, and take $f^*_\lambda \in \mathcal{I}^+(\lambda f - Af)$. Since $\|f^*_\lambda\| \leq 1$, by Theorem A.31, the set $\{f^*_\lambda\}$ has a weak*-accumulation point $f^* \in E^*$ as $\lambda \to \infty$. So, $\|f^*\| \leq 1$. Moreover, we have

$$\langle f, f^*_\lambda \rangle - \frac{1}{\lambda}\langle Af, f^*_\lambda \rangle = \left\| \left(f - \frac{1}{\lambda} Af \right)^+ \right\|.$$

Letting $\lambda \to \infty$ yields $\langle f, f^* \rangle = \|f^+\|$. So, $f^* \in \mathcal{I}^+(f)$.

On the other hand, by assumption,

$$\begin{aligned}
\lambda \|f^+\| &\leq \|(\lambda f - Af)^+\| \\
&= \langle \lambda f - Af, f^*_\lambda \rangle \\
&\leq \lambda \langle f^+, f^*_\lambda \rangle - \langle Af, f^*_\lambda \rangle \\
&\leq \lambda \|f^+\| - \langle Af, f^*_\lambda \rangle.
\end{aligned}$$

Thus, $\langle Af, f^*_\lambda \rangle \leq 0$ for all $\lambda > 0$, hence, $\langle Af, f^* \rangle \leq 0$. $\square$

Remark 11.13. On a (real) Hilbert lattice H the Riesz–Fréchet representation theorem, see Theorem A.28, shows that $\mathcal{I}^+(f) = \{f^+/\|f^+\|\}$ for every $f \in H$. As a consequence, a linear operator A on a Hilbert lattice is dispersive if and only if $(Af|f^+) \leq 0$ for every $f \in D(A)$.

We conclude this section by providing two interesting examples of dispersive operators.

Example 11.14.

a) First we consider a simple delay equation on the space $E = C([-1,0], \mathbb{R})$, where we define the operator

$$Af = f' \text{ with } D(A) = \{f \in C^1[-1,0] : f'(0) = Lf\}$$

for a positive linear operator $L : C[-1,0] \to \mathbb{R}$. The domain $D(A)$ can also be written as the kernel of a linear form

$$D(A) = \ker \varphi, \text{ where } \varphi : C^1[-1,0] \to \mathbb{R}, \ \varphi(f) = f'(0) - Lf.$$

This linear form is unbounded, hence $\ker \varphi$ is dense in X, see Proposition A.30.

We now show that $A - \|L\|$ is a dispersive operator. To see this, take $f \in E$ and $\tau_0 \in [-1,0]$ such that $f(\tau_0) = \|f^+\|_\infty$. Then, $\delta_{\tau_0} \in \mathcal{I}^+(f)$, see Example 11.11.a). Thus, showing that $A - \|L\|$ is dispersive is the same as showing that

$$\langle Af - \|L\|f, \delta_{\tau_0}\rangle = f'(\tau_0) - \|L\|f(\tau_0) \leq 0 \quad \text{for } f \in D(A). \tag{11.13}$$

Assume that $\tau_0 \in (-1,0)$. Then $f'(\tau_0) = 0$, and so (11.13) holds.

If now $\tau_0 = -1$, then $f'(-1) \leq 0$, and so (11.13) holds if $\tau_0 \in [-1,0)$. If $\tau_0 = 0$, the positivity of L yields

$$\langle Af - \|L\|f, \delta_{\tau_0}\rangle = f'(0) - \|L\|f(0)$$
$$= Lf - \|L\|\|f^+\|_\infty$$
$$\leq Lf^+ - \|L\|\|f^+\|_\infty \leq 0.$$

This proves the dispersiveness of $A - \|L\|$.

Further, let us prove that $\lambda - A$ is surjective for all $\lambda > \|L\|$. To see this, take any $g \in X$. We have to find $f \in D(A)$ such that $\lambda f - f' = g$. It is not difficult to verify that such a function is given by

$$f(\tau) = \frac{g(0) + Lh}{\lambda - L\varepsilon_\lambda}\varepsilon_\lambda(\tau) + h(\tau),$$

where

$$h(\tau) = \int_\tau^0 e^{\lambda(\tau-s)}g(s) \, ds \quad \text{and} \quad \varepsilon_\lambda(\tau) = e^{\lambda\tau}$$

for $\tau \in [-1,0]$.

Hence, by Theorem 11.10, the operator $A - \|L\|$ generates a positive contraction semigroup on E and, by Corollary 11.7, the operator A generates a positive C_0-semigroup $(T(t))_{t \geq 0}$ such that

$$\|T(t)\| \leq e^{\|L\|t}, \quad t \geq 0.$$

One can show that the obtained semigroup $(T(t))_{t \geq 0}$ satisfies the *translation property*, i.e., for $f \in X$

$$(T(t)f)(\tau) = \begin{cases} f(t + \tau) & \text{if } t + \tau \leq 0, \\ (T(t + \tau)f)(0) & \text{otherwise.} \end{cases}$$

Moreover, for any $f \in D(A)$ the function $u : [-1, \infty) \to \mathbb{C}$ defined by

$$u(t) = \begin{cases} f(t) & \text{if } t \in [-1, 0], \\ (T(t)f)(0) & \text{if } t > 0, \end{cases}$$

is the unique classical solution of the *delay equation*

$$\begin{cases} \dot{u}(t) = Lu_t, & t \geq 0, \\ u_0 = f, \end{cases}$$

where $u_t(s) = u(t + s)$, $s \in [-1, 0]$ is the *history function*. A more thorough treatment of delay equations follows in Chapter 15.

b) This example deals with uniformly elliptic second-order operators on L^2-spaces. Let $\Omega \subseteq \mathbb{R}^n$ be an open set. On $L^2(\Omega)$ we denote the standard inner product by

$$(f|g) = \int_\Omega f\bar{g}\, dx, \quad f, g \in L^2(\Omega).$$

We consider functions $a_{ij} \in L^\infty(\Omega)$ with $a_{ij} = a_{ji}$ and assume the *uniform ellipticity condition*: there is $\eta > 0$ such that

$$\sum_{i=1}^n \sum_{j=1}^n a_{ij}(x)\xi_i\bar{\xi}_j \geq \eta|\xi|^2, \quad x \in \mathbb{R}^n, \, \xi \in \mathbb{C}^n. \tag{11.14}$$

Define the second-order elliptic differential operator $A : D(A) \to L^2(\Omega)$ by

$$Af := \sum_{i,j=1}^n D_i(a_{ij}D_j f), \quad \text{with}$$

$$D(A) := \left\{ f \in H_0^1(\Omega) : \sum_{i,j=1}^n D_i(a_{ij}D_j f) \in L^2(\Omega) \right\}.$$

Here $D_j f$ denotes the partial distributional derivative of $f \in L^1_{\text{loc}}(\Omega)$ with respect to the variable x_j, see Appendix A.11. The space

$$\mathrm{H}^1(\Omega) := \left\{ f \in \mathrm{L}^2(\Omega) : D_j f \in \mathrm{L}^2(\Omega) \text{ for all } j = 1, \ldots, n \right\},$$

endowed with the inner product

$$(f|g)_{\mathrm{H}^1} := \int_\Omega f \bar{g} \, dx + \sum_{j=1}^n \int_\Omega D_j f D_j \bar{g} \, dx, \quad f, g \in \mathrm{H}^1(\Omega),$$

is a Hilbert space. The space $\mathrm{H}^1_0(\Omega)$ is the closure of the space of test functions $\mathrm{C}^\infty_c(\Omega)$ in $\mathrm{H}^1(\Omega)$.

For $f \in D(A)$, the ellipticity condition in (11.14) and Proposition A.48 yield

$$(Af|f^+) = -\sum_{i,j=1}^n \int_\Omega a_{ij} D_j f D_i f \chi_{[f \geq 0]} \, dx \leq -\eta \int_\Omega |\nabla f^+|^2 \, dx \leq 0.$$

Thus, by Remark 11.13, A is a dispersive operator and, by Theorem 11.10, it generates a positive C_0-semigroup of contractions on $\mathrm{L}^2(\Omega)$ if and only if the range condition is satisfied.

To show this range condition, we use that thanks to (11.14) and since $a_{ij} \in \mathrm{L}^\infty(\Omega)$, the space $\mathrm{H}^1_0(\Omega)$ can be seen as a Hilbert space endowed with the inner product

$$(g|\bar{f})_a := \sum_{i,j=1}^n \int_\Omega a_{ij} D_i g D_j f \, dx + \int_\Omega fg \, dx, \quad f, g \in \mathrm{H}^1_0(\Omega).$$

The norms $\| \cdot \|_a$ and $\| \cdot \|$, where $\| \cdot \|_a$ denotes the norm associated to $(\cdot|\cdot)_a$, are equivalent. From now on we consider $\mathrm{H}^1_0(\Omega)$ with the inner product $(\cdot|\cdot)_a$. Let us take $f \in \mathrm{L}^2(\Omega)$ and define the mapping

$$\varphi(g) := \int_\Omega fg \, dx \quad \text{for } g \in \mathrm{H}^1_0(\Omega).$$

Then, $\varphi \in (\mathrm{H}^1_0(\Omega))^*$. So, by the Riesz–Fréchet theorem, Theorem A.28, there exists a unique $h \in \mathrm{H}^1_0(\Omega)$ such that $\varphi(g) = (g|\bar{h})_a$ for all $g \in \mathrm{H}^1_0(\Omega)$. In particular, this holds for all $g \in \mathrm{C}^\infty_c(\Omega)$. Thus, $h - Ah = f$ holds (in the sense of distributions). Hence, $h \in D(A)$ and $h - Ah = f$. The denseness of $D(A)$ in $\mathrm{L}^2(\Omega)$ is obvious. Thus, by Theorem 11.10 the operator A generates a positive contraction C_0-semigroup.

11.4   Positive Minimum Principle

We investigate further the question how to decide whether a semigroup generator actually generates a positive semigroup. In this section we concentrate on the space of continuous functions, where a special behaviour can be observed. To this end, in this section we suppose that K is a compact Hausdorff space, and E is the real Banach lattice $E = \mathrm{C}(K)$. The following is a straightforward generalization of Theorem 10.33.

Theorem 11.15. *Suppose that $E = \mathrm{C}(K)$ and let A generate a strongly continuous semigroup $(T(t))_{t\geq 0}$ on E. Then the semigroup $(T(t))_{t\geq 0}$ is positive if and only if the following positive minimum principle holds:*

$$\begin{cases} \text{for all } f \in D(A) \text{ with } f \geq 0, \text{ the condition} \\ f(y) = 0 \text{ for some } y \in K \text{ implies } Af(y) \geq 0. \end{cases} \tag{11.15}$$

Proof. Assume first that $(T(t))_{t\geq 0}$ is positive and take $f \in D(A)$ with $f \geq 0$. If $f(y) = 0$ for some $y \in K$, then

$$(Af)(y) = \left(\lim_{h\downarrow 0} \frac{T(h)f - f}{h} \right)(y) = \lim_{h\downarrow 0} \frac{(T(h)f)(y)}{h} \geq 0.$$

For the other direction, suppose that (11.15) holds. By Corollary 11.4, we have to show that $R(\lambda, A) \geq 0$ for large real λ. Since every positive function can be approximated by strictly positive ones, it suffices to show this for $f \gg 0$. Introduce the numbers

$$\omega := \inf\{\lambda \in \mathbb{R} : [\lambda, \infty) \subset \rho(A)\}$$

and

$$\lambda_f := \inf\{\lambda > \omega : R(\mu, A)f \gg 0 \text{ for all } \mu \in (\lambda, \infty)\}.$$

Since $\mu R(\mu, A)f \to f$ as $\mu \to \infty$, we clearly have $\omega \leq \lambda_f < \infty$. We show that $\lambda_f = \omega$ for all $f \in E$ with $f \gg 0$.

Assume, by contradiction, that for $f \in E$ with $f \gg 0$ we have $\lambda_f > \omega$. Then $[\lambda_f, \infty) \subset \rho(A)$ and $R(\lambda_f, A)f \geq 0$, but $R(\lambda_f, A)f \gg 0$ does not hold (otherwise, this would contradict the definition of λ_f). This means that there is $y \in K$ such that $R(\lambda_f, A)f(y) = 0$. The positive minimum principle in (11.15) then implies that

$$AR(\lambda_f, A)f(y) \geq 0.$$

Therefore,

$$0 < f(y) = \lambda_f R(\lambda_f, A)f(y) - AR(\lambda_f, A)f(y) = -AR(\lambda_f, A)f(y) \leq 0,$$

a contradiction. Hence, $R(\lambda, A)f \gg 0$ for all $\lambda > \omega$ and for all $f \gg 0$. □

Example 11.16. We consider $E = C[0,1]$ and $Af = f''$ with Neumann boundary conditions, i.e., $D(A) = \{f \in C^2[0,1] : f'(0) = f'(1) = 0\}$. We show that the positive minimum principle is satisfied.

To this end, take $f \in D(A)$, $f \geq 0$, and $x \in [0,1]$ with $f(x) = 0$. We have two cases. If $x \in (0,1)$, then, because it is a minimum of the function f, we have that $f'' \geq 0$, which means exactly $(Af)(x) \geq 0$.

If x is on the boundary of the domain, for example if $x = 1$, then suppose by contradiction that $(Af)(1) = f''(1) < 0$. Then, by continuity, there is $\varepsilon > 0$ such that f' is strictly monotonically decreasing on $(1 - \varepsilon, 1]$. Since $f'(1) = 0$, this means that $f'(y) > 0$ for $y \in (1 - \varepsilon, 1)$. Hence, f is strictly monotonically growing on this interval, which means that $f(y) < 0 = f(1)$ for $y \in (1 - \varepsilon, 1)$, which contradicts to our assumption that $f \geq 0$. The other boundary point can be treated similarly.

11.5   Notes and Remarks

For most parts of this chapter we refer to the texts by Engel and Nagel [43, 44], Pazy [110], and Goldstein [54].

Theorem 11.1 was originally proved by Hille [65] and Yosida [156]. The Dyson–Phillips expansion in (11.6) was discovered by Dyson [37] studying quantum electrodynamics and rediscovered by Phillips in his pioneering work [112] on perturbation theory for operator semigroups.

Dispersive operators and Theorem 11.10 originate from Phillips [113]. For the fact, mentioned in Example 11.11, that in L^p-spaces with $1 < p < \infty$, the sets $\mathcal{I}^+(f)$ are singletons, we refer to Megginson [92, Example 5.1.4, Corollary 5.1.16].

The positivity minimum principle, Theorem 11.15 is due to W. Arendt and is taken from Nagel (ed.) [101, Theorem B-II.1.6].

11.6   Exercises

1. Consider the operators from Exercise 9.10.8. Which of them are generators?

2. Let $X = C_0(\mathbb{R})$ and let the operator A be given as

$$(Af)(x) := g(x)f(x),$$

where $g \in C(\mathbb{R})$ is a given function and with

$$D(A) := \{f \in C_0(\mathbb{R}) : gf \in C_0(\mathbb{R})\}.$$

Show that
$$\sigma(A) = \overline{\{f(x) \,:\, x \in \mathbb{R}\}}.$$

For which functions g is A a generator?

3. Prove that the generator of the Gaussian semigroup defined in Section 9.6 is given by

$$Af = f'', \quad f \in D(A) = \{f \in L^p(\mathbb{R}) : f'' \in L^p(\mathbb{R})\},$$

where f'' is defined in the distributional sense.

4. Let A be an operator satisfying the estimates (11.1). Without loss of generality (by considering $A - \omega$ instead of A) one can take $\omega = 0$ in (11.1). For every $\mu > 0$, define a new norm on X by

$$\|f\|_\mu := \sup_{k \geq 0} \|\mu^k R(\mu, A)^k f\|.$$

Show that:

a) $\|f\| \leq \|f\|_\mu \leq M \|f\|$, i.e., these new norms are all equivalent to $\|\cdot\|$.

b) $\|\mu R(\mu, A)\|_\mu \leq 1$.

c) $\|\lambda R(\lambda, A)\|_\mu \leq 1$ for all $0 < \lambda \leq \mu$.

d) $\|\lambda^k R(\lambda, A)^k f\| \leq \|\lambda^k R(\lambda, A)^k f\|_\mu \leq \|f\|_\mu$ for all $0 < \lambda \leq \mu$ and $k \in \mathbb{N}$.

e) $\|f\|_\lambda \leq \|f\|_\mu$ for $0 < \lambda \leq \mu$.

5. Using the notation of the previous exercise, show that for the norm

$$\|\!|f|\!\| := \sup_{\mu > 0} \|f\|_\mu$$

we have

a) $\|f\| \leq \|\!|f|\!\| \leq M \|f\|$.

b) $\|\lambda R(\lambda, A)\|\! \leq 1$ for all $\lambda > 0$.

6. Let $X = C[0,1]$ and consider the operator $Af = f''$ with domain

$$D(A) = \{f \in C^2[0,1] \,:\, f'(0) + \alpha f(0) = f'(1) + \beta f(1) = 0\},$$

for some $\alpha, \beta \in \mathbb{R}$. Show that A generates a positive semigroup.

7. Let A generate the C_0-semigroup $(T(t))_{t \geq 0}$ and $A + B$ the C_0-semigroup $(S(t))_{t \geq 0}$ for $B \in \mathcal{L}(X)$. Show that there is $K > 0$ such that

$$\|T(t) - S(t)\| \leq Kt$$

for $t \in [0,1]$.

Chapter 12

Spectral Theory for Positive Semigroups

We have discovered in the finite-dimensional case that exponential functions enjoy some rather special spectral properties. Such properties are, for example, that the spectrum of a semigroup operator is determined by the spectrum of its generator, or, that the stability of a semigroup is guaranteed whenever the spectrum of its generator lies in the left half-plane.

Unfortunately, since strongly continuous semigroups are not exactly exponential functions, these properties fail to hold in general. However, we will see that positivity has significant impact on the spectrum of the semigroup. We will show, for example, that the spectral bound is always an element of the spectrum of the generator of a positive semigroup and we will be able to make some more results analogous to the finite-dimensional case.

Throughout this chapter we suppose that E is a complex Banach lattice and X is a Banach space.

12.1 Asymptotic Stability of Semigroups

We are interested in the asymptotic behavior of the solution of the abstract Cauchy problem

$$\begin{cases} \dot{u}(t) = Au(t), & t \geq 0, \\ u(0) = f \in X, \end{cases}$$

where A is the generator of a C_0-semigroup $(T(t))_{t \geq 0}$ on X. Recall from Proposition 9.15 that the solution to this Cauchy problem is given by $u(t) = T(t)f$. In Chapter 9 we have also already defined the growth bound of the semigroup $(T(t))_{t \geq 0}$ as

$$\omega_0(T) := \inf\{\omega \in \mathbb{R} : \text{ there is } M = M_\omega \geq 1 \text{ with } \|T(t)\| \leq Me^{\omega t} \text{ for all } t \geq 0\}.$$

181

There is an important connection between the growth bound and the spectral radius of semigroup operators.

Proposition 12.1. *Let $(T(t))_{t\geq 0}$ be a C_0-semigroup on X.*

a) *We have that*

$$\omega_0(T) = \lim_{t\to\infty} \frac{\log\|T(t)\|}{t} = \inf_{t>0} \frac{\log\|T(t)\|}{t}. \tag{12.1}$$

b) *For every $t \geq 0$, the spectral radius $\mathrm{r}(T(t))$ of the operator $T(t)$ satisfies*

$$\mathrm{r}(T(t)) = \mathrm{e}^{t\omega_0(T)}.$$

Proof. a) By Exercise 1,

$$\lim_{t\to\infty} \frac{\log\|T(t)\|}{t} = \inf_{t>0} \frac{\log\|T(t)\|}{t}.$$

Setting

$$\eta := \inf_{t>0} \frac{\log\|T(t)\|}{t} = \lim_{t\to\infty} \frac{\log\|T(t)\|}{t},$$

we obtain $\mathrm{e}^{\eta t} \leq \|T(t)\|$ for all $t \geq 0$. So, by the definition of $\omega_0(T)$, we infer that $\eta \leq \omega_0(T)$. Take now $\omega > \eta$. Then there is a $\tau > 0$ such that

$$\frac{\log\|T(t)\|}{t} \leq \omega, \quad \text{for all } t \geq \tau.$$

Hence, $\|T(t)\| \leq \mathrm{e}^{\omega t}$ for all $t \geq \tau$. Since the function $t \mapsto T(t)$ is bounded on $[0, \tau]$, we see that $\|T(t)\| \leq M\mathrm{e}^{\omega t}$ for all $t \geq 0$ and some constant $M \geq 1$. This implies that $\omega_0(T) \leq \eta$ and therefore $\omega_0(T) = \eta$.

b) Since

$$\mathrm{r}(T(t)) = \lim_{k\to\infty} \|T(kt)\|^{1/k},$$

we obtain for $t > 0$ that

$$\mathrm{r}(T(t)) = \lim_{k\to\infty} \mathrm{e}^{t(kt)^{-1}\log\|T(kt)\|} = \mathrm{e}^{t\omega_0(T)}. \qquad \square$$

As in finite dimensions (compare with Definition 4.6) we define the *spectral bound* of A by

$$\mathrm{s}(A) := \sup\{\mathrm{Re}\,\lambda : \lambda \in \sigma(A)\}.$$

Motivated by the finite-dimensional case, see Corollary 4.8, one may ask whether for a generator A of a C_0-semigroup $(T(t))_{t\geq 0}$ on X we have $\omega_0(T) = \mathrm{s}(A)$. We will, however, see later that this equality is in general not even true for positive C_0-semigroups on a Banach lattice E.

We introduce now different stability concepts.

Definition 12.2. A C_0-semigroup $(T(t))_{t\geq 0}$ on X is called

a) *uniformly exponentially stable* if $\omega_0(T) < 0$,

b) *strongly stable* if $\lim_{t\to\infty} \|T(t)f\| = 0$ for every $f \in X$.

It is clear that a) implies b). The converse does not hold in general, as the following example shows.

Example 12.3. Let us consider the shift semigroup on $L^p(\mathbb{R}_+)$, $1 \leq p < \infty$, defined by

$$(T(t)f)(s) := f(t+s), \quad t \geq 0, \ \text{a.e. } s \in \mathbb{R}_+.$$

Then $(T(t))_{t\geq 0}$ is a C_0-semigroup of contraction operators on $L^p(\mathbb{R}_+)$ satisfying $\lim_{t\to\infty} T(t)f = 0$ for any $f \in L^p(\mathbb{R}_+)$ since

$$\|T(t)f\|_p^p = \int_t^\infty |f(s)|^p \, ds, \quad t \geq 0, \ f \in L^p(\mathbb{R}_+).$$

On the other hand, by considering the function

$$f_t(s) = \chi_{(t,t+1)}(s) = \begin{cases} 1 & \text{if } s \in (t, t+1), \\ 0 & \text{otherwise,} \end{cases}$$

we have $\|T(t)f_t\|_p = 1$. So, since $(T(t))_{t\geq 0}$ is a semigroup of contractions, we deduce that $\|T(t)\| = 1$.

The definition of the growth bound and Proposition 12.1 yield the following characterization of uniform exponential stability. Compare with Theorem 4.12 in the finite-dimensional case.

Proposition 12.4. *For a C_0-semigroup $(T(t))_{t\geq 0}$ on X, the following assertions are equivalent.*

(i) $\omega_0(T) < 0$, *i.e.,* $(T(t))_{t\geq 0}$ *is uniformly exponentially stable.*

(ii) $\lim_{t\to\infty} \|T(t)\| = 0$.

(iii) $\|T(t_0)\| < 1$ *for some* $t_0 > 0$.

(iv) $r(T(t_1)) < 1$ *for some* $t_1 > 0$.

Proof. The implications (i) $\implies$ (ii) $\implies$ (iii) $\implies$ (iv) are straightforward, while (iv) $\implies$ (i) is an immediate consequence of Proposition 12.1.b). $\qquad\square$

It is clear that if $\omega_0(T) < 0$, then there are constants $\varepsilon > 0$ and $M \geq 1$ such that

$$\|T(t)\| \leq Me^{-\varepsilon t}, \quad t \geq 0.$$

Hence, for every $p \in [1, \infty)$, $\int_0^\infty \|T(t)f\|^p \, dt < \infty$ for all $f \in X$. The following result due to Datko shows that the converse is also true.

Theorem 12.5 (Datko). *A C_0-semigroup $(T(t))_{t\geq 0}$ on X is uniformly exponentially stable if and only if for some (and hence for every) $p \in [1,\infty)$*

$$\int_0^\infty \|T(t)f\|^p \, \mathrm{d}t < \infty$$

for all $f \in X$.

Proof. We only have to prove the sufficiency. By Proposition 12.4, it suffices to prove that $\lim_{t\to\infty}\|T(t)\| = 0$. We note first that the set

$$\{\mathcal{T}_k f : k \in \mathbb{N}\} \subset L^p(\mathbb{R}_+, X)$$

is bounded for each $f \in X$, where $\mathcal{T}_k f := \chi_{[0,k]}(\cdot)T(\cdot)f$. The uniform boundedness principle (see Theorem A.15) implies the existence of $C > 0$ such that

$$\int_0^t \|T(s)f\|^p \, \mathrm{d}s \leq C^p \|f\|^p$$

holds for all $f \in X$, $p \in [1,\infty)$ and $t \geq 0$. Since there are constants $M, \omega \in \mathbb{R}_+$ with $\|T(t)\| \leq Me^{\omega t}$, $t \geq 0$, we obtain

$$\frac{1 - e^{-p\omega t}}{p\omega}\|T(t)f\|^p = \int_0^t e^{-p\omega s}\|T(s)T(t-s)f\|^p \, \mathrm{d}s$$

$$\leq M^p \int_0^t \|T(t-s)f\|^p \, \mathrm{d}s$$

$$\leq M^p C^p \|f\|^p$$

for all $f \in X$ and $t \geq 0$. Hence,

$$\|T(t)f\|^p \leq \frac{p\omega}{1 - e^{-p\omega}} M^p C^p \|f\|^p$$

for all $f \in X$ and $t \geq 1$. Thus, there exists a constant $L > 0$ with $\|T(t)\| \leq L$ for all $t \geq 0$, therefore

$$t\|T(t)f\|^p = \int_0^t \|T(t-s)T(s)f\|^p \, \mathrm{d}s$$

$$\leq L^p \int_0^t \|T(s)f\|^p \, \mathrm{d}s$$

$$\leq L^p C^p \|f\|^p$$

for all $f \in X$ and $t \geq 0$. Thus

$$\|T(t)\| \leq LCt^{-\frac{1}{p}}$$

for $t > 0$, which implies $\lim_{t\to\infty}\|T(t)\| = 0$. $\square$

12.2 The Spectral Bound for Positive Semigroups

In this section we characterize the spectral bound $s(A)$ of the generator A of a positive C_0-semigroup $(T(t))_{t\geq 0}$ on a complex Banach lattice E. We will see that $s(A)$ is always contained in $\sigma(A)$ provided that $\sigma(A) \neq \emptyset$. Compare this result with Perron's theorem (see Theorem 5.6) for positive matrices.

To this end we first improve the integral representation formula for the resolvent in the case of positive semigroups (see Proposition 9.33). This requires the following auxiliary result.

Lemma 12.6. *If for some* $\lambda \in \mathbb{C}$, $R_\lambda f := \lim_{t\to\infty} \int_0^t e^{-\lambda s} T(s) f \, ds$ *exists for all* $f \in E$, *then* $\lambda \in \rho(A)$ *and* $R_\lambda f = R(\lambda, A)f$ *for all* $f \in E$.

Proof. For $f \in E$ and $t > 0$ we have

$$\frac{1}{t}(T(t) - I)R_\lambda f = \frac{1}{t} \int_0^\infty e^{-\lambda s}(T(t+s) - T(s))f \, ds$$

$$= \frac{1}{t}\left((e^{\lambda t} - 1) \int_0^\infty e^{-\lambda s} T(s) f \, ds - e^{\lambda t} \int_0^t e^{-\lambda s} T(s) f \, ds \right).$$

Taking the limit $t \to 0^+$ we obtain $R_\lambda f \in D(A)$ and

$$AR_\lambda f = \lambda R_\lambda f - f,$$

i.e.,

$$(\lambda - A)R_\lambda f = f.$$

On the other hand, for $f \in D(A)$, by the closedness of A, we obtain

$$R_\lambda(\lambda - A)f = \lim_{t\to\infty} \int_0^t e^{-\lambda s} T(s)(\lambda - A)f \, ds$$

$$= \lim_{t\to\infty} (\lambda - A) \int_0^t e^{-\lambda s} T(s) f \, ds$$

$$= (\lambda - A)R_\lambda f = f.$$

So, R_λ defines a two-sided inverse of $\lambda - A$, and hence R_λ is closed. Thus, by the closed graph theorem, $R_\lambda \in \mathcal{L}(E)$. Therefore, $\lambda \in \rho(A)$ and $R_\lambda = R(\lambda, A)$. $\square$

Theorem 12.7. *Let A be the generator of a positive C_0-semigroup $(T(t))_{t\geq 0}$ on E. For λ with $\operatorname{Re} \lambda > s(A)$ we have*

$$R(\lambda, A)f = \lim_{t\to\infty} \int_0^t e^{-\lambda s} T(s) f \, ds, \quad f \in E.$$

Moreover, $\int_0^t e^{-\lambda s} T(s) \, ds$ converges to $R(\lambda, A)$ with respect to the operator norm as $t \to \infty$.

Proof. Let $\lambda_0 > \omega_0(T)$ be fixed. We recall from Proposition 9.33 that

$$R(\lambda_0, A)f = \int_0^\infty e^{-\lambda_0 t} T(t) f \, \mathrm{d}t$$

and that

$$R(\lambda_0, A)^{k+1} f = \frac{1}{k!} \int_0^\infty t^k e^{-\lambda_0 t} T(t) f \, \mathrm{d}t$$

holds for $k \in \mathbb{N}$ and $f \in E$. Let $\mu \in (s(A), \lambda_0)$, $f \in E_+$, and $f^* \in E_+^*$. By Corollary 9.30, $\frac{1}{\lambda_0 - \mu} > r(R(\lambda_0, A))$ and hence,

$$\langle R(\mu, A) f, f^* \rangle = \sum_{k=0}^\infty (\lambda_0 - \mu)^k \langle R(\lambda_0, A)^{k+1} f, f^* \rangle$$

$$= \sum_{k=0}^\infty \int_0^\infty \frac{1}{k!} [(\lambda_0 - \mu)s]^k e^{-\lambda_0 s} \langle T(s) f, f^* \rangle \, \mathrm{d}s$$

$$= \int_0^\infty \left(\sum_{k=0}^\infty \frac{1}{k!} [(\lambda_0 - \mu)s]^k \right) e^{-\lambda_0 s} \langle T(s) f, f^* \rangle \, \mathrm{d}s$$

$$= \int_0^\infty e^{(\lambda_0 - \mu)s} e^{-\lambda_0 s} \langle T(s) f, f^* \rangle \, \mathrm{d}s$$

$$= \int_0^\infty e^{-\mu s} \langle T(s) f, f^* \rangle \, \mathrm{d}s$$

$$= \lim_{t \to \infty} \left\langle \int_0^t e^{-\mu s} T(s) f \, \mathrm{d}s, f^* \right\rangle.$$

The equality above remains valid for all $f^* \in E^*$ since any $f^* \in E^*$ can be decomposed into real and imaginary components and these in turn into their positive and negative parts. Hence, for $f \in E_+$, $\left(\int_0^t e^{-\mu s} T(s) f \, \mathrm{d}s \right)_{t \geq 0}$ converges weakly to $R(\mu, A) f$ as $t \to \infty$. Since $f \in E_+$, the positivity of the semigroup $(T(t))_{t \geq 0}$ implies that $\left(\int_0^t e^{-\mu s} T(s) f \, \mathrm{d}s \right)_{t \geq 0}$ is monotone increasing and so, by Proposition 10.9, we have strong convergence. Thus

$$\lim_{t \to \infty} \int_0^t e^{-\mu s} T(s) f \, \mathrm{d}s = R(\mu, A) f$$

for all $f \in E_+$ and hence for all $f \in E$. If $\lambda = \mu + i\nu$ with $\mu, \nu \in \mathbb{R}$ and $\mu > s(A)$, then for any $0 \leq r < t$, $f \in E$ and $f^* \in E^*$, we have

$$\left| \left\langle \int_r^t e^{-\lambda s} T(s) f \, \mathrm{d}s, f^* \right\rangle \right| \leq \int_r^t e^{-\mu s} \langle T(s) |f|, |f^*| \rangle \, \mathrm{d}s,$$

since $\langle |f|, |f^*| \rangle = \sup_{|g| \leq |f|} |\langle g, |f^*| \rangle|$. Hence,

$$\left\| \int_r^t e^{-\lambda s} T(s) f \, \mathrm{d}s \right\| \leq \left\| \int_r^t e^{-\mu s} T(s) |f| \, \mathrm{d}s \right\|,$$

which implies that

$$\lim_{t \to \infty} \int_0^t e^{-\lambda s} T(s) f \, ds \text{ exists for all } f \in E.$$

Then, using Lemma 12.6, we obtain $\lambda \in \rho(A)$ and

$$R(\lambda, A)f = \int_0^\infty e^{-\lambda t} T(t) f \, dt$$

for all $f \in E$. It remains to prove that $\left(\int_0^t e^{-\lambda s} T(s) \, ds \right)$ converges in the operator norm as $t \to \infty$. We fix $\mu \in (s(A), \operatorname{Re} \lambda)$. As we have seen above, the function

$$\psi_{f, f^*} : s \longmapsto e^{-\mu s} \langle T(s) f, f^* \rangle$$

belongs to $L^1(\mathbb{R}_+)$ for all $f \in E$ and $f^* \in E^*$. The closed graph theorem implies that the bilinear form

$$b : E \times E^* \longrightarrow L^1(\mathbb{R}_+), \quad (f, f^*) \longmapsto \psi_{f, f^*}$$

is separately continuous and hence continuous. Thus, there exists a constant $M \geq 0$ such that

$$\int_0^\infty e^{-\mu s} |\langle T(s) f, f^* \rangle| \, ds \leq M \|f\| \|f^*\|, \quad f \in E, f^* \in E^*.$$

For $0 \leq t < r$ and $\varepsilon := \operatorname{Re} \lambda - \mu$ we have

$$\left| \int_t^r e^{-\lambda s} \langle T(s) f, f^* \rangle \, ds \right| \leq \int_t^r e^{-(\operatorname{Re} \lambda - \mu) s} e^{-\mu s} |\langle T(s) f, f^* \rangle| \, ds$$
$$\leq e^{-\varepsilon t} \int_t^r e^{-\mu s} |\langle T(s) f, f^* \rangle| \, ds$$
$$\leq e^{-\varepsilon t} M \|f\| \|f^*\|.$$

Hence, $\left\| \int_t^r e^{-\lambda s} T(s) \, ds \right\| \leq M e^{-\varepsilon t}$ and this implies that the function

$$t \longmapsto \int_0^t e^{-\lambda s} T(s) \, ds$$

satisfies the Cauchy convergence criterion in $\mathcal{L}(E)$ as $t \to \infty$.

Thus, $\left(\int_0^\infty e^{-\lambda s} T(s) \, ds \right)$ converges in the operator norm. $\qquad \square$

As an immediate consequence we obtain the following

Corollary 12.8. *Let A be the generator of a positive C_0-semigroup $(T(t))_{t \geq 0}$ on E. If $\operatorname{Re} \lambda > s(A)$, then*

$$|R(\lambda, A)f| \leq R(\operatorname{Re} \lambda, A)|f| \quad \text{for all } f \in E.$$

Another important corollary is the following, previously announced result.

Corollary 12.9. *If A is the generator of a positive C_0-semigroup $(T(t))_{t \geq 0}$ on E and $s(A) > -\infty$, then $s(A) \in \sigma(A)$.*

Proof. Assume that $s(A) \in \rho(A)$. First we show that

$$M := \sup_{\mu > s(A)} \|R(\mu, A)\| < \infty.$$

Applying Corollary 12.8, we have

$$|R(\lambda, A)f| \leq R(\operatorname{Re}\lambda, A)|f| \quad \text{for all } \operatorname{Re}\lambda > s(A), \ f \in E.$$

Thus

$$\|R(\lambda, A)\| \leq M \quad \text{for all } \operatorname{Re}\lambda > s(A).$$

Since

$$\|R(\lambda, A)\| \geq \frac{1}{\operatorname{dist}(\lambda, \sigma(A))}$$

for $\lambda \in \rho(A)$ (see Corollary 9.30), we infer that

$$\{\lambda \in \mathbb{C} : \operatorname{Re}\lambda = s(A)\} \subseteq \rho(A)$$

and

$$\|R(\lambda, A)\| \leq M \text{ for all } \lambda \text{ with } \operatorname{Re}\lambda = s(A).$$

Thus, since $\rho(A)$ is open,

$$\{\lambda \in \mathbb{C} : |\operatorname{Re}\lambda - s(A)| < M^{-1}\} \subseteq \rho(A).$$

This contradicts the definition of $s(A)$. $\square$

We also obtain a relation between $s(A)$ and the positivity of the resolvent.

Corollary 12.10. *Suppose that A generates a positive C_0-semigroup $(T(t))_{t \geq 0}$ on E and $\lambda_0 \in \rho(A)$. Then the following assertions hold.*

a) *$R(\lambda_0, A)$ is positive if and only if $\lambda_0 > s(A)$.*
b) *If $\lambda > s(A)$, then $r(R(\lambda, A)) = \frac{1}{\lambda - s(A)}$.*

Proof. a) Assume first that $R(\lambda_0, A) \geq 0$. So, one has

$$R(\lambda_0, A)E_{\mathbb{R}} \subset E_{\mathbb{R}},$$

where

$$E_{\mathbb{R}} := \{\operatorname{Re} f : f \in E\}.$$

Let $f \in E_{\mathbb{R}} \setminus \{0\}$. Then $g = R(\lambda_0, A)f \in E_{\mathbb{R}}$ and so $Ag \in E_{\mathbb{R}}$. So, it follows from $\lambda_0 g - Ag = f$ that $\lambda_0 g \in E_{\mathbb{R}}$ and hence $\lambda_0 \in \mathbb{R}$. On the other hand, Theorem 12.7 implies that $R(\lambda, A) \geq 0$ for all $\lambda > \max(\lambda_0, s(A))$, and hence

$$R(\lambda_0, A) = R(\lambda, A) + (\lambda - \lambda_0)R(\lambda, A)R(\lambda_0, A)$$
$$\geq R(\lambda, A) \geq 0$$

for all $\lambda > \max(\lambda_0, s(A))$. Therefore,

$$(\lambda - s(A))^{-1} = r(R(\lambda, A)) \leq \|R(\lambda, A)\| \leq \|R(\lambda_0, A)\|$$

for all $\lambda > \max(\lambda_0, s(A))$. But this is only true if $\lambda_0 > s(A)$.

The converse follows from Theorem 12.7.

b) This is a simple consequence of Corollary 12.9 and Proposition 9.29. $\square$

We give now a statement on the spectral properties of positive perturbations of semigroups.

Proposition 12.11. *Let A be the generator of a positive C_0-semigroup $(T(t))_{t\geq 0}$ on the Banach lattice E and let $B \in \mathcal{L}(E)$ be positive. Then the following assertions hold.*

a) *$A + B$ generates a positive C_0-semigroup $(S(t))_{t\geq 0}$ with $0 \leq T(t) \leq S(t)$.*
b) *We have $s(A) \leq s(A + B)$ and $R(\lambda, A) \leq R(\lambda, A + B)$ for $\lambda > s(A + B)$.*

Proof. a) Recall from Theorem 11.5 that $A + B$ generates a C_0-semigroup $(S(t))_{t\geq 0}$, which is positive by Corollary 11.7. Moreover, the Dyson–Phillips expansion formula (11.6) yields $S(t) \geq T(t) \geq 0$.

b) Applying now Theorem 12.7 we see that

$$R(\lambda, A + B) \geq R(\lambda, A)$$

for $\lambda > \max\{s(A), s(A + B)\}$. Hence, for such λ we also have

$$\|R(\lambda, A)\| \leq \|R(\lambda, A + B)\|.$$

But by Corollary 12.9 we have $s(A) \in \sigma(A)$, and so

$$\lim_{\lambda \downarrow s(A)} \|R(\lambda, A)\| = \infty.$$

We conclude that $s(A + B) \geq s(A)$. $\square$

We say that A is a *resolvent positive operator* if there is $\mu \in \mathbb{R}$ such that $(\mu, \infty) \subset \rho(A)$ and $R(\lambda, A) \geq 0$ for all $\lambda > \mu$. Recall that, if A generates a positive C_0-semigroup, then A is a resolvent positive operator, see Corollary 11.4. There are, however, resolvent positive operators that are not generators of a C_0-semigroup. For a nontrivial example see Example 13.17.

Remark 12.12.

a) It can be proved that the statement in Corollary 12.9 remains valid if one assumes only that A is a resolvent positive operator, see Exercise 4. Therefore, if $R(\lambda_0, A) \geq 0$ for some $\lambda_0 \in \rho(A)$, then $\lambda_0 \in \mathbb{R}$ and $\lambda_0 > s(A)$, whenever A is a resolvent positive operator.

b) As an immediate consequence of Corollary 12.10 we obtain that

$$s(A) = \inf\{\lambda \in \rho(A) : R(\lambda, A) \geq 0\}$$

for the generator A of a positive C_0-semigroup on a Banach lattice E.

c) If $E = C(K)$, where K is a compact Hausdorff space, then $s(A) > -\infty$. In fact, we know from Lemma 11.2 that $\lim_{\lambda \to \infty} \lambda R(\lambda, A)f = f$ for all $f \in E$. In particular, we can find $\lambda_0 \in \mathbb{R}$ sufficiently large such that

$$\lambda_0 R(\lambda_0, A)\mathbb{1} \geq \frac{1}{2}\mathbb{1}.$$

Since $R(\lambda_0, A) \geq 0$, we infer that

$$R(\lambda_0, A)^k \mathbb{1} \geq \left(\frac{1}{2\lambda_0}\right)^k \mathbb{1} \quad \text{for all } k \in \mathbb{N}.$$

Thus,

$$\mathrm{r}(R(\lambda_0, A)) = \lim_{k \to \infty} \|R(\lambda_0, A)^k\|^{1/k} \geq \frac{1}{2\lambda_0} > 0,$$

and hence $\sigma(A) \neq \emptyset$.

The spectrum of a generator of a positive C_0-semigroup can in general be empty, as the following examples show.

Example 12.13.

a) On $E := C_0[0,1] := \{f \in C[0,1) : f(1) = 0\}$ we consider the nilpotent C_0-semigroup $(T(t))_{t \geq 0}$ given by

$$(T(t)f)(x) = \begin{cases} f(x+t) & \text{if } x+t < 1, \\ 0 & \text{otherwise,} \end{cases}$$

for $t \geq 0$, $x \in [0,1]$ and $f \in E$. Then, $T(t) = 0$ for $t \geq 1$ and hence $\sigma(T(t)) = \{0\}$. So by the spectral inclusion theorem (see Corollary 9.32), $\sigma(A) = \emptyset$.

b) Let $E := C_0(0,\infty) := \{f \in C(\mathbb{R}_+) : \lim_{x \to \infty} f(x) = 0\}$. On E we define the C_0-semigroup $(T(t))_{t \geq 0}$ by

$$(T(t)f)(x) := e^{-\frac{t^2}{2} - xt} f(x+t), \quad x, t \geq 0 \text{ and } f \in E.$$

Then the generator A of $(T(t))_{t \geq 0}$ on E is given by

$$(Af)(x) = f'(x) - xf(x), \quad x \geq 0, \text{ and}$$

$$f \in D(A) = \{f \in E : f \in C^1(\mathbb{R}_+) \text{ and } Af \in E\}.$$

One proves that $\sigma(A) = \emptyset$, see Exercise 2.

However, for generators of positive C_0-groups the spectrum is always non-empty.

Corollary 12.14. *If A generates a positive C_0-group on a Banach lattice E, then*

$$\sigma(A) \neq \emptyset.$$

Proof. Assume that $\sigma(A) = \emptyset$. By Theorem 12.7, $R(\lambda, A) \geq 0$ for all $\lambda \in \mathbb{R}$. Again, one can apply the same theorem to $-A$ and obtain $R(\lambda, -A) \geq 0$ for all $\lambda \in \mathbb{R}$. But $R(\lambda, -A) = -R(-\lambda, A) \leq 0$ for all $\lambda \in \mathbb{R}$, and hence $R(\lambda, -A) = 0$ for all $\lambda \in \mathbb{R}$. This contradicts the fact that $E \neq \{0\}$. $\square$

We end this section by proving the existence of a positive eigenfunction. We will see in Chapter 14 that such an eigenfunction is unique (up to a scalar factor) if A generates an irreducible positive C_0-semigroup. Compare this to Theorem 5.6 in the finite-dimensional case.

Theorem 12.15 (Krein–Rutman). *Let A be a resolvent positive operator on E with compact resolvent such that $s(A) > -\infty$. Then there exists $0 \lneqq u \in D(A)$ such that $Au = s(A)u$.*

Proof. Replacing A with $A - s(A)$, one can assume without loss of generality that $s(A) = 0$. Since, by Exercise 4,

$$s(A) \in \sigma(A),$$

Corollary 9.30 implies that $\|R(\lambda_k, A)\| \to \infty$ as $k \to \infty$ for every $\lambda_k > 0$ with $\lim_{k \to \infty} \lambda_k = s(A) = 0$. By the uniform boundedness principle (see Theorem A.15), there is $f \in E$ such that

$$\lim_{k \to \infty} \|R(\lambda_k, A)f\| = \infty.$$

Since $|R(\lambda_k, A)f| \leq R(\lambda_k, A)|f|$, one can assume that $f \gneqq 0$. Take

$$u_k = \frac{R(\lambda_k, A)f}{\|R(\lambda_k, A)f\|}.$$

Then $0 \leq u_k \in D(A)$ with $\|u_k\| = 1$ and

$$\lim_{k \to \infty} (\lambda_k u_k - Au_k) = \lim_{k \to \infty} \|R(\lambda_k, A)f\|^{-1}f = 0.$$

Thus, (u_k) is bounded in the graph norm. Using the compactness of the resolvent of A we have that the embedding $D(A) \hookrightarrow E$ is compact and so we can assume that $\lim_{k \to \infty} u_k = u$ exists in E, taking an appropriate subsequence if necessary. Then $\|u\| = 1$, $u \geq 0$, and by the closedness of A we obtain $u \in D(A)$ and $Au = 0$. $\square$

12.3   The Identity $\omega_0(T) = s(A)$ for Positive Semigroups

As we have seen, for operator semigroups the spectral bound is not always equal to the growth bound. It turns out, however, that for positive semigroups on some special Banach lattices like the space of continuous functions or Lebesgue spaces, the spectral bound equals the growth bound. These are wonderful results where the special geometry of the spaces comes into play.

We proved in the finite-dimensional setting that $s(A) = \omega_0(T)$, see Corollary 4.8. On the other hand, Proposition 9.33 implies that $s(A) \leq \omega_0(T)$ holds, whenever A generates a C_0-semigroup $(T(t))_{t\geq 0}$ on a Banach space X. Let us show that in the infinite-dimensional case in general $s(A) \neq \omega_0(T)$, even for positive C_0-semigroups.

Example 12.16. Consider the Banach lattice $E := C_0(\mathbb{R}_+) \cap L^1(\mathbb{R}_+, e^s \, ds)$ endowed with the norm

$$\|f\| := \sup_{s\geq 0} |f(s)| + \int_0^\infty |f(s)| e^s \, ds =: \|f\|_\infty + \|f\|_1.$$

On E we define the translation semigroup

$$(T(t)f)(s) = f(s+t), \quad t, s \geq 0.$$

Then $(T(t))_{t\geq 0}$ is a positive C_0-semigroup. Its generator is given by

$$Af = f' \text{ for } f \in D(A) = \{f \in E : f \in C^1(\mathbb{R}_+) \text{ and } f' \in E\}.$$

Note that $\|T(t)\| = 1$ for all $t \geq 0$, hence $\omega_0(T) = 0$.

On the other hand, the function $\varepsilon_\lambda(s) := e^{\lambda s}$ is an eigenfunction for A associated with λ provided that $\operatorname{Re} \lambda < -1$. Hence,

$$\{\lambda \in \mathbb{C} : \operatorname{Re} \lambda \leq -1\} \subseteq \sigma(A).$$

Moreover, for $\lambda \in \mathbb{C}$ with $\operatorname{Re} \lambda > -1$ one sees that

$$\| \cdot \|_1 - \lim_{N\to\infty} \int_0^N e^{-\lambda s} T(s)f \, ds \quad \text{and} \quad \| \cdot \|_\infty - \lim_{N\to\infty} \int_0^N e^{-\lambda s} T(s)f \, ds$$

exist, because $\|T(s)f\|_1 \leq e^{-s}\|f\|_1$ for all $s \geq 0$ and $\int_0^\infty e^s |f(s)| \, ds < \infty$. Therefore, $\int_0^\infty e^{-\lambda s} T(s)f \, ds$ exists in E for all $f \in E$. Thus, by Lemma 12.6, $\lambda \in \rho(A)$. It follows that

$$\sigma(A) = \{\lambda \in \mathbb{C} : \operatorname{Re} \lambda \leq -1\}, \quad \text{whence } s(A) = -1.$$

Another example of such a situation is given in Exercise 3.

We now look for sufficient conditions implying the equality $\omega_0(T) = s(A)$ when A generates a positive C_0-semigroup $(T(t))_{t\geq 0}$ on E.

Theorem 12.17. *Let A be the generator of a positive C_0-semigroup $(T(t))_{t\geq 0}$ on a Banach lattice E. Then $\omega_0(T) = s(A)$ holds in the following cases.*

a) *The space E is an AL-space, i.e., the norm satisfies $\|f + g\| = \|f\| + \|g\|$ for all $f, g \in E_+$.*

b) *The space E is $C_0(\Omega)$, where Ω is a locally compact Hausdorff space.*

Proof. a) For $\lambda > s(A)$ and $f \in E_+$ we obtain from Theorem 12.7 that

$$\|R(\lambda, A)f\| = \left\| \int_0^\infty e^{-\lambda s} T(s)f \, ds \right\| = \int_0^\infty e^{-\lambda s} \|T(s)f\| \, ds,$$

where the second equality follows from the fact that the norm is additive on the positive cone. Hence,

$$\int_0^\infty \|(e^{-\lambda s} T(s))f\| \, ds < \infty$$

for all $f \in E$. So, by Theorem 12.5, we have $\omega_0(T) - \lambda < 0$, and thus

$$\omega_0(T) \leq s(A).$$

b) Is follows immediately that $\|f \vee g\| = \|f\| \vee \|g\|$ for all $f, g \in E_+$. Then, for $\gamma, \nu \in E_+^*$, we have

$$\langle f, \gamma \rangle + \langle g, \nu \rangle \leq \langle f \vee g, \gamma + \nu \rangle|$$
$$\leq \|\gamma + \nu\| \|f \vee g\|$$
$$= \|\gamma + \nu\|(\|f\| \vee \|g\|), \quad f, g \in E_+.$$

Hence, $\langle f, \gamma \rangle + \langle g, \nu \rangle \leq \|\gamma + \nu\|$ for all $f, g \in E_+$ with $\|f\| = \|g\| = 1$. Proposition 10.22.a) implies that $\|\gamma\| + \|\nu\| \leq \|\gamma + \nu\|$ and so

$$\|\gamma\| + \|\nu\| = \|\gamma + \nu\|, \quad \gamma, \nu \in E_+^*.$$

This implies that E^* is an *AL*-space. If we set $F := \overline{D(A^*)}$, then it follows from Exercise 5 that F is a closed ideal and hence also an *AL*-space. On F we consider the positive C_0-semigroup $(S(t))_{t\geq 0}$ given by

$$S(t) := T(t)^*|_F \quad \text{for } t \geq 0,$$

and we denote by B its generator. Then, by Proposition 9.38, B is the part of A^* in F, i.e.,

$$D(B) = \{\nu \in D(A^*) : A^*\nu \in F\} \text{ and } B\nu = A^*\nu \text{ for } \nu \in D(B).$$

Moreover, one can show that

$$\sigma(B) = \sigma(A^*) = \sigma(A),$$

see Proposition 9.40.

Consequently, $s(B) = s(A)$ holds. Since B is the generator of the positive C_0-semigroup $(S(t))_{t \geq 0}$ on the AL-space F, it follows from a) that $s(B) = \omega_0(S)$. Now, it suffices to prove that

$$\omega_0(S) = \omega_0(T).$$

The inequality $\omega_0(S) \leq \omega_0(T)$ is straightforward from the definition of $(S(t))_{t \geq 0}$. Let $\omega > \omega_0(S)$, $f \in E$, and $\nu \in F$. Then we have

$$|\langle T(t)f, \nu \rangle| = |\langle f, S(t)\nu \rangle| \leq M\|f\|e^{\omega t}\|\nu\|$$

for $t \geq 0$ and some constant $M \geq 1$. On the other hand, since the Yosida approximations yield $f = \lim_{\lambda \to \infty} \lambda R(\lambda, A)f$ for all $f \in E$, we have

$$c := \limsup_{\lambda \to \infty} \lambda \|R(\lambda, A)\| < \infty.$$

Therefore

$$
\begin{aligned}
|\langle T(t)f, \gamma \rangle| &= \lim_{\lambda \to \infty} |\langle \lambda R(\lambda, A)T(t)f, \gamma \rangle| \\
&= \lim_{\lambda \to \infty} |\langle T(t)f, \lambda R(\lambda, A^*)\gamma \rangle| \\
&\leq M\|f\|e^{\omega t} \limsup_{\lambda \to \infty} \lambda \|R(\lambda, A)^*\gamma\| \\
&\leq Mce^{\omega t}\|f\|\|\gamma\|, \qquad \gamma \in E^*.
\end{aligned}
$$

Consequently, $\|T(t)\| \leq Mce^{\omega t}$ for all $t \geq 0$, and hence $\omega_0(T) \leq \omega$ for all $\omega > \omega_0(S)$. Thus, we have shown that

$$\omega_0(S) = \omega_0(T). \qquad \square$$

Remark 12.18. We will present another result of this type in the case when E is also a Hilbert space in Corollary 15.11.

12.4   Notes and Remarks

The spectral and stability theory of semigroups is a broad subject and well documented in the literature. We refer to the monographs by Engel and Nagel [43, 44], van Neerven [103], or Arendt et al. [6]. Datko's theorem originates from Datko [28]. Theorem 12.15 is a variation of the famous Krein–Rutman theorem, see [77].

The content of Theorem 12.17 is due to Derndinger [29]. It remains true also in L^p-spaces due to a result by Weis [153]. For an elegant proof of Weis' Theorem, see Arendt et al. [6, Theorem 5.3.6].

12.5 Exercises

1. Let the mapping $\zeta : [0, \infty) \to \mathbb{R}$ be bounded on compact intervals and subadditive, i.e., $\zeta(t+s) \le \zeta(t) + \zeta(s)$ for all t, $s \ge 0$. Prove that

$$\inf_{t>0} \frac{\zeta(t)}{t} = \lim_{t \to \infty} \frac{\zeta(t)}{t}.$$

2. On $E := C_0([0, \infty))$ consider the family of operators

$$(T(t)f)(x) := e^{-\frac{t^2}{2} - xt} f(x+t), \quad x, t \ge 0 \text{ and } f \in E.$$

   a) Show that $(T(t))_{t \ge 0}$ is a positive C_0-semigroup on E.

   b) Prove that its generator is given by

$$(Af)(x) = f'(x) - xf(x), \quad x \ge 0, \quad \text{for}$$
$$f \in D(A) = \{f \in E : f \in C^1(\mathbb{R}_+) \text{ and } Af \in E\}.$$

   c) Prove that $\sigma(A) = \emptyset$.

3. Let $1 < p < q < \infty$ and $E := L^p[1, \infty) \cap L^q[1, \infty)$ the Banach lattice endowed with the norm

$$\|f\| := \|f\|_p + \|f\|_q, \quad f \in E.$$

   Consider the family of operators

$$T(t)f(s) = f(se^t) \quad s \ge 1, t \ge 0.$$

   a) Show that $(T(t))_{t \ge 0}$ is a positive C_0-semigroup on E with generator

$$(Af)(s) = sf'(s), \quad s \ge 1, \text{ for}$$
$$f \in D(A) = \{f \in E : f \text{ absolutely continuous and } Af \in E\}.$$

   b) Prove that $s(A) = -\frac{1}{p} < -\frac{1}{q} = \omega_0(T)$.

4. Let A be a resolvent positive operator on a Banach lattice E with $s(A) > -\infty$. Show that $s(A) \in \sigma(A)$.

5. Let A be a resolvent positive operator on a Banach lattice with order continuous norm E. Prove that $\overline{D(A)}$ is an ideal in E.

Chapter 13

Unbounded Positive Perturbations

For two unbounded linear operators A and B on a Banach space X it is not always evident how to define in a reasonable way their sum $A + B$. In order to avoid such a discussion we will present some standard perturbation results in the situation where the operator A generates a C_0-semigroup on a Banach space X and the perturbing operator B satisfies $D(A) \subseteq D(B)$. In this case the sum $A + B$ will be defined on the dense set $D(A)$.

The simplest case where B is bounded was already considered in Section 11.2. Here we consider unbounded operators B and focus on results where the positivity of the semigroup and the perturbation plays an important role.

13.1 Unbounded Dispersive Perturbations

We start with dispersive and A-bounded perturbations of the generator A of a positive contraction C_0-semigroup on a Banach lattice.

First we state the definition of A-boundedness.

Definition 13.1. Let A be a closed operator on a Banach space X. A linear operator B with domain $D(B)$ is called A-bounded if $D(A) \subseteq D(B)$ and there exist $a, b > 0$ such that for all $f \in D(A)$ the inequality

$$\|Bf\| \leq a\|Af\| + b\|f\| \tag{13.1}$$

holds. The A-bound of B is defined by

$$a_0 := \inf\{a \geq 0 : \text{ there exists } b \geq 0 \text{ such that inequality (13.1) holds }\}.$$

The following is not difficult to prove and we leave it as an exercise.

Lemma 13.2. Let A be a closed operator with $\rho(A) \neq \emptyset$. Then the operator B is A-bounded if and only if $BR(\lambda, A) \in \mathcal{L}(X)$ for some/all $\lambda \in \rho(A)$.

Now we can state a fundamental perturbation result for positive contraction semigroups.

Theorem 13.3. *Let A be the generator of a positive contraction C_0-semigroup on a Banach lattice E and B a dispersive and A-bounded operator with A-bound $a_0 < 1$. Then $A+B$, defined on $D(A)$, generates a positive contraction C_0-semigroup on E.*

Proof. Since B is dispersive, by Proposition 11.12, for every $f \in D(A)$ there is an $f^* \in \mathcal{I}^+(f)$ such that $\mathrm{Re}\langle Bf, f^* \rangle \le 0$. By Exercise 4, $\mathrm{Re}\langle Af + Bf, f^* \rangle \le 0$, which means that $A + B$ is a dispersive operator.

a) Assume first that $a_0 < \frac{1}{2}$. By Theorem 11.10, it suffices to prove that for some $\lambda_0 > 0$ we have $\mathrm{im}(\lambda_0 - A - B) = E$. Now, for any $\lambda > 0$ we infer that

$$\|BR(\lambda, A)f\| \le a\|AR(\lambda, A)f\| + b\|R(\lambda, A)f\|$$

$$\le a\|\lambda R(\lambda, A)f - f\| + \frac{b}{\lambda}\|f\|$$

$$\le \left(2a + \frac{b}{\lambda}\right)\|f\| \quad \text{for all } f \in E \text{ and some } a, b > 0.$$

Since $a_0 < \frac{1}{2}$, one has

$$\|BR(\lambda, A)\| \le 2a + \frac{b}{\lambda},$$

with $a < \frac{1}{2}$. Thus, for $\lambda > \frac{b}{1-2a}$ one obtains $\|BR(\lambda, A)\| < 1$ and so $\lambda \in \rho(A + B)$ and $R(\lambda, A + B) = R(\lambda, A)(I - BR(\lambda, A))^{-1}$ by Neumann's series.

b) Consider now the general case when $a_0 < 1$. For any $\alpha \in [0, 1]$ we define operators

$$C_\alpha = A + \alpha B \quad \text{with} \quad D(C_\alpha) = D(A).$$

Then for $x \in D(A)$ we obtain

$$\|Bf\| \le a\|Af\| + b\|f\|$$

$$\le a\left(\|C_\alpha f\| + \alpha\|Bf\|\right) + b\|f\|$$

$$\le a\left(\|C_\alpha f\| + \|Bf\|\right) + b\|f\|.$$

Hence,

$$\|Bf\| \le \frac{a}{1 - a}\|C_\alpha f\| + \frac{b}{1 - a}\|f\| \quad \text{for all } \alpha \in [0, 1].$$

Let $k \in \mathbb{N}$ such that $\frac{a}{k(1-a)} < \frac{1}{2}$. Then for any $\alpha \in [0, 1]$, $\frac{1}{k}B$ is a C_α-bounded operator with C_α-bound $a_0 < \frac{1}{2}$. From the calculations above we have that

$$C_\alpha + \frac{1}{k}B = A + \left(\alpha + \frac{1}{k}\right)B$$

generates a positive contraction C_0-semigroup on E provided that C_α generates a positive contraction C_0-semigroup. This is the case for $A + \frac{1}{k}B$. So, we have the generation result for $A + \frac{2}{k}B$. Iterating the process, we finally obtain that

$$\left(A + \frac{k-1}{k}B\right) + \frac{1}{k}B = A + B$$

generates a positive contraction C_0-semigroup on E. $\square$

A simple application of the perturbation theorem above is given in the following example.

Example 13.4. On $L^2([0,1])$ we consider the one-dimensional second-order elliptic operator with Dirichlet boundary conditions

$$D(A) := \{f \in H^2([0,1]) : f(0) = f(1) = 0\}$$
$$Af := f'', \quad f \in D(A).$$

Here $H^2([0,1])$ is the Sobolev space

$$H^2([0,1]) = \{f \in H^1([0,1]) : f' \in H^1([0,1])\},$$

where the derivatives are in sense of distributions and H^1 is defined in Appendix A.11. It is known that $f \in H_0^1([0,1])$ if and only if $f(0) = f(1) = 0$ and $f \in H^1([0,1])$, see Theorem A.46. Hence,

$$D(A) = \{f \in H_0^1([0,1]) : f'' \in L^2([0,1])\}.$$

By Example 11.14.b) we know that A generates a positive contraction C_0-semigroup on $L^2([0,1])$.

Now let $Bf := f'$ for $f \in D(B) := H^1([0,1])$. Then, $D(A) \subset D(B)$ and for any $\varepsilon > 0$ there is a constant $C(\varepsilon) > 0$ such that

$$\|f'\|_2 \le \varepsilon \|f''\|_2 + C(\varepsilon)\|f\|_2 \quad \text{for all } f \in H^2([0,1]),$$

see Proposition A.47. Hence, B is A-bounded with A-bound equal to 0. We show that B is also dispersive. Using A.48, we compute

$$(Bf|f^+) = \int_0^1 f'(s)f^+(s)\,ds = \int_0^1 (f^+)'(s)f^+(s)\,ds$$
$$= \frac{1}{2}\int_0^1 (f^2)'(s)\,ds = \frac{1}{2}\left(f^2(1) - f^2(0)\right) = 0$$

for all $f \in D(B)$. By Remark 11.13, this implies that B is dispersive. By Theorem 13.3, we deduce that the operator

$$Af + Bf = f'' + f', \quad f \in D(A),$$

generates a positive contraction C_0-semigroup on $L^2([0,1])$.

13.2   Miyadera Perturbations

Next we introduce the so-called Miyadera perturbations. We will see that such
perturbations are useful for positive perturbations of positive semigroups on L^1-
spaces, or for delay equations, to mention a few of the many applications. Throughout this section we suppose that A generates a C_0-semigroup $(T(t))_{t \geq 0}$ on a Banach
space X satisfying $\|T(t)\| \leq M\mathrm{e}^{\omega t}$ for all $t \geq 0$ and some constants $M \geq 1$ and
$\omega \in \mathbb{R}$.

Definition 13.5. A linear operator B is called a *Miyadera perturbation* of A if B
is A-bounded and there are $\alpha \in (0, \infty)$ and $\gamma \in [0, 1)$ such that

$$\int_0^\alpha \|BT(t)f\| \ \mathrm{d}t \leq \gamma \|f\| \quad \text{for all } f \in D(A). \tag{13.2}$$

For the Miyadera perturbations we can prove a perturbation result analogous
to the one for bounded perturbations, cf. Proposition 11.6.

Theorem 13.6. *If B is a Miyadera perturbation of A, then $A + B$ with domain $D(A)$
generates a C_0-semigroup $(S(t))_{t \geq 0}$ on X. Moreover, the semigroup $(S(t))_{t \geq 0}$ is
given by the Dyson–Phillips series*

$$S(t) = \sum_{k=1}^\infty S_k(t),$$

where the operators $S_k(t) \in \mathcal{L}(X)$ satisfy

$$S_0(t) := T(t) \quad and \quad S_{k+1}(t)f := \int_0^t S_k(t-s)BT(s)f \ \mathrm{d}s \tag{13.3}$$

*for $t \geq 0$, $f \in D(A)$, and for all $k \in \mathbb{N}$. Further, the semigroup $S(t)$ satisfies
Duhamel's equation*

$$S(t)f = T(t)f + \int_0^t S(t-s)BT(s)f \ \mathrm{d}s, \quad t \geq 0, f \in D(A). \tag{13.4}$$

Proof. First we assume that

$$\|T(t)\| \leq M\mathrm{e}^{-\eta t}$$

for all $t \geq 0$ and some $\eta > 0$. Indeed, if B is a Miyadera perturbation of A, then
B is a Miyadera perturbation of $A - \lambda$ for any $\lambda \geq 0$. We define

$$S_1(t)f := \int_0^t T(t-s)BT(s)f \ \mathrm{d}s, \quad f \in D(A).$$

Using the notation $[t/\alpha]$ for the integer part of t/α for any $t \geq \alpha$, we obtain

$$\int_0^t \|BT(s)f\|\, ds \leq \sum_{k=0}^{[t/\alpha]} \int_{k\alpha}^{(k+1)\alpha} \|BT(s)f\|\, ds$$

$$= \sum_{k=0}^{[t/\alpha]} \int_0^\alpha \|BT(s)T(k\alpha)f\|\, ds$$

$$= \int_0^\alpha \|BT(s)f\|\, ds + \sum_{k=1}^{[t/\alpha]} \int_0^\alpha \|BT(s)T(k\alpha)f\|\, ds$$

$$\leq \gamma\|f\| + \gamma\|f\| \sum_{k=1}^\infty \|T(k\alpha)\|$$

$$\leq \gamma \left(1 + \frac{Me^{-\eta\alpha}}{1 - e^{-\eta\alpha}}\right) \|f\| \tag{13.5}$$

for any $f \in D(A)$ and $t > \alpha$. The estimate (13.5) holds also for $t \in [0, \alpha]$ since B is a Miyadera perturbation of A and $\gamma < \tilde{\gamma}$, where

$$\tilde{\gamma} = \gamma \left(1 + \frac{Me^{-\eta\alpha}}{1 - e^{-\eta\alpha}}\right).$$

Then

$$\|S_1(t)f\| \leq M \int_0^t \|BT(s)f\|\, ds \leq M\tilde{\gamma}\|f\|$$

for all $t \geq 0$ and $f \in D(A)$. So, by the denseness of $D(A)$ in X, $S_1(t)$ can be extended uniquely to a bounded linear operator on X and satisfies

$$\|S_1(t)\| \leq M\tilde{\gamma}, \quad t \geq 0.$$

By induction we see that each of the operators $S_k(t)$ defined in (13.3) can be extended uniquely to a bounded linear operator on X satisfying

$$\|S_k(t)\| \leq M\tilde{\gamma}^k, \quad k \in \mathbb{N}, t \geq 0. \tag{13.6}$$

Since $\gamma < 1$, we can assume that $\tilde{\gamma} < 1$ by choosing η sufficiently large. Thus the series

$$S(t) := \sum_{k=0}^\infty S_k(t), \quad t \geq 0,$$

converges in $\mathcal{L}(X)$ uniformly on $[0, \infty)$ and satisfies

$$\|S(t)\| \leq \frac{M}{1 - \tilde{\gamma}}, \quad t \geq 0.$$

The semigroup property $S(t+s) = S(t)S(s)$, $t, s \geq 0$, can be obtained in the same way as in the proof of Proposition 11.6.

To obtain the strong continuity of the function $t \mapsto S(t)$ we first note that for each $f \in D(A)$, the mapping $s \mapsto BT(s)f$ is continuous, and thus the local boundedness of the function $t \mapsto T(t)$ implies that $t \mapsto S_1(t)f$ is continuous. So, by (13.6) and the denseness of $D(A)$, we have the strong continuity of $t \mapsto S_1(t)$. Using (13.6) again we deduce by induction that $t \mapsto S_k(t)$ is strongly continuous for all $k \in \mathbb{N}$. Thus, because the series $\sum_{k=0}^{\infty} S_k(t)$ converges uniformly for $t \in [0, \infty)$, the function $t \mapsto S(t)$ is strongly continuous and hence a C_0-semigroup on X.

Moreover, the definition of $S_k(t)$ implies that $S(t)$ satisfies Duhamel's equation (13.4).

Let us denote by G the generator of the semigroup $(S(t))_{t\geq 0}$. Using equation (13.4), Fubini's theorem (see Theorem A.24), and a change of variables, we compute

$$\int_0^\infty e^{-\lambda t} S(t)f \, dt = \int_0^\infty e^{-\lambda t} T(t)f \, dt + \int_0^\infty e^{-\lambda t} \int_0^t S(t-s)BT(s)f \, ds \, dt$$

$$= \int_0^\infty e^{-\lambda t} T(t)f \, dt + \int_0^\infty e^{-\lambda s} \int_0^\infty e^{-\lambda t} S(t)BT(s)f \, dt \, ds$$

for $\lambda > 0$ and $f \in D(A)$. The A-boundedness of B yields

$$R(\lambda, G)f = R(\lambda, A)f + R(\lambda, G)B \int_0^\infty e^{-\lambda s} T(s)f \, ds$$

$$= R(\lambda, A)f + R(\lambda, G)BR(\lambda, A)f$$

for $f \in D(A)$. Thus, using the denseness of $D(A)$, we see that

$$R(\lambda, G)(I - BR(\lambda, A)) = R(\lambda, A).$$

But we know from (13.5) and the A-boundedness of B that

$$\|BR(\lambda, A)f\| = \left\| \int_0^\infty e^{-\lambda t} BT(t)f \, dt \right\|$$

$$\leq \int_0^\infty \|BT(t)f\| \, dt \leq \tilde{\gamma}\|f\|$$

for all $f \in D(A)$. Hence

$$\|BR(\lambda, A)\| < \tilde{\gamma} < 1$$

yielding $\lambda \in \rho(A + B)$ and

$$R(\lambda, G) = R(\lambda, A)(I - BR(\lambda, A))^{-1} = R(\lambda, A + B).$$

This proves that $G = A + B$. $\square$

13.3 Positive Perturbations in L^1

In this section we consider positive semigroups perturbed by positive unbounded operators. In the L^1-setting we deduce from Miyadera's theorem very interesting perturbation results with elegant proofs.

Throughout the section we assume that (Ω, μ) is a σ-finite measure space and that the operator A is the generator of a positive C_0-semigroup on $E := L^1(\Omega, \mu)$. We begin our investigations by showing that in this setting there is a large class of Miyadera perturbations.

Proposition 13.7. *Let A be the generator of a positive C_0-semigroup $(T(t))_{t \geq 0}$ on $E := L^1(\Omega, \mu)$, $B : D(A) \to E$ a positive operator, and $C : D(A) \to E$ a linear operator satisfying $|Cf| \leq Bf$ for any $f \in D(A)_+$. If there exists $\lambda > s(A)$ such that $\|BR(\lambda, A)\| < 1$, then $\|CR(\lambda, A)\| < 1$, the operator $A + C$ generates a C_0-semigroup, and $A + B$ generates a positive C_0-semigroup on E.*

Proof. First note that for any $f \in E$ we have

$$
\begin{aligned}
|CR(\lambda, A)f| &= |CR(\lambda, A)(f^+ - f^-)| \\
&= |CR(\lambda, A)f^+ - CR(\lambda, A)f^-| \\
&\leq |CR(\lambda, A)f^+| + |CR(\lambda, A)f^-| \\
&\leq BR(\lambda, A)(f^+ + f^-) = BR(\lambda, A)|f|.
\end{aligned}
$$

Note that $BR(\lambda, A) : E \to E$ is a positive operator and hence, by Theorem 10.20, also bounded. Therefore, we have

$$
\|CR(\lambda, A)\| \leq \|BR(\lambda, A)\|. \tag{13.7}
$$

So, the operators B and C are both A-bounded, and $\|CR(\lambda, A)\| < 1$ whenever the inequality $\|BR(\lambda, A)\| < 1$ holds. Since the norm is additive on the positive cone, we obtain

$$
\begin{aligned}
\int_0^\alpha \|Ce^{-\lambda t}T(t)f\|_1 \, dt &\leq \int_0^\alpha \|Be^{-\lambda t}T(t)f\|_1 \, dt \\
&= \left\| B \int_0^\alpha e^{-\lambda t}T(t)f \, dt \right\|_1 \\
&\leq \left\| B \int_0^\infty e^{-\lambda t}T(t)f \, dt \right\|_1 \\
&= \|BR(\lambda, A)f\|_1 \\
&\leq \|BR(\lambda, A)\| \|f\|_1 =: \gamma \|f\|_1
\end{aligned}
$$

for all $f \in D(A)_+$ and any $\alpha > 0$. For $f \in D(A)$ and $k \in \mathbb{N}$ sufficiently large, we put $f_k^\pm := kR(k, A)f^\pm$. Then

$$
\lim_{k \to \infty} \|A(f_k^+ - f_k^-) - Af\|_1 = 0 \quad \text{and} \quad \lim_{k \to \infty} \|f_k^\pm - f^\pm\|_1 = 0.
$$

This implies that

$$\int_0^\alpha \|Ce^{-\lambda t}T(t)(f_k^+ - f_k^-)\|_1 \, dt \leq \gamma \left(\|f_k^+\|_1 + \|f_k^-\|_1 \right)$$

and

$$\int_0^\alpha \|Ce^{-\lambda t}T(t)f\|_1 \, dt \leq \gamma \left(\|f^+\|_1 + \|f^-\|_1 \right) = \gamma \|f\|_1.$$

Finally, since $\gamma = \|BR(\lambda, A)\| < 1$, we see that C is a Miyadera perturbation of $A - \lambda$. So, by Theorem 13.6, the operator $A + C - \lambda$, and hence also $A + C$, generates a C_0-semigroup on E.

Finally, observe that, since both $(T(t))_{t \geq 0}$ and B are positive, the positivity of the semigroup $(S(t))_{t \geq 0}$ generated by $A + B$ follows from the positivity of the terms $S_k(\cdot)$ in the Dyson–Phillips series given in Theorem 13.6. $\square$

In the following example we present an application of this result to Schrödinger operators on $L^1(\mathbb{R}^n)$.

Example 13.8. On $L^1(\mathbb{R}^n)$ we consider the Laplacian Δ with domain

$$D(\Delta) = \{f \in L^1(\mathbb{R}^n) \ : \ \Delta f \in L^1(\mathbb{R}^n)\}.$$

By Remark 9.24 we know that the corresponding semigroup is given by

$$T(t)f(x) = \int_{\mathbb{R}^n} f(y)G_t(x - y)\mathrm{d}y, \qquad x \in \mathbb{R}^n, \, t > 0,$$

where

$$G_t(x - y) := (4\pi t)^{-n/2} e^{-\frac{|x-y|^2}{4t}}, \qquad x, y \in \mathbb{R}^n, \, t > 0.$$

Let $V \in L^r(\mathbb{R}^n)$ with $r > \max\{1, n/2\}$. For $f \in L^1(\mathbb{R}^n)$ and $\lambda > 0$, we have by Fubini's theorem (see Theorem A.24) and Hölder's inequality (cf. Lemma A.13),

$$\|VR(\lambda, \Delta)f\|_1 = \left\| V \int_0^\infty e^{-\lambda t}T(t)f \, dt \right\|_1$$

$$\leq \int_{\mathbb{R}^n} |V(x)| \int_0^\infty e^{-\lambda t} \int_{\mathbb{R}^n} G_t(x - y)|f(y)|\mathrm{d}y \, dt \, dx$$

$$= \int_{\mathbb{R}^n} |f(y)| \int_0^\infty e^{-\lambda t} \int_{\mathbb{R}^n} G_t(x - y)|V(x)| \, dx \, dt\mathrm{d}y$$

$$\leq \|f\|_1 \|V\|_r \int_0^\infty e^{-\lambda t}\|G_t\|_{r'} \, dt,$$

with $\frac{1}{r} + \frac{1}{r'} = 1$. One can verify easily that $\|G_t\|_{r'} = ct^{-n/2r}$, hence $D(\Delta) \subset D(V)$ and

$$\|VR(\lambda, \Delta)\| \leq c\|V\|_r \int_0^\infty e^{-\lambda t}t^{-n/2r} \, dt = c\|V\|_r \lambda^{(n/2r)-1} \int_0^\infty e^{-s}s^{-n/2r}\mathrm{d}s.$$

Thus, the inequality $\frac{n}{2r} - 1 < 0$ implies that

$$\|VR(\lambda, \Delta)\| < 1$$

for $\lambda > 0$ large enough. By Proposition 13.7, $\Delta + V$ generates a C_0-semigroup on $L^1(\mathbb{R}^n)$.

Motivated by the following lemma we may ask if the condition $\|BR(\lambda, A)\| < 1$ in Proposition 13.7 can be replaced by $r(BR(\lambda, A)) < 1$.

Lemma 13.9. *Let Λ be a positive operator on $E := L^1(\Omega, \mu)$ with $r(\Lambda) < 1$. Then there exists an equivalent norm $\| \cdot \|_\Lambda$, which is additive on the positive cone of E, such that $\|\Lambda\|_\Lambda < 1$.*

Proof. Since $r(\Lambda) < 1$, there exist $m_0 \in \mathbb{N}$ and $\nu \in (0, 1)$ such that $\|\Lambda^m\| \leq \nu^m$ for all $m \geq m_0$. By taking $\nu < \kappa < 1$ we define

$$\|f\|_\Lambda := \sum_{m=0}^\infty \frac{\|\Lambda^m f\|_1}{\kappa^m}, \quad f \in E.$$

Then, by the positivity of Λ, one can see that $\| \cdot \|_\Lambda$ is a norm on E which is additive on the positive cone of E and, obviously, $\|f\|_1 \leq \|f\|_\Lambda$ for any $f \in E$. On the other hand,

$$\|f\|_\Lambda = \sum_{m=0}^\infty \frac{\|\Lambda^m f\|_1}{\kappa^m}$$

$$= \sum_{m=0}^{m_0} \frac{\|\Lambda^m f\|_1}{\kappa^m} + \sum_{m=m_0+1}^\infty \frac{\|\Lambda^m f\|_1}{\kappa^m}$$

$$\leq \left(\sum_{m=0}^{m_0} \frac{\|\Lambda^m\|}{\kappa^m} + \sum_{m=m_0+1}^\infty \left(\frac{\nu}{\kappa}\right)^m \right) \|f\|_1 = M\|f\|_1$$

for any $f \in E$. Moreover, for any $f \in E$, we have

$$\|\Lambda f\|_\Lambda = \sum_{m=0}^\infty \frac{\|\Lambda^{m+1} f\|_1}{\kappa^m} = \kappa \sum_{m=0}^\infty \frac{\|\Lambda^{m+1} f\|_1}{\kappa^{m+1}} \leq \kappa \sum_{m=0}^\infty \frac{\|\Lambda^m f\|_1}{\kappa^m}.$$

Thus

$$\|\Lambda f\|_\Lambda \leq \kappa \|f\|_\Lambda, \quad f \in E. \qquad \square$$

Remark 13.10. It follows directly that if Λ satisfies $|\Lambda f| = \Lambda|f|$ for any $f \in E$, then $(E, \| \cdot \|_\Lambda)$ is Banach lattice. The condition above is satisfied, for example, when Λ is a positive multiplication operator.

Furthermore, the condition $r(BR(\lambda, A)) < 1$ is equivalent to positivity of the resolvent of the operator $A + B$.

Proposition 13.11. *Let A be the generator of a positive C_0-semigroup on E, $\lambda >$ $s(A)$, and $B : D(A) \to E$ a positive linear operator. Then the following assertions are equivalent.*

(i) $\lambda \in \rho(A + B)$ and $R(\lambda, A + B) \geq 0$.

(ii) $r(BR(\lambda, A)) < 1$.

Proof. Let us first note that, by Corollary 12.10, we have $R(\lambda, A) \geq 0$. So, the operator $BR(\lambda, A) : E \to E$ is positive, and hence, by Theorem 10.20, also bounded.

Take $\lambda > s(A)$ such that $\lambda \in \rho(A + B)$ and $R(\lambda, A + B) \geq 0$. Since

$$\lambda - A - B = (I - BR(\lambda, A))(\lambda - A), \tag{13.8}$$

we deduce that $1 \in \rho(BR(\lambda, A))$ and

$$(I - BR(\lambda, A))^{-1} = (\lambda - A)R(\lambda, A + B) = I + BR(\lambda, A + B) \geq 0.$$

By Lemma 10.25, we obtain $r(BR(\lambda, A)) < 1$.

The converse follows from relation (13.8) and the Neumann series (see (A.4)),

$$(I - BR(\lambda, A))^{-1} = \sum_{k=0}^{\infty}(BR(\lambda, A))^{k} \geq 0. \qquad \square$$

Remark 13.12. One can see that Proposition 13.11 remains true for a general Banach lattice with order continuous norm. We mention here that a Banach lattice E has *order continuous norm* if every monotone order bounded net of E is convergent.

We end this section by the following perturbation result due to W. Desch which generalizes Proposition 13.7.

Theorem 13.13. *Let A be the generator of a positive C_0-semigroup on the Banach lattice $E := \mathrm{L}^1(\Omega, \mu)$, $B : D(A) \to E$ a positive linear operator, and $C : D(A) \to E$ a linear operator with $|Cf| \leq Bf$ for any $f \in D(A)_+$. If for some $\lambda > s(A)$ the resolvent $R(\lambda, A + B)$ is positive, then $A + C$ generates a C_0-semigroup on E.*

Proof. We prove first that $A + B$ generates a positive C_0-semigroup. Proposition 13.11 implies that $r(BR(\lambda, A)) < 1$ and

$$R(\lambda, A + B) = R(\lambda, A) \sum_{k=0}^{\infty}(BR(\lambda, A))^{k}.$$

On the other hand, $BR(\lambda, A + B) : E \to E$ is positive and hence a bounded operator on E, by Theorem 10.20. So, there is an $m \in \mathbb{N}$ such that

$$\|BR(\lambda, A + B)\| < m.$$

From

$$0 \leq BR(\lambda, A) \leq BR\left(\lambda, A + \frac{j}{m}B\right) \leq BR(\lambda, A + B)$$

we obtain for any $j = 0, \ldots, m - 1$, that

$$\left\|\frac{1}{m}BR\left(\lambda, A + \frac{j}{m}B\right)\right\| < 1.$$

Applying Proposition 13.7 to A and $m^{-1}B$ we obtain that $A + m^{-1}B$ generates a positive C_0-semigroup. Taking $A + m^{-1}B$ instead of A and applying Proposition 13.7 again, one sees that $A + m^{-1}B + m^{-1}B = A + 2m^{-1}B$ generates a positive C_0-semigroup. Repeating this process we obtain that $A + \frac{m-1}{m}B + \frac{1}{m}B = A + B$ generates a positive C_0-semigroup on E.

We need to prove that $A + C$ is a generator. To this end take $\mu > \max\{s(A), s(A + B)\}$. Then, as in the proof of Proposition 13.7, we have

$$|CR(\mu, A)f| \leq BR(\mu, A)|f|, \quad f \in E,$$

and by iteration we see that

$$|(CR(\mu, A))^k f| \leq (BR(\mu, A))^k |f|, \quad f \in E, \ k \in \mathbb{N}. \tag{13.9}$$

Since $\mu > s(A + B)$, by Corollary 12.10 and Proposition 13.11, we have that $r(BR(\mu, A)) < 1$. Hence, using inequality (13.9) we obtain that $r(CR(\mu, A)) < 1$ and for any $f \in E$

$$|R(\mu, A + C)f| = \left|R(\mu, A)\sum_{k=0}^{\infty}(CR(\mu, A))^k f\right|$$

$$\leq R(\mu, A)\sum_{k=0}^{\infty}(BR(\mu, A))^k |f|$$

$$= R(\mu, A + B)|f|.$$

Iterating this one obtains

$$|R(\mu, A + C)^k f| \leq R(\mu, A + B)^k |f|, \quad f \in E, \ k \in \mathbb{N}. \tag{13.10}$$

Since, by the Hille–Yosida theorem (see Theorem 11.1), there is an $\omega \in \mathbb{R}$ such that

$$\sup_{\mu > \omega, k \in \mathbb{N}} \|[(\mu - \omega)R(\mu, A + B)]^k\| < \infty,$$

Inequality (13.10) implies that

$$\sup_{\mu > \omega, k \in \mathbb{N}} \|[(\mu - \omega)R(\mu, A + C)]^k\| < \infty.$$

So, again by the Hille–Yosida theorem, $A + C$ generates a C_0-semigroup. $\qquad\square$

Remark 13.14. Denote by $(S(t))_{t\geq 0}$ and $(U(t))_{t\geq 0}$ the semigroups from Theorem 13.13 generated by $A + B$ and $A + C$, respectively. Setting $\mu = \frac{k}{t}$ for $t > 0$ and taking $k \in \mathbb{N}$ sufficiently large yields the following relation between the two semigroups.

$$|U(t)f| = \lim_{k\to\infty} |(I - (t/k)(A + C))^{-k} f|$$

$$\leq \lim_{k\to\infty} (I - (t/k)(A + B))^{-k} |f|$$

$$= S(t)|f|, \quad f \in E.$$

As a consequence we obtain the following domination result for Schrödinger semigroups on $\mathrm{L}^1(\mathbb{R}^n)$.

Example 13.15. Denote by $(S_V(t))_{t\geq 0}$ the semigroup generated by $\Delta + V$ on $\mathrm{L}^1(\mathbb{R}^n)$ discussed in Example 13.8, where $V \in \mathrm{L}^r(\mathbb{R}^n)$ with $r > \max\{1, n/2\}$. Then,

$$|S_V(t)f| \leq S_{|V|}(t)|f|, \quad f \in \mathrm{L}^1(\mathbb{R}^n),\, t \geq 0.$$

Remark 13.16. Looking at the proof of Theorem 13.13 we see that the perturbation result remains valid in any Banach lattice E if we consider perturbations of finite rank. Specifically, a linear operator $C : D(A) \to E$ is called a *finite-rank perturbation* if there exist $\varphi_i \in \operatorname{span} D(A)_+^*$, $g_i \in E$, $i = 1, \dots, k$ such that

$$Cf = \sum_{i=1}^{k} \varphi_i(f) g_i, \quad f \in D(A). \tag{13.11}$$

To see this, take for simplicity $Bf = \varphi(f)g$, $f \in D(A)$ for some $\varphi \in D(A)_+^*$ and $g \in E_+$. Then the claim follows from the estimate

$$\int_0^\alpha \|B e^{-\lambda t} T(t) f\|\, \mathrm{d}t = \int_0^\alpha e^{-\lambda t} \varphi(T(t)f)\, \mathrm{d}t \|g\|$$

$$\leq \varphi(R(\lambda, A)f) \|g\|$$

$$= \|B R(\lambda, A) f\|$$

for any $f \in D(A)_+$ and $\lambda > \mathrm{s}(A)$.

We conclude with an example showing that Theorem 13.13 is, however, not true in general when the state space is not an L^1-space.

Example 13.17. Let $X := \mathrm{C}_0(0, 1] := \{f \in \mathrm{C}([0, 1]) : f(0) = 0\}$ and $Lf = -f'$ for $f \in D(L)$, where

$$D(L) = \left\{ f \in \mathrm{C}^1([0, 1]) : f'(0) = f(0) = 0 \right\}.$$

By Exercise 3, L generates a positive contraction C_0-semigroup on X with spectral bound $\mathrm{s}(L) = -\infty$. Let $C : D(L) \to X$ be given by

$$Cf(s) = \begin{cases} \frac{1}{s} f(s) & \text{if } s \in (0, 1], \\ 0 & \text{if } s = 0. \end{cases}$$

Then $A + B$ is resolvent positive, but does not generate a C_0-semigroup on $E :=$ $X \times X$, where

$$A := \begin{pmatrix} L & 0 \\ 0 & L \end{pmatrix} \text{ and } B := \begin{pmatrix} 0 & 0 \\ C & 0 \end{pmatrix}.$$

To see this we compute the resolvent of $A + B$ as

$$R(\lambda, A + B) = \begin{pmatrix} R(\lambda, L) & 0 \\ R(\lambda, L)CR(\lambda, L) & R(\lambda, L) \end{pmatrix}$$

for $\lambda \in \mathbb{R}$, where $R(\lambda, L)f(s) = e^{-\lambda s} \int_0^s e^{\lambda r} f(r) \, dr$. So $A + B$ is resolvent positive. For $f \in X$ we compute

$$
\begin{aligned}
R(\lambda, L)CR(\lambda, L)f(s) &= e^{-\lambda s} \int_0^s e^{\lambda r} CR(\lambda, L)f(r) \, dr \\
&= e^{-\lambda s} \int_0^s e^{\lambda r} \frac{1}{r} e^{-\lambda r} \int_0^r e^{\lambda y} f(y) \, dy \, dr \\
&= e^{-\lambda s} \int_0^s e^{\lambda y} f(y) \int_y^s \frac{1}{r} \, dr \, dy \\
&= \int_0^s e^{-\lambda(s-y)} f(y) \log(s/y) \, dy \\
&= \int_0^s e^{-\lambda t} W(t) f(s) \, dt,
\end{aligned}
$$

where

$$W(t)f(s) := \begin{cases} \log(s/(s-t))f(s-t) & \text{if } s > t, \\ 0 & \text{otherwise.} \end{cases} \tag{13.12}$$

If $A + B$ is the generator of a C_0-semigroup $(S(t))_{t \geq 0}$ on E, then, by the uniqueness of the Laplace transform, the semigroup is the form

$$S(t)(f, g) = \begin{pmatrix} T(t)f & 0 \\ W(t)f & T(t)g \end{pmatrix}, \quad (f, g) \in E.$$

But this is not possible because the operator defined in (13.12) is not bounded on $C_0((0, 1])$.

13.4   Notes and Remarks

The study of unbounded perturbations of C_0-semigroups started first in Hilbert spaces. For this we refer to the monographs by Kato [73] and by Reed and Simon [119, Sec. X.2].

Our Theorem 13.3 is a dispersive version of a result due to Gustafson [61]. Theorem 13.6 is taken from Voigt [148], where it is shown that the perturbed

semigroup is given by a Dyson–Phillips series for small times. The convergence of
the Dyson–Phillips expansion for all times is due to Rhandi [120]. Lemma 13.9
originates in Mokhtar–Kharroubi [97, Lemma 8.3] and Proposition 13.11 is due to
Voigt [152]. Theorem 13.13 was originally proved by Desch, but the proof using
Miyadera's perturbation theorem is taken from Voigt [152]. Extension to finite-
rank perturbations can be found in Arendt and Rhandi [7]. Another important
application of the Miyadera perturbation theorem is for the delay semigroup, see
Chapter 15.

13.5   Exercises

1. Show that if $\rho(A) \neq \emptyset$, then B is A-bounded if and only if $BR(\lambda, A) \in \mathcal{L}(X)$
   for some/all $\lambda \in \rho(A)$.

2. Prove that every AL-space has an order continuous norm, see Remark 13.12.

3. Show that the operator L defined in Example 13.17 generates a positive
   contraction C_0-semigroup on $C_0(0, 1]$ and $s(L) = -\infty$.

4. Prove that if A generates a positive contraction C_0-semigroup on a Banach
   lattice E, then for every $f \in D(A)$ and for every $f^* \in \mathcal{I}^+(f)$,

$$\mathrm{Re}\langle Af, f^* \rangle \leq 0.$$

5. Let A be the generator of a positive C_0-semigroup on a Banach lattice E.
   Define $B : D(A) \to E$ as $Bf = \varphi(f)g$ for a given $\varphi \in D(A)^*_+$ and $g \in E$.

   a) Verify that $r(BR(\lambda, A)) \leq |\varphi(R(\lambda, A)g)|$ for any $\lambda > s(A)$.

   b) Prove that $A + B$ generates a C_0-semigroup on E.

   c) Let $E = \mathrm{L}^p(0, 1)$, $1 \leq p < \infty$, and let $Af = -f'$ with

$$D(A) = \{f \in \mathrm{W}^{1,p}(0, 1) : f(0) = 0\}.$$

   Let μ be a bounded positive measure on $[0, 1]$, $0 \leq g \in E$ and define

$$Bf = \left(\int_0^1 f(s)\, d\mu(s) \right) g,$$

   for $f \in E$. Show that A and $A + B$ generate positive C_0-semigroups
   on E.

6. Finish the proof of Theorem 13.13 for finite-rank perturbations of the type as
   in (13.11) in arbitrary Banach lattices. This was sketched in Remark 13.16.

Part III

Advanced Topics and Applications

Chapter 14

Advanced Spectral Theory and Asymptotics

In this chapter we continue our investigation of spectral properties of positive C_0-semigroups on Banach lattices and show how the Perron–Frobenius theory can be generalized to the infinite-dimensional setting. We also list some important properties of irreducible semigroups. We will see that many results valid for positive matrix semigroups continue to hold also in infinite dimensions.

Our main goal is to describe the asymptotic behavior of a semigroup (such as asymptotic periodicity or balanced exponential growth) via the spectral properties of its generator.

14.1 Spectral Decomposition

First we define and discuss spectral projections and spectral decompositions for an unbounded closed operator. Recall that in finite dimensions we have constructed a functional calculus using spectral projections corresponding to the eigenvalues (cf. Theorem 2.11). As already mentioned in Section 2.4, these projections can be obtained by Dunford's integral representation. We now start with such a representation in the case of bounded operators.

Let $T \in \mathcal{L}(X)$, where X is a Banach space. For a function f holomorphic on a neighborhood of $\overline{W}$ for some open neighborhood W of $\sigma(T)$ with a smooth, positively oriented boundary ∂W^+, we define

$$f(T) := \frac{1}{2\pi i} \int_{\partial W^+} f(\lambda) R(\lambda, T) \, d\lambda.$$

As in the finite-dimensional situation, the map $f \mapsto f(T)$ is linear and multiplicative, and for $g(z) := z^k$, $k \in \mathbb{N}$, one obtains $g(T) = T^k$.

Assume now that the spectrum $\sigma(T)$ can be decomposed as

$$\sigma(T) = \sigma_1 \cup \sigma_2, \tag{14.1}$$

where σ_1, σ_2 are closed and disjoint sets. The *spectral projection* P_i of T belonging to σ_i is defined to be $\chi_i(T)$, where χ_i is the characteristic function of a neighborhood W_i of σ_i such that $\overline{W_i} \cap \sigma(T) = \sigma_i$ (compare with relation (2.5) in the finite-dimensional case). Hence, P_i can be written as

$$P_i := \frac{1}{2\pi i} \int_{\gamma_i} R(\lambda, T) \, d\lambda, \tag{14.2}$$

where γ_i is a smooth curve in $\rho(T)$ enclosing σ_i. These projections commute with T and yield the *spectral decomposition*

$$X = X_1 \oplus X_2$$

with the T-invariant spaces $X_1 := \operatorname{im} P_1 = \ker P_2$, $X_2 := \operatorname{im} P_2 = \ker P_1$. The restrictions $T_i \in \mathcal{L}(X_i)$ of T to X_i satisfy

$$\sigma(T_i) = \sigma_i, \quad i = 1, 2,$$

a property that characterizes the above decomposition of X and T (again recall corresponding results in finite dimensions, e.g., Theorem 2.9).

For an unbounded operator A and an arbitrary decomposition of the spectrum $\sigma(A)$ into disjoint closed sets, it is not always possible to find an associated spectral decomposition. However, the spectral mapping theorem for the resolvent allows us to construct such decompositions if one of the subsets is compact.

Proposition 14.1. *Let $A : D(A) \subset X \to X$ be a closed operator such that its spectrum $\sigma(A)$ can be decomposed into the disjoint union of two closed subsets σ_c and σ_u, i.e.,*

$$\sigma(A) = \sigma_c \cup \sigma_u.$$

If σ_c is compact, then there exists a spectral decomposition $X = X_c \oplus X_u$ for A in the following sense.

a) *The restriction $A_c := A|_{X_c}$ is bounded on the Banach space X_c.*
b) *$D(A) = X_c \oplus D(A_u)$, where A_u is the part of A in X_u, i.e., $A_u := A|_{X_u}$, $A_u f := Af$ for $f \in D(A_u) := \{g \in X_u \cap D(A) : Ag \in X_u\}$.*
c) *The operator A decomposes as $A = A_c \oplus A_u$.*
d) *$\sigma(A_c) = \sigma_c$ and $\sigma(A_u) = \sigma_u$.*

Proof. Supposing that A is unbounded and taking $\lambda_0 \in \rho(A)$, we see that $0 \in \sigma(R(\lambda_0, A))$ and, by Proposition 9.29, we obtain

$$\sigma\big(R(\lambda_0, A)\big) = \underbrace{\left\{\frac{1}{\lambda_0 - \mu} : \mu \in \sigma_c\right\}}_{\tau_c} \cup \underbrace{\left\{\frac{1}{\lambda_0 - \mu} : \mu \in \sigma_u\right\}}_{\tau_u} \cup \{0\}, \tag{14.3}$$

where τ_c, τ_u are compact and disjoint subsets of $\mathbb{C}$. (If σ_c is not compact, 0 is in the closure of τ_c.) Let now P be the spectral projection for $R(\lambda_0, A)$ associated to the decomposition in (14.3) and put $X_c := \operatorname{im} P$ and $X_u := \ker P$. Since $R(\lambda_0, A)$ and P commute, we have $R(\lambda_0, A)X_c \subseteq X_c$, hence

$$\lambda_0 \in \rho(A_c) \quad \text{and} \quad R(\lambda_0, A_c) = R(\lambda_0, A)|_{X_c}. \tag{14.4}$$

Moreover, we know that $\sigma(R(\lambda_0, A_c)) = \tau_c \not\ni 0$. Therefore, $A_c = \lambda_0 - R(\lambda_0, A_c)^{-1}$ is bounded on X_c, and we obtain a).

To verify b), observe that by similar arguments as above we obtain

$$\lambda_0 \in \rho(A_u) \quad \text{and} \quad R(\lambda_0, A_u) = R(\lambda_0, A)|_{X_u}. \tag{14.5}$$

Combining this with (14.4) yields

$$\begin{aligned} X_c + D(A_u) &= R(\lambda_0, A_c)X_c + R(\lambda_0, A_u)X_u \\ &\subseteq D(A) = R(\lambda_0, A)(X_c + X_u) \\ &\subseteq R(\lambda_0, A_c)X_c + R(\lambda_0, A_u)X_u \\ &= X_c + D(A_u), \end{aligned}$$

implying that $D(A) = X_c + D(A_u)$. This proves b), while c) follows from a) and b). Finally, d) is a consequence of Proposition 9.29 and (14.3), (14.4), and (14.5). $\qquad \square$

A particularly important case of the above decomposition occurs when $\sigma_c = \{\mu\}$ consists of a single point. This means that μ is isolated in $\sigma(A)$ and therefore the holomorphic function $\rho(A) \ni \lambda \mapsto R(\lambda, A) \in \mathcal{L}(X)$ can be expanded in a Laurent series

$$R(\lambda, A) = \sum_{k=-\infty}^{\infty} (\lambda - \mu)^k U_k$$

for $0 < |\lambda - \mu| < \delta$ and some sufficiently small $\delta > 0$. The coefficients U_k of this series are bounded operators given by the formulas

$$U_k = \frac{1}{2\pi i} \int_\gamma \frac{R(\lambda, A)}{(\lambda - \mu)^{k+1}} \, d\lambda, \quad k \in \mathbb{Z}, \tag{14.6}$$

where γ is, for example, the positively oriented boundary of the disc with radius $\frac{\delta}{2}$ centered at μ. The coefficient U_{-1} is called the *residue* of $R(\cdot, A)$ at μ. From formula (14.6) one deduces

$$U_{k+1} = (A - \mu)^k U_{-1} \tag{14.7}$$

and the identity

$$U_{-(k+1)} \cdot U_{-(\ell+1)} = U_{-(k+\ell+1)} \tag{14.8}$$

for $k, \ell \geq 0$. Indeed,

$$\frac{1}{2\pi i} \int_\gamma (\lambda - \mu)^k (\lambda - \mu)^\ell R(\lambda, A) \, d\lambda$$

$$= \left(\frac{1}{2\pi i} \int_\gamma (\lambda - \mu)^k R(\lambda, A) \, d\lambda \right) \cdot \left(\frac{1}{2\pi i} \int_\gamma (\lambda - \mu)^\ell R(\lambda, A) \, d\lambda \right)$$

can be proved as in the case of a bounded operator A since the proof only uses the resolvent equation and the residue theorem.

If there exists $k > 0$ such that $U_{-k} \neq 0$, while $U_{-\ell} = 0$ for all $\ell > k$, then the spectral value μ is called a *pole of $R(\cdot, A)$ of order k* (compare with Remark 2.17). In view of (14.8), this is true if and only if $U_{-k} \neq 0$ and $U_{-(k+1)} = 0$. Moreover, we obtain U_{-k} as

$$U_{-k} = \lim_{\lambda \to \mu} (\lambda - \mu)^k R(\lambda, A). \tag{14.9}$$

The dimension of the spectral subspace $\operatorname{im} P$ is called the *algebraic multiplicity* m_{a} of μ, while $m_{\mathrm{g}} := \dim \ker(\mu - A)$ is its *geometric multiplicity*. One can show that the following relation holds:

$$m_{\mathrm{g}} + k - 1 \leq m_{\mathrm{a}} \leq m_{\mathrm{g}} \cdot k. \tag{14.10}$$

In the case $m_{\mathrm{a}} = 1$, we call μ an *algebraically simple pole*. We also denote by

$$\operatorname{Pol}(A) := \{\mu \in \mathbb{C} : \mu \text{ is a pole of } R(\cdot, A)\}. \tag{14.11}$$

The following result shows that, as in the case of a bounded operator, the spectral projection of A belonging to an isolated point $\mu \in \sigma(A)$ is the residue of $R(\cdot, A)$ at μ.

Proposition 14.2. *Let A be a closed linear operator having nonempty resolvent set $\rho(A)$ and take some $\lambda_0 \in \rho(A)$. Then $\mu \in \mathbb{C}$ is an isolated point of $\sigma(A)$ if and only if $(\lambda_0 - \mu)^{-1}$ is isolated in $\sigma(R(\lambda_0, A))$. In this case, the residues and the orders of the poles of $R(\cdot, A)$ at μ and of $R(\cdot, R(\lambda_0, A))$ at $(\lambda_0 - \mu)^{-1}$ coincide.*

Proof. The first claim follows easily from Proposition 9.29 and the fact that the map $z \mapsto (\lambda_0 - z)^{-1}$ is homeomorphic between $\mathbb{C} \setminus \{\lambda_0\}$ and $\mathbb{C} \setminus \{0\}$.

In order to prove the assertion concerning the residues, we choose a positively oriented circle $\gamma \subset \rho(A)$ with center μ such that λ_0 lies in the exterior of γ. Then the residue P of $R(\cdot, A)$ at μ is given by

$$P = \frac{1}{2\pi i} \int_\gamma R(\lambda, A) \, d\lambda$$

$$= \frac{1}{2\pi i} \int_\gamma \frac{R\big((\lambda_0 - \lambda)^{-1}, R(\lambda_0, A)\big)}{(\lambda_0 - \lambda)^2} \, d\lambda - \frac{1}{2\pi i} \int_\gamma \frac{d\lambda}{(\lambda_0 - \lambda)}$$

$$= \frac{1}{2\pi i} \int_\gamma \frac{R\big((\lambda_0 - \lambda)^{-1}, R(\lambda_0, A)\big)}{(\lambda_0 - \lambda)^2} \, d\lambda,$$

where we used the identity

$$R(\lambda, A) = \frac{R((\lambda_0 - \lambda)^{-1}, R(\lambda_0, A))}{(\lambda_0 - \lambda)^2} - \frac{1}{(\lambda_0 - \lambda)}$$

and Cauchy's integral theorem. The substitution $z := (\lambda_0 - \lambda)^{-1}$ then yields a path $\tilde{\gamma}$ around $(\lambda_0 - \mu)^{-1}$, and we obtain

$$P = \frac{1}{2\pi i} \int_{\tilde{\gamma}} R(z, R(\lambda_0, A)) \, dz,$$

which is the residue of $R(\cdot, R(\lambda_0, A))$ at $(\lambda_0 - \mu)^{-1}$.

The final assertion concerning the pole orders is obtained as follows. By the same calculations as above we see that for $k \in \mathbb{N}$

$$\frac{1}{2\pi i} \int_{\gamma} (\lambda - \mu)^{k-1} R(\lambda, A) \, d\lambda = \frac{1}{2\pi i} \int_{\tilde{\gamma}} \left(\lambda_0 - \mu - \frac{1}{z} \right)^{k-1} R(z, R(\lambda_0, A)) \, dz.$$

Since $\lambda_0 - \mu - \frac{1}{z} = \left(\frac{\lambda_0 - \mu}{z}\right)\left(z - \frac{1}{\lambda_0 - \mu}\right)$, by the multiplicativity of the functional calculus for $R(\lambda_0, A)$ the last integral can be interpreted as

$$\left((\lambda_0 - \mu)(\lambda_0 - A)\right)^{k-1} V_{-k},$$

where V_{-k} denotes the $-k$th coefficient in the Laurent expansion of $R(\cdot, R(\lambda_0, A))$ at $(\lambda_0 - \mu)^{-1}$. Hence, we obtain for the coefficients U_{-k} of the Laurent expansion of $R(\cdot, A)$ at μ

$$U_{-k} = \left((\lambda_0 - \mu)(\lambda_0 - A)\right)^{k-1} V_{-k}, \qquad k \in \mathbb{N},$$

and therefore

$$V_{-k} = \left((\lambda_0 - \mu)^{-1} R(\lambda_0, A)\right)^{k-1} U_{-k}, \qquad k \in \mathbb{N},$$

which proves the assertion. $\qquad\qquad\qquad\qquad\qquad\qquad\qquad\qquad\qquad\qquad\square$

We continue by further refining the spectral decomposition. First recall that the *essential spectrum* of a bounded operator $T \in \mathcal{L}(X)$ is the spectrum of $T + \mathcal{K}(X)$ in the Calkin algebra $\mathcal{C}(X) := \mathcal{L}(X)/\mathcal{K}(X)$, where $\mathcal{K}(X)$ denotes the ideal of compact operators. Accordingly, the *essential spectral radius* is

$$r_{ess}(S) := r(S + \mathcal{K}(X)),$$

see also Appendix A.9.

Analogously, we define the *essential growth bound* $\omega_{ess}(T)$ of a C_0-semigroup $(T(t))_{t \geq 0}$ as the growth bound of the quotient semigroup $(T(t) + \mathcal{K}(X))_{t \geq 0}$ on $\mathcal{C}(X)$, i.e.,

$$\omega_{ess}(T) := \inf\{\omega \in \mathbb{R} : \exists M > 0 \text{ such that } \|T(t)\|_{ess} \leq M e^{\omega t} \text{ for all } t \geq 0\},$$

where $\| \cdot \|_{\mathrm{ess}}$ is the quotient norm in $\mathcal{C}(X)$. Then, as in Proposition 12.1, one can see that

$$\omega_{\mathrm{ess}}(T) = \frac{\log r_{\mathrm{ess}}(T(t_0))}{t_0} = \lim_{t \to \infty} \frac{\log \|T(t)\|_{\mathrm{ess}}}{t} \qquad (14.12)$$

holds for all $t_0 > 0$. The following result gives the relationship between $\omega_{\mathrm{ess}}(T)$ and $\omega_0(T)$.

Proposition 14.3. *Let $(T(t))_{t \geq 0}$ be a C_0-semigroup with generator A on a Banach space X. Then*

$$\omega_0(T) = \max\{\mathrm{s}(A), \omega_{\mathrm{ess}}(T)\}.$$

Proof. If $\omega_{\mathrm{ess}}(T) < \omega_0(T)$, then $r_{\mathrm{ess}}(T(1)) < r(T(1))$. Let $\lambda \in \sigma(T(1))$ such that $|\lambda| = r(T(1))$. Then by Proposition A.34, λ is an eigenvalue of $T(1)$ and by the spectral mapping theorem for the point spectrum, Theorem A.33, there is a $\lambda_1 \in \sigma_{\mathrm{p}}(A)$ with $e^{\lambda_1} = \lambda$. Therefore, $\mathrm{Re}\,\lambda_1 = \omega_0(T)$, and thus $\omega_0(T) = \mathrm{s}(A)$. □

We are finally able to give an infinite-dimensional analogue of the formula for the matrix exponential function given in (2.9). As in finite dimensions, this will be an important tool to study the asymptotic behavior of the semigroup.

Theorem 14.4. *Let A be the generator of a C_0-semigroup $(T(t))_{t \geq 0}$ on a Banach space X such that $\omega_{\mathrm{ess}}(T) < 0$. Then the following assertions hold.*

a) *The set $\sigma_+ := \{\lambda \in \sigma(A) : \mathrm{Re}\,\lambda \geq 0\}$ is finite (or empty) and consists of poles of $R(\cdot, A)$ of finite algebraic multiplicity.*

b) *Let $\sigma_+ := \{\lambda_1, \ldots, \lambda_m\}$ where λ_j is a pole of order k_j with the corresponding spectral projection P_j, $j = 1, \ldots, m$. Then $T(t) = T_1(t) + \cdots + T_m(t) + R(t)$, where*

$$T_j(t) := e^{\lambda_j t} \sum_{k=0}^{k_j - 1} \frac{t^k}{k!} (A - \lambda_j)^k P_j, \quad j = 1, \ldots, m, \text{ and } t \geq 0,$$

and

$$\|R(t)\| \leq M e^{-\varepsilon t}, \quad \text{for some } \varepsilon > 0, \ M \geq 1, \text{ and all } t \geq 0.$$

Proof. a) Let $t_0 > 0$. Since $\omega_{\mathrm{ess}}(T) < 0$, (14.12) shows that $r_{\mathrm{ess}}(T(t_0)) < 1$. So, by Proposition A.34, every $\lambda \in \sigma(T(t_0))$ with $|\lambda| \geq 1$ is an isolated point. The set

$$\sigma_{\mathrm{c}} := \sigma(T(t_0)) \cap \{z \in \mathbb{C} : |z| \geq 1\}$$

is thus finite and consists of the points $\{\lambda_1, \ldots, \lambda_m\}$.

Set $\sigma_{\mathrm{u}} := \sigma(T(t_0)) \setminus \sigma_{\mathrm{c}}$. Then $\sigma(T(t_0))$ is the disjoint union of the closed sets σ_{c} and σ_{u} with σ_{c} compact, and we can apply Proposition 14.1, yielding the spectral decomposition

$$X = \mathrm{im}\,P_{\mathrm{c}} \oplus \ker P_{\mathrm{c}} =: X_{\mathrm{c}} \oplus X_{\mathrm{u}}$$

with the associated spectral projection P_c. Since σ_c is finite and any of its elements is a pole of $R(\cdot, T(t_0))$, we deduce that X_c is finite-dimensional. To this decomposition we associate semigroups $T_c(\cdot) := T(\cdot)|_{X_c}$ and $T_u(\cdot) := T(\cdot)|_{X_u}$, and the corresponding generators are, respectively, $A_c := A|_{X_c} \in \mathcal{L}(X_c)$ and A_u the part of A in X_u. Moreover, $\sigma(A_c) = \sigma_c$ and $\sigma(A_u) = \sigma_u$.

Since X_c is finite-dimensional, $\sigma(A_c)$ is finite and $A = A_c \oplus A_u$. Moreover, every element of σ_c is a pole of $R(\cdot, A) = R(\cdot, A_c) \oplus R(\cdot, A_u)$. Thus, the spectral mapping theorem (see Theorems 2.20 and 2.28) yields

$$\sigma(A_c) = \{\lambda_1, \ldots, \lambda_m\} \quad \text{and} \quad \sigma(T_c(t)) = \{e^{\lambda_1 t}, \ldots, e^{\lambda_m t}\}.$$

In particular,

$$\sigma_c = \sigma(T_c(t_0)) \subset \{z \in \mathbb{C} : |z| \geq 1\},$$

and hence $\operatorname{Re} \lambda_j \geq 0$ for all $j = 1, \ldots, m$.

Next, we show that $(T_u(t))_{t \geq 0}$ is uniformly exponentially stable. By the spectral decomposition, we know that $\sigma(T_u(t_0)) = \sigma_u \subset \{z \in \mathbb{C} : |z| < 1\}$. So, $r(T_u(t_0)) < 1$ and by Proposition 12.1 we obtain $\omega_0(T_u) < 0$, which also implies $s(A_u) < 0$. This proves a).

b) In order to verify this, we define the spectral projection $P := \sum_{j=1}^m P_j$ of A corresponding to the spectral set $\{\lambda_1, \ldots, \lambda_m\}$, i.e., $P = P_c$. We decompose now the semigroup $(T(t))_{t \geq 0}$ as

$$T(t) = T(t)P_1 + \cdots + T(t)P_m + T(t)(I - P),$$

where each restricted semigroup $T(\cdot)P_j$ has generator $A|_{\operatorname{im} P_j}$. Since $\operatorname{im} P_j$ is finite-dimensional and $(A - \lambda_j)^{k_j} P_j = 0$, we can use Theorem 2.11 and get

$$T_j(t) := T(t)P_j = e^{\lambda_j t} \sum_{k=0}^{k_j - 1} \frac{t^k}{k!} (A - \lambda_j)^k P_j, \quad t \geq 0.$$

To show the last assertion, it suffices to note that

$$R(t) = T(t)(I - P) = T_u(t)(I - P_c) = T_u(t)$$

and $\omega_0(T_u) < 0$. This ends the proof of the theorem. $\qquad\square$

Semigroups satisfying $\omega_{\mathrm{ess}}(T) < 0$ are also called *quasi-compact* semigroups (for an explanation of this name see Exercise 1). They include uniformly exponentially stable semigroups and eventually compact semigroups.

14.2 Periodic Semigroups

In this section we characterize periodic semigroups in terms of their spectrum. As we shall see later in this chapter, this class of semigroups plays an important role for the asymptotics of general semigroups.

Definition 14.5. A C_0-semigroup $(T(t))_{t\geq 0}$ on a Banach space X is called *periodic* if there is $t_0 > 0$ such that $T(t_0) = I$. In this case its *period* is defined as the smallest $\tau > 0$ such that $T(\tau) = I$.

Since, for every $k \in \mathbb{N}$ and $0 \leq t \leq k\tau$, $T(t)T(k\tau - t) = I$, we readily see that a periodic semigroup always extends to a group.

As in finite dimensions, we can characterize periodic semigroups in terms of their spectrum, compare with Theorem 4.12.c) in the finite-dimensional situation.

Theorem 14.6. *For a C_0-semigroup $(T(t))_{t\geq 0}$ with generator A on a Banach space X the following assertions are equivalent.*

(i) *$(T(t))_{t\geq 0}$ is a periodic semigroup.*

(ii) *$\sigma(A) = \sigma_p(A) \subset 2\pi i\alpha\mathbb{Z}$ for some $\alpha > 0$ and the corresponding eigenvectors span a dense subspace of X.*

Proof. (ii) $\implies$ (i): First observe that for any $\lambda \in \sigma_p(A)$ and a corresponding eigenvector $f \in D(A)$, Corollary 9.32 yields

$$T(t)f = e^{\lambda t}f, \quad t \geq 0. \tag{14.13}$$

Thus, taking $\lambda = 2\pi i k\alpha \in \sigma_p(A)$ we deduce that $T(t)f = e^{2\pi i k\alpha t}f$ for all $t \geq 0$. Since these eigenvectors span a dense subspace of X, we obtain that $(T(t))_{t\geq 0}$ is periodic with period $\tau \leq \frac{1}{\alpha}$.

(i) $\implies$ (ii): Let τ be the period of $(T(t))_{t\geq 0}$ and $\lambda \neq \frac{2\pi ki}{\tau}$, $k \in \mathbb{Z}$. From Lemma 9.31 we infer that $\lambda \in \rho(A)$ and

$$R(\lambda, A) = \frac{1}{1 - e^{-\lambda\tau}} \int_0^\tau e^{-\lambda s}T(s)\ ds. \tag{14.14}$$

So the resolvent is a meromorphic function having poles only at (some) $\mu_k = \frac{2\pi ki}{\tau}$, $k \in \mathbb{Z}$, of order less than or equal to one. Using formula (14.14) and the residue theorem one obtains the residues in μ_k as

$$P_k = \frac{1}{\tau} \int_0^\tau e^{-\mu_k s}T(s)\ ds, \quad k \in \mathbb{Z}. \tag{14.15}$$

Now we prove that $\overline{\text{span}}\bigcup_{k\in\mathbb{Z}} P_k X = X$. More precisely, we prove that

$$f = \sum_{k=-\infty}^{+\infty} P_k f \quad \text{for all } f \in D(A), \tag{14.16}$$

which clearly implies the assertion, since $D(A)$ is dense in X.

Setting $g = Af$, we have $P_k g = P_k A f = \frac{2\pi k i}{\tau} P_k f$. This implies $P_k f = \frac{\tau}{2\pi k i} P_k g$. Hence, by applying the Cauchy–Schwarz inequality (A.3), we obtain

$$\left| \sum_{k \in F} \langle P_k f, f^* \rangle \right| = \left| \sum_{k \in F} \frac{\tau}{2\pi k i} \langle P_k g, f^* \rangle \right|$$

$$\leq \frac{\tau}{2\pi} \left(\sum_{k \in F} \frac{1}{k^2} \right)^{1/2} \left(\sum_{k \in F} |\langle P_k g, f^* \rangle|^2 \right)^{1/2}$$

$$\leq \frac{\tau}{2\pi} \left(\sum_{k \in F} \frac{1}{k^2} \right)^{1/2} \left(\frac{1}{\tau} \int_0^\tau |\langle T(s)g, f^* \rangle|^2 \, ds \right)^{1/2}$$

$$\leq \frac{\tau}{2\pi} \left(\sum_{k \in F} \frac{1}{k^2} \right)^{1/2} \|f^*\| \underbrace{\left(\frac{1}{\tau} \int_0^\tau \|T(s)g\|^2 \, ds \right)^{1/2}}_{C}$$

$$= \frac{C\tau}{2\pi} \left(\sum_{k \in F} \frac{1}{k^2} \right)^{1/2} \|f^*\|$$

for any finite subset $F \subset \mathbb{Z}$. Thus,

$$\left\| \sum_{k \in F} P_k f \right\| \leq \frac{C\tau}{2\pi} \left(\sum_{k \in F} \frac{1}{k^2} \right)^{1/2}$$

for any finite subset $F \subset \mathbb{Z}$. This gives the convergence of $\sum_{k \in \mathbb{Z}} P_k f$ for all $f \in D(A)$.

On the other hand, using the relation (14.13) with $\lambda = \mu_m$ and the corresponding eigenvector $P_m f$, we obtain

$$P_k P_m f = \frac{1}{\tau} \int_0^\tau e^{-\mu_k s} T(s) P_m f \, ds = \frac{1}{\tau} \int_0^\tau e^{(\mu_m - \mu_k)s} P_m f \, ds = 0,$$

if $k \neq m$. From this we see that, for any $f^* \in X^*$, the Fourier coefficients of the functions $s \mapsto \langle T(s)(\sum_{k \in \mathbb{Z}} P_k f), f^* \rangle$ and $s \mapsto \langle T(s)f, f^* \rangle$ coincide. So, the two functions are equal and, in particular,

$$\left\langle \sum_{k \in \mathbb{Z}} P_k f, f^* \right\rangle = \left\langle T(0) \left(\sum_{k \in \mathbb{Z}} P_k f \right), f^* \right\rangle = \langle T(0)f, f^* \rangle = \langle f, f^* \rangle.$$

This proves that $f = \sum_{k \in \mathbb{Z}} P_k f$. $\qquad\square$

The calculations in the proof above yield the following expansion formula for a periodic semigroup and its generator.

Corollary 14.7. *Let $(T(t))_{t\geq 0}$ be a periodic C_0-semigroup with period τ and generator A on a Banach space X. Then*

$$T(t)f = \sum_{-\infty}^{+\infty} e^{\mu_k t} P_k f \quad for \ f \in D(A) \quad and$$

$$Af = \sum_{-\infty}^{+\infty} \mu_k P_k f \quad for \ f \in D(A^2),$$

where P_k are the residues of $R(\cdot, A)$ at $\mu_k := \frac{2\pi i k}{\tau}$ given in (14.15).

Proof. One has to apply expansion (14.16) to $T(t)f$ and Af instead of f, respectively, and use the identities $AP_k = \mu_k P_k$ and $T(t)P_k = e^{\mu_k t}P_k$, see (14.13). $\square$

Example 14.8. Let $\Gamma := \{z \in \mathbb{C} : |z| = 1\}$ denote the unit circle and $\tau > 0$. On $X := L^p(\Gamma)$, $1 \leq p < \infty$, we define

$$R_\tau(t)f(z) := f\left(z e^{(2\pi i/\tau)t}\right), \quad z \in \Gamma, t \in \mathbb{R}.$$

Then $R_\tau(\cdot)$ defines a periodic C_0-group with period τ. Moreover, one can prove that its generator is given by

$$D(A) = \{f \in X : f \text{ absolutely continuous}, f' \in X\}$$
$$Af(z) = \frac{2\pi i}{\tau} z f'(z), \quad f \in D(A),$$

with

$$\sigma(A) = \frac{2\pi i}{\tau}\mathbb{Z} \quad and \quad P_k f(z) = \frac{z^k}{2\pi i}\int_\Gamma f(u) u^{-(k+1)} \, du.$$

14.3   Irreducible Semigroups

We now return to positive semigroups. The concept of irreducibility of bounded operators on a Banach lattice was already introduced in Definition 10.26. Let us restate it for operators forming a C_0-semigroup.

Definition 14.9. A positive C_0-semigroup $(T(t))_{t\geq 0}$ on a Banach lattice E is called *irreducible* if $\{0\}$ and E are the only closed ideals that are invariant under all the operators $T(t)$, $t \geq 0$.

The following result gives two properties equivalent to irreducibility that are sometimes easier to verify.

Proposition 14.10. *Let $(T(t))_{t\geq 0}$ be a positive C_0-semigroup on a Banach lattice E with generator A. The following assertions are equivalent.*

   (i) *$(T(t))_{t\geq 0}$ is irreducible.*

(ii) *For some (and then for every)* $\lambda > s(A)$, *there is no* $R(\lambda, A)$*-invariant closed ideal except* $\{0\}$ *and* E.

(iii) *For some (and then for every)* $\mu > s(A)$ *and for every* $f > 0$, $R(\mu, A)f$ *is a quasi-interior point of* E_+.

Proof. We prove first that if for some $\mu > s(A)$ there is no $R(\mu, A)$-invariant closed ideal except $\{0\}$ and E, then this holds for every $\lambda > s(A)$. Let I be a closed ideal of E such that $R(\mu, A)I \subset I$ for $\mu > s(A)$. The inequality

$$0 \le R(\lambda, A) \le R(\mu, A),$$

holding for all $\lambda \ge \mu$, and the definition of ideals imply that $R(\lambda, A)I \subset I$.

On the other hand, for $\mu - \frac{1}{r(R(\mu, A))} < \lambda < \mu$, we have $R(\lambda, A)I \subset I$, since

$$R(\lambda, A) = R(\mu, A) \sum_{k=0}^{\infty} ((\mu - \lambda)R(\mu, A))^k$$

(see (9.10)) and $R(\mu, A)I \subset I$.

Iteration of the argument establishes that $R(\lambda, A)I \subset I$ for every $\lambda > s(A)$. This proves the above claim.

(i) $\Longrightarrow$ (ii): Let $I \ne \{0\}$ be a closed ideal of E such that $R(\lambda, A)I \subset I$ for some (and then for every) $\lambda > s(A)$. By the approximation formula (see the proof of Theorem 11.1),

$$T(t)f = \lim_{k \to \infty} e^{tA_k} f, \quad f \in E,$$

where $A_k = kAR(k, A) \in \mathcal{L}(E)$ are the Yosida approximants, we obtain that $T(t)I \subset I$ for all $t > 0$, hence $I = E$.

(ii) $\Longrightarrow$ (i): This follows from Theorem 12.7.

(ii) $\Longrightarrow$ (iii): Let $\lambda > s(A)$, $0 \ne f \in E_+$ and consider the ideal generated by $R(\lambda, A)f$, i.e.,

$$E_{R(\lambda, A)f} := \bigcup_{k \in \mathbb{N}} [-kR(\lambda, A)f, kR(\lambda, A)f].$$

We infer from the resolvent equation that for $g \in E_{R(\lambda, A)f}$,

$$|R(\mu, A)g| \le R(\mu, A)|g| \le kR(\mu, A)R(\lambda, A)f \le \frac{k}{\mu - \lambda} R(\lambda, A)f$$

for $\mu > \lambda$. Hence, $R(\mu, A)g \in E_{R(\lambda, A)f}$ for any $g \in E_{R(\lambda, A)f}$ and $\mu > \lambda$. Thus, we see that $\overline{E_{R(\lambda, A)f}}$ is a nontrivial $R(\mu, A)$-invariant closed ideal and hence equals E. This means that $R(\lambda, A)f$ is a quasi-interior point of E_+.

(iii) $\Longrightarrow$ (ii): Let $I \ne \{0\}$ be an $R(\mu, A)$-invariant closed ideal for some $\mu > s(A)$, and let $0 \ne f \in E_+ \cap I$. It follows that for any $g \in E_{R(\mu, A)f}$ we have $|g| \le nR(\mu, A)f$ for some $n \in \mathbb{N}$ and hence $g \in I$. This implies that $E_{R(\mu, A)f} \subset I$ and, furthermore, $E = \overline{E_{R(\mu, A)f}} = I$. $\qquad\square$

Example 14.11. From the characterization of closed ideals given in Propositions 10.13, 10.14, and 10.15 (see also Examples 10.16) we obtain the following characterization of irreducible semigroups in certain function spaces.

a) Let $E := L^p(\Omega, \mu)$, $1 \le p < \infty$, and let $(T(t))_{t \ge 0}$ be a positive C_0-semigroup on E with generator A. Then, $(T(t))_{t \ge 0}$ is irreducible if and only if

$$0 \lneq f \in E \implies (R(\lambda, A)f)(s) > 0 \text{ for a.e. } s \in \Omega \text{ and some } \lambda > s(A).$$

b) Let $E := C_0(\Omega)$, where Ω is locally compact Hausdorff space, and let $(T(t))_{t \ge 0}$ be a positive C_0-semigroup on E with generator A. Then $(T(t))_{t \ge 0}$ is irreducible if and only if

$$0 \lneq f \in E \implies (R(\lambda, A)f)(s) > 0 \text{ for all } s \in \Omega \text{ and some } \lambda > s(A).$$

We collect here some properties of irreducible C_0-semigroups. Many of them resemble properties already observed in finite dimensions. The most import one is a generalization of the Perron–Frobenius theorem, Theorem 5.13 (see also the same result for matrix semigroups given in Theorem 7.6).

Proposition 14.12. *Assume that A is the generator of an irreducible C_0-semigroup $(T(t))_{t \ge 0}$ on a Banach lattice E. Then the following assertions hold.*

a) *Every positive eigenvector of A is a quasi-interior point.*
b) *Every positive eigenvector of A^* is strictly positive.*
c) *If $\ker(s(A) - A^*)$ contains a positive element, then $\dim \ker(s(A) - A) \le 1$.*
d) *If $s(A)$ is a pole of the resolvent, then it has algebraic (and geometric) multiplicity equal to 1. The corresponding residue has the form*

$$P_{s(A)} = u^* \otimes f,$$

where $f \in E$ is a strictly positive eigenvector of A, $u^* \in E^*$ is a strictly positive eigenvector of A^*, and $\langle f, u^* \rangle = 1$.

Proof. a) Let f be a positive eigenvector of A and λ its corresponding eigenvalue. Since $\lambda f = Af = \lim_{t \to 0^+} \frac{1}{t}(T(t)f - f)$, we have $\lambda \in \mathbb{R}$. We also have

$$f = (\mu - \lambda)R(\mu, A)f \quad \text{for } \mu > s(A) > \lambda.$$

Thus a) follows from Proposition 14.10.

b) Let f^* be a positive eigenvector of A^* and λ its corresponding eigenvalue. By the same argument as above, $\lambda \in \mathbb{R}$ and, by Corollary 9.32, $T(t)^* f^* = e^{\lambda t} f^*$ for $t \ge 0$. Hence,

$$\langle |T(t)u|, f^* \rangle \le \langle T(t)|u|, f^* \rangle = \langle |u|, e^{\lambda t} f^* \rangle, \quad u \in E, \, t \ge 0.$$

Thus $I := \{u \in E : \langle |u|, f^* \rangle = 0\}$ is a $(T(t))_{t \geq 0}$ invariant closed ideal. Since $f^* \neq 0$, we have $I \subsetneq E$, and so by irreducibility we obtain $I = \{0\}$. Therefore $f^* > 0$.

c) For $0 \lneqq f^* \in \ker(s(A) - A^*)$ we see from b) that f^* is strictly positive. Assume that $\ker(s(A) - A) \neq \{0\}$ and define the rescaled positive semigroup as

$$T_{-s(A)}(t)g := e^{-s(A)t} T(t)g,$$

see also Exercise 9.10.4. Then for $f \in \ker(s(A) - A)$ we have by Corollary 9.32 that $T_{-s(A)}(t)f = f$ and hence, by Lemma 10.18,

$$|f| = |T_{-s(A)}(t)f| \leq T_{-s(A)}(t)|f|, \quad t \geq 0.$$

Thus, for $t \geq 0$,

$$\begin{aligned} \langle |f|, f^* \rangle &\leq \langle T_{-s(A)}(t)|f|, f^* \rangle \\ &= \langle |f|, f^* \rangle. \end{aligned}$$

This implies that $\langle T_{-s(A)}(t)|f| - |f|, f^* \rangle = 0$, and since $f^* > 0$, we obtain $T_{-s(A)}(t)|f| = |f|$ for $t \geq 0$. Therefore,

$$|f| \in \ker(s(A) - A).$$

By Lemma 10.18, we also have $(T_{-s(A)}(t)f)^+ \leq T_{-s(A)}(t)f^+$ and $(T_{-s(A)}(t)f)^- \leq T_{-s(A)}(t)f^-$. By the same arguments as above, we obtain $f^+ \in \ker(s(A) - A)$ and $f^- \in \ker(s(A) - A)$. This implies that $F := E_{\mathbb{R}} \cap \ker(s(A) - A)$ is a real sublattice of E. For $f \in F$ we consider the ideal E_{f^+} (resp. E_{f^-}) generated by f^+ (resp. f^-). Then E_{f^+} and E_{f^-} are $T_{-s(A)}(t)$-invariant for all $t \geq 0$. Since E_{f^+} and E_{f^-} are orthogonal, see Proposition 10.4, the irreducibility of $(T_{-s(A)}(t))_{t \geq 0}$ implies that either $f^+ = 0$ or $f^- = 0$. Consequently, F is totally ordered and, by Lemma 10.10, we have

$$\dim F = \dim \ker(s(A) - A) = 1.$$

d) We claim first that, if $s(A)$ is a pole of the resolvent, then there is an eigenvector $0 \lneqq f \in E$ of A corresponding to $s(A)$. Indeed, let k be the pole order of $s(A)$ and

$$U_{-k} = \lim_{\lambda \to s(A)^+} (\lambda - s(A))^k R(\lambda, A),$$

see (14.9). Then $U_{-k} \neq 0$ and $U_{-(k+1)} = 0$. Moreover, by Corollary 12.10, we have $U_{-k} \geq 0$. Hence, there is $0 \leq g \in E$ with $f := U_{-k}g \gneqq 0$. By the relation $U_{-(k+1)} = (A - s(A))U_{-k} = 0$, we obtain $(A - s(A))f = 0$. This proves the claim.

We can now use a) to obtain $\overline{E_f} = E$. By taking the adjoint $U^*_{-(k+1)}$ of $U_{-(k+1)}$ and by the same computation as before, one deduces that there is $0 \lneqq f^* \in \ker(s(A) - A^*)$. So by c) we have $\dim \ker(s(A) - A) = 1$.

Assume now that $k \geq 2$. Then we have

$$
\begin{aligned}
\langle f, f^* \rangle &= \langle U_{-k}g, f^* \rangle \\
&= \langle g, U_{-k}^* f^* \rangle \\
&= \langle g, U_{-(k-1)}^* (A^* - s(A)) f^* \rangle = 0.
\end{aligned}
$$

Since $\overline{E_f} = E$, we infer that $\langle g, f^* \rangle = 0$ for all $g \in E_+$. This contradicts the assertion b), hence $k = 1$. From the inequality $m_{\mathrm{g}} + k - 1 \leq m_{\mathrm{a}} \leq m_{\mathrm{g}}k$, see (14.10), we further obtain

$$
m_{\mathrm{a}} = m_{\mathrm{g}} = \dim P_{\mathrm{s}(A)}E = \dim \ker(\mathrm{s}(A) - A) = 1,
$$

and

$$
P_{\mathrm{s}(A)}E = \ker(\mathrm{s}(A) - A),
$$

where we recall that $P_{\mathrm{s}(A)} = U_{-1}$.

We now show the last part of assertion d). To this end, let

$$
0 \lneqq f \in \ker(\mathrm{s}(A) - A).
$$

Without loss of generality we suppose that $\|f\| = 1$. Then $P_{\mathrm{s}(A)}E = \mathrm{span}\{f\}$, i.e., for every $g \in E$ there is a $\lambda \in \mathbb{C}$ such that $P_{\mathrm{s}(A)}g = \lambda f$. By the Hahn–Banach theorem (see Exercise 10.9.6), there exists

$$
0 \leq g^* \in (\ker(\mathrm{s}(A) - A))^* \text{ with } \|g^*\| = 1 \text{ and } \langle f, g^* \rangle = \|f\| = 1.
$$

Hence,

$$
\langle P_{\mathrm{s}(A)}g, g^* \rangle = \lambda = \langle g, P_{\mathrm{s}(A)}^* g^* \rangle.
$$

Putting $u^* := P_{\mathrm{s}(A)}^* g^* \geq 0$, we obtain $P_{\mathrm{s}(A)} = u^* \otimes f$ and $\langle f, u^* \rangle = \langle P_{\mathrm{s}(A)}f, g^* \rangle = \langle f, g^* \rangle = 1$. Moreover, $0 \lneqq u^* \in P_{\mathrm{s}(A)}^* E^* \subseteq \ker(\mathrm{s}(A) - A^*)$, so $u^* > 0$ by b).

In the proof of c) we have seen that for every $g \in \ker(\mathrm{s}(A) - A)$ we have either $g^+ = 0$, or $g^- = 0$. So, we may assume that our eigenvector f is strictly positive. This ends the proof of the proposition. $\qquad\square$

Now we study the boundary spectrum of irreducible semigroups on Banach lattices. The results resemble the properties of imprimitive matrices obtained in Chapter 5.

Before going on, we need some auxiliary results on the structure of Banach lattices and their quasi-interior points. The following result, due to Kakutani, shows that for every $e \in E_+$ the generated ideal satisfies $E_e \cong \mathrm{C}(K)$ for some compact Hausdorff space K. Here, E_e is equipped with the norm

$$
\|f\|_e := \inf\{\lambda > 0 : f \in \lambda[-e, e]\}, \quad f \in E_e.
$$

We recall that $T \in \mathcal{L}(E, F)$ is called a *lattice homomorphism* if $|Tf| = T|f|$ for every $f \in E$, where F is a complex Banach lattice (see Definition 10.19).

Theorem 14.13 (Kakutani). *Let $e \in E_+$ and let E_e be the ideal generated by $\{e\}$. Further, take $B := \{f^* \in (E_e)^*_+ : \langle e, f^* \rangle = 1\}$ and denote by K the set of all extreme points of B. Then K is $\sigma(E^*, E)$-compact and the mapping*

$$U_e : E_e \ni f \longmapsto \varphi_f \in C(K), \quad \varphi_f(f^*) = \langle f, f^* \rangle, f^* \in K,$$

is an isometric lattice isomorphism.

Now, if $|h|$ is a quasi-interior point of E_+, then $E_{|h|}$ is a dense subspace of E, isomorphic to a space of continuous functions $C(K)$ on some K. Let $U_{|h|}$ be the lattice isomorphism obtained from Kakutani's theorem and let $\widetilde{h} := U_{|h|}h$. Then $|\widetilde{h}| = U_{|h|}|h| = \mathbb{1}$. Consider the operator

$$\widetilde{S}_0 : C(K) \longrightarrow C(K), \quad f \longmapsto (\text{sign}\,\widetilde{h})f := \frac{\widetilde{h}}{|\widetilde{h}|}f = \widetilde{h}f, \qquad (14.17)$$

and put $S_h := U_{|h|}^{-1}\widetilde{S}_0 U_{|h|}$. Then S_h is a linear mapping from $E_{|h|}$ into itself satisfying

a) $S_h\overline{h} = |h|$, where $\overline{h} = \text{Re}\,h - i\,\text{Im}\,h$,

b) $|S_h f| \leq |f|$ for every $f \in E_{|h|}$.

Since b) implies the continuity of S_h for the norm induced by E and $|h|$ is a quasi-interior point of E_+, S_h can be uniquely extended to E. This extension is also denoted by S_h and is called the *signum operator* with respect to h.

In the following we generalize Wielandt's lemma (see Lemma 5.18) and its consequences to irreducible semigroups on Banach lattices.

Lemma 14.14. *Let E be a Banach lattice and $|h|$ a quasi-interior point of E_+. Suppose that for $T, R \in \mathcal{L}(E)$ we have $Rh = h$, $T|h| = |h|$, and $|Rg| \leq T|g|$ for all $g \in E$. Then $T = S_h^{-1}RS_h$, where S_h is the signum operator.*

Proof. First observe that for $g \in E_+$ we have

$$Tg = T|g| \geq |Rg| \geq 0,$$

so T is a positive operator. Since $T|h| = |h|$, the ideal $E_{|h|}$ is T-and R-invariant. Consider the operators $\widetilde{T} := U_{|h|}TU_{|h|}^{-1}$ and $\widetilde{R} := U_{|h|}RU_{|h|}^{-1}$, and put $\widetilde{h} := U_{|h|}h$. We then have

$$\widetilde{R}\widetilde{h} = \widetilde{h}, \quad \widetilde{T}\mathbb{1} = \mathbb{1}, \quad \text{and} \quad |\widetilde{R}f| \leq \widetilde{T}|f| \text{ for all } f \in C(K). \qquad (14.18)$$

Define $T_1 := \widetilde{S}_0^{-1}\widetilde{R}\widetilde{S}_0$, where $\widetilde{S}_0$ is the multiplication operator by $\widetilde{h}$ on $C(K)$ defined in (14.17). By (14.18), we have

$$T_1\mathbb{1} = \mathbb{1} \quad \text{and}$$
$$|T_1 f| = |\widetilde{S}_0^{-1}\widetilde{R}\widetilde{S}_0 f| = |\widetilde{R}\widetilde{S}_0 f| \leq \widetilde{T}|\widetilde{S}_0 f| = \widetilde{T}|f| \qquad (14.19)$$

for all $f \in C(K)$. Hence, $\|T_1\| \leq \|\widetilde{T}\| = \|\widetilde{T}\mathbb{1}\|_\infty = 1$. So by Lemma 10.27, T_1 is a positive operator and (14.19) implies that $0 \leq T_1 \leq \widetilde{T}$. Therefore,

$$\|\widetilde{T} - T_1\| = \|(\widetilde{T} - T_1)\mathbb{1}\|_\infty = 0,$$

thus $T_1 = \widetilde{T}$ and hence $T = S_h^{-1} R S_h$. $\square$

The following result describes the eigenvalues of an irreducible semigroup which are contained in the boundary spectrum $\sigma_b(A) = \{\lambda \in \sigma(A) : \mathrm{Re}\,\lambda = s(A)\}$, where A is the corresponding generator. Compare this with Theorem 5.19 for finite imprimitive matrices.

Proposition 14.15. *Let $(T(t))_{t \geq 0}$ be an irreducible C_0-semigroup with generator A on a Banach lattice E. Assume that $s(A) = 0$ and there is $0 \lneq f^* \in D(A^*)$ with $A^* f^* = 0$. If $\sigma_p(A) \cap i\mathbb{R} \neq \emptyset$, then the following assertions hold.*

a) *For $0 \neq h \in D(A)$ and $\alpha \in \mathbb{R}$ with $Ah = i\alpha h$, $|h| \in \ker A$ is a quasi-interior point,*

$$S_h(D(A)) = D(A), \quad and \quad S_h^{-1} A S_h = A + i\alpha,$$

where S_h is the signum operator defined above.

b) *$\dim \ker(\lambda - A) = 1$ for every $\lambda \in \sigma_p(A) \cap i\mathbb{R}$.*

c) *$\sigma_p(A) \cap i\mathbb{R}$ is an additive subgroup of $i\mathbb{R}$.*

d) *0 is the only eigenvalue of A admitting a positive eigenvector.*

Proof. We first remark that by Proposition 14.12.b) we have $f^* > 0$, and by Corollary 9.32, $f^* = T(t)^* f^*$ for all $t \geq 0$.

a) Assume that $Ah = i\alpha h$ for some $0 \neq h \in D(A)$ and $\alpha \in \mathbb{R}$. Then, by Corollary 9.32, $T(t)h = e^{i\alpha t}h$ and hence $|h| = |T(t)h| \leq T(t)|h|$. This implies that

$$T(t)|h| - |h| \geq 0$$

for every $t \geq 0$. On the other hand,

$$\langle T(t)|h| - |h|, f^* \rangle = \langle |h|, T(t)^* f^* \rangle - \langle |h|, f^* \rangle = 0$$

for all $t \geq 0$. Since $f^* > 0$, we obtain $T(t)|h| = |h|$ for every $t \geq 0$, which implies that $A|h| = 0$. So, by Proposition 14.12.a), the vector $|h|$ is a quasi-interior point. If we set

$$T_\alpha(t) := e^{-i\alpha t} T(t)$$

for $t \geq 0$, then $T(t)$ and $T_\alpha(t)$ satisfy the assumptions of Lemma 14.14 and hence

$$T(t) = S_h^{-1} T_\alpha(t) S_h, \quad t \geq 0.$$

Therefore, $S_h(D(A)) = D(A)$ and $A = S_h^{-1}(A - i\alpha)S_h$.

b) The calculations in the proof of a) imply that $\ker A \neq \{0\}$ and $\dim \ker(i\alpha - A) = \dim \ker A$, so Proposition 14.12.c) yields the claim.

c) Let $\alpha, \beta \in \mathbb{R}$ be such that $Ah = i\alpha h$ and $Ag = i\beta g$ for some $0 \neq h, g \in D(A)$. By a), we have

$$S_h^{-1}AS_h = A + i\alpha \quad \text{and} \quad S_g^{-1}AS_g = A + i\beta.$$

Thus

$$A + i(\alpha + \beta) = S_h(A + i\beta)S_h^{-1} = S_h S_g^{-1} A S_g S_h^{-1},$$

which implies that

$$\ker(A + i(\alpha + \beta)) = S_h S_g^{-1} \ker A \neq \{0\}.$$

Therefore $i(\alpha + \beta) \in \sigma_p(A) \cap i\mathbb{R}$.

d) If $Af = \lambda f$, where $0 \lneqq f \in D(A)$, then

$$\lambda\langle f, f^*\rangle = \langle Af, f^*\rangle = \langle f, A^* f^*\rangle = 0.$$

Since $f^* > 0$, we see that $\langle f, f^*\rangle > 0$. Hence, $\lambda = 0$. $\qquad\square$

The following result, which we recall without proof, states that the boundary spectrum of the generator A of an irreducible C_0-semigroup $(T(t))_{t\geq 0}$ on a Banach lattice E is always contained in the point spectrum $\sigma_p(A)$ if $s(A)$ is a pole of $R(\cdot, A)$.

Lemma 14.16. *Let A be the generator of an irreducible C_0-semigroup $(T(t))_{t\geq 0}$ on a Banach lattice E. If $s(A)$ is a pole of $R(\cdot, A)$, then $\sigma_b(A) \subset \sigma_p(A)$.*

As a consequence, we obtain the following description of the boundary spectrum of irreducible semigroups.

Theorem 14.17. *Let $(T(t))_{t\geq 0}$ be an irreducible C_0-semigroup with generator A on a Banach lattice E and assume that $s(A)$ is a pole of the resolvent. Then there is $\alpha \geq 0$ such that*

$$\sigma_b(A) = s(A) + i\alpha\mathbb{Z}.$$

Moreover, $\sigma_b(A)$ consists of simple poles.

Proof. Without loss of generality we may suppose that $s(A) = 0$. It can be shown that $\sigma_b(A) \subseteq \sigma_p(A)$, see Lemma 14.16. Hence

$$\sigma_b(A) = \sigma_p(A) \cap i\mathbb{R}.$$

Proposition 14.12.d) yields the existence of a positive eigenvector $f^* \in D(A^*)$ corresponding to the eigenvalue $s(A) = 0$. Proposition 14.15.c) implies that $\sigma_b(A)$ is a subgroup of $(i\mathbb{R}, +)$. Since $\sigma_b(A)$ is closed and $s(A) - 0$ is an isolated point, we have

$$\sigma_b(A) = i\alpha\mathbb{Z}$$

for some $\alpha \geq 0$. Proposition 14.12.d) implies that 0 is a simple pole and by Proposition 14.15.a) we have, for $\lambda \in \rho(A)$,

$$R(\lambda + ik\alpha, A) = S_h^k R(\lambda, A) S_h^{-k}$$

for all $k \in \mathbb{Z}$. Therefore, $ik\alpha$ is a simple pole for each $k \in \mathbb{Z}$. This ends the proof of the theorem. $\qquad\qquad\qquad\qquad\qquad\qquad\qquad\qquad\qquad\qquad\qquad\qquad\qquad\square$

14.4   Asymptotic Behavior

In many concrete examples one can observe some regularity in the long-term behavior of the orbits of a semigroup. We will encounter two types of such behavior that are interesting for applications: balanced exponential growth, and asymptotic periodicity.

Let us start with the first kind of behavior. We say that a semigroup $(T(t))_{t \geq 0}$ with a generator A possesses a *balanced exponential growth* if there are a rank-one projection P and constants $\varepsilon > 0$ and $M \geq 1$ such that

$$\|e^{-s(A)t}T(t) - P\| \leq Me^{-\varepsilon t} \quad \text{for all } t \geq 0.$$

We will present an example of such a semigroup in Chapter 17, see also Exercise 2. Using our spectral results, we can prove such behavior for certain class of irreducible semigroups.

Theorem 14.18. *Let $(T(t))_{t \geq 0}$ be an irreducible C_0-semigroup with the generator A on a Banach lattice E. If $\omega_{\mathrm{ess}}(T) < \omega_0(T)$, then there exist a quasi-interior point $0 \leq f \in E$ and $0 < f^* \in E^*$ with $\langle f, f^* \rangle = 1$ such that*

$$\|e^{-s(A)t}T(t) - f^* \otimes f\| \leq Me^{-\varepsilon t} \quad \text{for all } t \geq 0$$

and appropriate constants $M \geq 1$ and $\varepsilon > 0$.

Proof. Since $\omega_{\mathrm{ess}}(T) < \omega_0(T)$, Proposition 14.3 implies that $s(A) = \omega_0(T)$. On the other hand, $\omega_{\mathrm{ess}}(T) < \omega_0(T)$ implies that $r_{\mathrm{ess}}(T(1)) < r(T(1))$. Hence, by Proposition A.34, $r(T(1))$ is a pole of the resolvent of $T(1)$ and thus $\omega_0(T) = s(A)$ is a pole of $R(\cdot, A)$.

Now, by Theorem 14.17, there exists $\alpha \geq 0$ such that $\sigma_{\mathrm{b}}(A) = s(A) + i\alpha\mathbb{Z}$ and therefore $\sigma_{\mathrm{b}}(A - \omega_0(T)) = i\alpha\mathbb{Z}$, where $A - \omega_0(T)$ is the generator of the rescaled semigroup

$$T_{-\omega_0(T)}(t) := e^{-\omega_0(T)t}T(t), \quad t \geq 0.$$

Since

$$\omega_{\mathrm{ess}}\left(T_{-\omega_0(T)}\right) = \omega_{\mathrm{ess}}(T) - \omega_0(T) < 0 \quad \text{and} \quad \omega_0\left(T_{-\omega_0(T)}\right) = 0,$$

we have, by Theorem 14.4, that the set

$$\{\lambda \in \sigma(A - \omega_0(T)) : \operatorname{Re}\lambda \geq 0\} = \{\lambda \in \sigma(A - \omega_0(T)) : \operatorname{Re}\lambda = 0\} = \sigma_{\mathrm{b}}(A - \omega_0(T))$$

is finite. Therefore, $\sigma_b(A - \omega_0(T)) = \{0\}$. The theorem is now proved by applying Theorem 14.4 and Proposition 14.12 to the rescaled semigroup $(T_{-\omega_0(T)}(t))_{t \geq 0}$. $\qquad\square$

Without the quasi-compactness assumption for the rescaled semigroup, i.e., $\omega_{\text{ess}}(T) < \omega_0(T)$, one obtains that the semigroup $(T(t))_{t \geq 0}$ behaves in the long run like a rotation group. Here we assume $s(A) > -\infty$. So, by considering the rescaled semigroup $(e^{-s(A)t}T(t))_{t \geq 0}$ instead of $(T(t))_{t \geq 0}$, one may without loss of generality assume $s(A) = 0$. Compare the following theorem with Definition 4.14 and Theorem 4.15 for the finite-dimensional case.

Theorem 14.19. *Let $(T(t))_{t \geq 0}$ be a bounded and irreducible C_0-semigroup with the generator A on a Banach lattice $E := L^p(\Omega, \mu)$, $1 \leq p < \infty$. If $s(A) = 0$ is a pole of $R(\cdot, A)$ and there is $\xi \in \mathbb{R}$ such that $i\xi \in \sigma(A)$, then there exists a positive projection P commuting with $(T(t))_{t \geq 0}$ such that the following holds.*

a) *We have*

$$E = \operatorname{im} P \oplus \ker P, \quad T(t) = T_r(t) \oplus T_s(t), \quad t \geq 0, \quad and \quad A = A_r \oplus A_s,$$

corresponding to the decomposition $\sigma(A) = \sigma_r \cup \sigma_s$, where $\sigma_r = i\alpha\mathbb{Z}$ and $\sigma_s = \sigma(A) \setminus \sigma_r$ for some $\alpha > 0$.

b) *The subspace $\operatorname{im} P$ is a closed sublattice of E and $(T_r(t))_{t \geq 0}$ is a periodic and irreducible C_0-semigroup on $\operatorname{im} P$.*

c) *For every $f \in E$ we have*

$$\lim_{t \to \infty} \|T(t)f - T_r(t)f\| = 0.$$

Proof. First observe that by Theorem 14.17 we have

$$\sigma_b(A) = \sigma(A) \cap i\mathbb{R} = i\alpha\mathbb{Z} \quad \text{for some } \alpha > 0. \tag{14.20}$$

Next, from Proposition 14.15.a) and its proof, we see that there is a quasi-interior point $h \in E_+$ which is also a fixed point of $T(t)$. Hence, $|T(t)f| \leq T(t)h = h$ for all $f \in [-h, h]$. Since h is a quasi-interior point and order intervals in E are weakly compact, $(T(t))_{t \geq 0}$ is relatively weakly compact. Thus, by the Jacobs–de Leeuw–Glicksberg splitting theorem, see Theorem A.39, there is a projection $P \in \mathcal{L}(E)$ commuting with $T(t)$ such that $E = \operatorname{im} P \oplus \ker P$. Moreover,

$$\operatorname{im} P = \overline{\operatorname{span}}\{f \in D(A) : \exists k \in \mathbb{Z} \text{ such that } Af = i\alpha k f\}$$

and

$$\ker P = \{f \in E : 0 \text{ belongs to the weak closure of } \{T(t)f : t \geq 0\}\}.$$

Furthermore, from (14.20) and Proposition A.40, it follows that

$$\ker P = \{f \in E : \lim_{t \to \infty} \|T(t)f\| = 0\}. \tag{14.21}$$

Since P commutes with each $T(t)$, it splits $(T(t))_{t \geq 0}$ into $(T_{\mathrm{r}}(t))_{t \geq 0}$ on $\mathrm{im}\, P$ and $(T_{\mathrm{s}}(t))_{t \geq 0}$ on $\ker P$. Moreover, by Corollary 9.32, $T\left(\frac{2\pi}{\alpha}\right) f = f$ for all $f \in D(A)$ such that $Af = i\alpha k f$ for some $k \in \mathbb{Z}$. Hence, $T_{\mathrm{r}}\left(\frac{2\pi}{\alpha}\right) = I$ and $T_{\mathrm{r}}(\cdot)$ is a periodic C_0-semigroup on $\mathrm{im}\, P$.

Theorem A.39 tells us that P belongs to the weak closure of $(T(t))_{t \geq 0}$. Since $(T(t))_{t \geq 0}$ is irreducible, we see that $Pf \gneq 0$ whenever $f \gneq 0$. So, by Lemma A.41, $\mathrm{im}\, P$ is a closed sublattice of E. This and the irreducibility of $(T(t))_{t \geq 0}$ imply that $(T_{\mathrm{r}}(t))_{t \geq 0}$ is a periodic and irreducible C_0-semigroup on $\mathrm{im}\, P$. Denote its generator by A_{r}. Then, by Theorem 14.6 and equation (14.14), we have $\sigma(A_{\mathrm{r}}) = i\alpha\mathbb{Z}$.

The family $T_{\mathrm{s}}(t) := T(t)|_{\ker P}$, $t \geq 0$, defines a C_0-semigroup on $\ker P$. We denote its generator by A_{s}. Then, by the spectral decomposition, we have $\sigma(A_{\mathrm{s}}) = \sigma(A) \setminus i\alpha\mathbb{Z}$. This ends the proof of assertions a) and b). Assertion c) follows from (14.21). $\qquad\square$

Remark 14.20.

a) Denote by G the closure of the set $\{T_r(t) : t \geq 0\}$ in the weak operator topology. Using abstract results from harmonic analysis and Theorem 14.19, one can prove that $\mathrm{im}\, P$ is lattice isomorphic to an $(R_\tau(t))_{t \in \mathbb{R}}$-invariant Banach function space $\mathcal{C}$ on G satisfying $C(G) \subset \mathcal{C} \subset \mathrm{L}^1(G, m)$ such that $(T_r(t))_{t \in \mathbb{R}}$ is similar to the group induced by $(R_\tau(t))_{t \in \mathbb{R}}$ on $\mathcal{C}$. Here m is the Haar measure on G and $(R_\tau(t))_{t \in \mathbb{R}}$ is the rotation group defined in Example 14.8 with period $\tau = \frac{2\pi}{\alpha}$. Moreover, if $E = \mathrm{L}^1(\Omega, \mu)$, then $\mathcal{C}$ can be identified with $L^1(G)$.

b) It can be seen that Theorem 14.19 holds if E is any Banach lattice with *order continuous norm*, see Remark 13.12 for the definition. In fact, it holds that a Banach lattice E has order continuous norm if and only if every order interval in E is weakly compact. This gives the weak compactness needed in the proof above.

c) One obtains from the proof above that $\mathrm{s}(A_{\mathrm{s}}) < 0$.

An example of a C_0-semigroup that behaves asymptotically periodic will be presented in Chapter 18.

14.5   Notes and Remarks

For the general spectral theory of operators we refer to monographs by Kato [73] or by Gohberg, Goldberg and Kaashoek [53]. More on spectral theory of irreducible semigroups can be found in the monograph edited by Nagel [101, Section B-III.3].

Kakutani's theorem, Theorem 14.13, originates from Kakutani [70]. We cited it from Meyer-Nieberg [95, Theorem 2.1.3], where you can also find a proof.

Concerning Lemma 14.14 and signum operators we refer to Nagel (ed.) [101, Chapter B-III]. For the proof of Lemma 14.16 see [101, p. 315].

Remark 14.20.a) is connected to abstract Halmos–von Neumann type theorems. We refer to Schaefer [126, Section III.10] for the corresponding abstract results. See also Keicher and Nagel [74]. For Remark 14.20.b) see Meyer-Nieberg [95, Theorem 2.4.2].

14.6 Exercises

1. Let $(T(t))_{t\geq 0}$ be a C_0-semigroup on a Banach space X. Prove that the following assertions are equivalent.

   (i) $\omega_{\mathrm{ess}}(T) < 0$.

   (ii) $\|T(t_0) - K\| < 1$ for some $t_0 > 0$ and $K \in \mathcal{L}(X)$ compact.

2. On the Banach space $C(K)$ with $K = [-\infty, 0]$ consider the operator

   $$Af = f' + mf$$
   $$D(A) = \{f \in C(K) : f \text{ is differentiable, } f' \in C(K) \text{ and } f'(0) = Lf\},$$

   where $m \in C(K)$ is real-valued and $L : C(K) \to \mathbb{R}$ a continuous linear form.

   a) Show that A generates a C_0-semigroup $(T(t))_{t\geq 0}$ on $C(K)$.

   b) Prove that $(T(t))_{t\geq 0}$ is given by

   $$T(t)f(s) = e^{\int_s^0 m(\nu)d\nu}\left(e^{(s+t)m(0)}f(0) + \int_0^{t+s} e^{\tau m(0)}LT(s+t-\tau)fd\tau\right)$$

   for $s + t > 0$ and

   $$T(t)f(s) = e^{\int_s^{t+s} m(\tau)d\tau}f(t+s)$$

   for $s + t \leq 0$.

   c) Using Exercise 1, prove that $\omega_{\mathrm{ess}}(T) < 0$ provided that $m(-\infty) < 0$.

   d) Describe the asymptotic behavior of $(T(t))_{t\geq 0}$.

3. Consider the transport operator

   $$D(A) = \left\{f \in L^1(I \times V) : v\frac{\partial f}{\partial x} \in L^1(I \times V) \text{ and } \begin{cases} f(0, v) = 0 \text{ if } v > 0, \\ f(1, v) = 0 \text{ if } v < 0, \end{cases}\right\},$$
   $$(Af)(x, v) = -v\frac{\partial f}{\partial x}(x, v),$$

   where $I = [0, 1]$ and $V = \{v \in \mathbb{R} : 1 \leq |v| \leq 2\}$. Prove that A generates a reducible C_0-semigroup on $L^1(I \times V)$.

4. On the Banach lattice $C([0, 1])$ consider the Laplace operator with Neumann boundary conditions:

   $$(Af)(x) = f''(x), \quad x \in [0, 1],$$
   $$f \in D(A) = \{f \in C^2([0, 1]) : f'(0) = f'(1) = 0\}.$$

   Prove that A generates an irreducible C_0-semigroup on $C[0, 1]$.

5. On $E = L^1([-1, 0])$ and for $0 \leq g \in L^\infty[-1, 0]$ define the operator

$$Af := f', \quad D(A) = \left\{ f \in E : f' \in E \text{ and } f(0) = \int_{-1}^0 f(s)g(s) \, ds \right\}.$$

a) Show that A generates a positive C_0-semigroup $(T(t))_{t \geq 0}$ on E.

b) Prove that $(T(t))_{t \geq 0}$ is reducible if and only if there exists $\varepsilon > 0$ such that g vanishes a.e. on $[-1, -1 + \varepsilon]$.

Chapter 15

Positivity and Delay Equations

We have seen that, in general, the growth bound $\omega_0(T)$ of a C_0-semigroup $(T(t))_{t\geq 0}$ and the spectral bound $s(A)$ of its generator A do not coincide, even if positivity is assumed. It turns out that in Hilbert spaces a deeper analysis is possible using the boundedness of the resolvent. This has the consequence that for a positive semigroup $(T(t))_{t\geq 0}$ on a Hilbert space the equality $s(A) = \omega_0(T)$ holds. This is the most important result of Section 15.2.

We start with an abstract approach to delay equations and apply the above-mentioned result to positive delay equations in Hilbert spaces. This yields a nice characterization of stability.

Throughout this chapter, we assume the reader to be familiar with the basics of measure theory, vector-valued integrals, and the vector-valued Fourier transform. We collected the needed results in Appendix A.10.

15.1 Abstract Delay Equations

Let X be a Banach space and let B generate a strongly continuous semigroup $(T(t))_{t\geq 0}$ on X. Suppose that we are given an operator-valued function of bounded variation

$$\eta : [-1, 0] \longrightarrow \mathcal{L}(X).$$

We consider the abstract delay equation of the form[12]

$$\begin{cases} \dot{u}(t) = Bu(t) + \displaystyle\int_{-1}^{0} \mathrm{d}\eta(s)u(t+s), & s \geq 0, \\ u(s) = \varphi(s), & s \in [-1, 0], \end{cases} \tag{15.1}$$

where $\varphi : [-1, 0] \to X$ is some given function.

For a simple situation see Example 11.14.

---

[12]Note that $\eta(s)$ is an operator for every s, and we write operators to the left when acting on vectors. This is the reason for the unusual form of the integral in equation (15.1).

Remark 15.1. For simplicity we assume that the delay takes place on the interval $[-1, 0]$. All what follows remains valid if we consider other finite delay intervals.

To write (15.1) as an abstract Cauchy problem, we have to choose a function space. As usual, the initial value function guides us to choose this space. A common choice is to suppose that $\varphi \in C([-1, 0], X)$. This, however, is not practical for applications to partial differential equations with delay. Hence, we choose[13] $\varphi \in L^p([-1, 0], X)$ for some $1 \le p < \infty$.

Example 15.2. If we take $C \in \mathcal{L}(X)$ and consider

$$\eta(s) = \begin{cases} C & \text{if } s = -1, \\ 0 & \text{if } s \in (-1, 0], \end{cases}$$

then we obtain

$$\int_{-1}^0 d\eta(s) u(t+s) = Cu(t-1),$$

which leads us to the standard delay equation

$$\dot{u}(t) = Bu(t) + Cu(t-1). \tag{15.2}$$

Consider now the special case $B = 0$. Suppose that the function $\varphi \in L^p([-1, 0], X)$ is even continuous. Then the formula

$$u(t) = \varphi(0) + \int_0^t C\varphi(s-1)\, ds$$

gives us the solution for $t \in (0, 1]$. This shows that in order to obtain a solution, it does not suffice to specify $\varphi \in L^p([-1, 0], X)$ (which is an equivalence class of functions). We also need the value $\varphi(0)$.

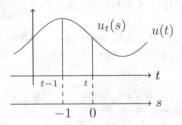

Figure 15.1: The history function $u_t(\cdot)$.

Now we can indeed rewrite the delay equation (15.1) as an abstract Cauchy problem. To do so, we first introduce the *history function*

$$u_t : [-1, 0] \longrightarrow X, \qquad u_t(s) := u(t+s),$$

---

[13]Bochner integrals are needed to define the spaces $L^p([-1, 0], X)$. The necessary ingredients are collected in Appendix A.10.

see Figure 15.1, and the *delay operator*

$$\Phi\varphi := \int_{-1}^{0} d\eta(s)\varphi(s).$$

For $\varphi = u_t$ this yields

$$\int_{-1}^{0} d\eta(s)u(t+s) = \Phi u_t.$$

The abstract delay equation (15.1) thus takes the form

$$\begin{cases} \dot{u}(t) = Bu(t) + \Phi u_t, & t \geq 0, \\ u(0) = f \in X, \\ u_0 = \varphi \in L^p([-1,0], X). \end{cases} \tag{15.3}$$

Next we introduce the product space $M_p = X \times L^p([-1,0], X)$ and the function

$$\mathcal{U} : t \longmapsto \left(\begin{smallmatrix} u(t) \\ u_t \end{smallmatrix}\right) \in M_p.$$

Further, on M_p we define the operator

$$A := \begin{pmatrix} B & \Phi \\ 0 & D \end{pmatrix}$$

with domain

$$D(A) := \left\{ \begin{pmatrix} f \\ \varphi \end{pmatrix} \in D(B) \times W^{1,p}([-1,0], X) : \varphi(0) = f \right\}.$$

Here D denotes the operators of weak differentiation on $L^p([-1,0], X)$. With all these spaces and operators we transform (15.1) into the abstract Cauchy problem

$$\begin{cases} \dot{\mathcal{U}}(t) = A\mathcal{U}(t), & t \geq 0, \\ \mathcal{U}(0) = \begin{pmatrix} f \\ \varphi \end{pmatrix} \end{cases} \tag{15.4}$$

on M_p. It is not to difficult to show that equation (15.4) is equivalent to the original Cauchy problem, see also Example 11.14. We accept this fact here and go on analysing the operator A.

To prove that this operator matrix generates a C_0-semigroup, we split it as

$$A := \begin{pmatrix} B & \Phi \\ 0 & D \end{pmatrix} = \begin{pmatrix} B & 0 \\ 0 & D \end{pmatrix} + \begin{pmatrix} 0 & \Phi \\ 0 & 0 \end{pmatrix} =: A_0 + A_1 \tag{15.5}$$

and apply a perturbation argument. To this end we need the following.

Proposition 15.3. *Let B generate the C_0-semigroup $(T(t))_{t\geq 0}$ on X. Then the operator A_0 defined in (15.5) with domain $D(A_0) := D(A)$ generates a C_0-semigroup given by the formula*

$$\mathcal{T}_0(t) := \begin{pmatrix} T(t) & 0 \\ T_t & L(t) \end{pmatrix}, \tag{15.6}$$

where $(L(t))_{t\geq 0}$ is the nilpotent left-shift semigroup[14] on $\mathrm{L}^p([-1,0],X)$ with generator D_0, and $T_t : X \to \mathrm{L}^p([-1,0],X)$ is defined by

$$(T_t\, f)(\tau) := \begin{cases} T(t+\tau)f & \text{if } -t < \tau \leq 0, \\ 0 & \text{if } -1 \leq \tau \leq -t. \end{cases}$$

Proof. First we show that $(\mathcal{T}_0(t))_{t\geq 0}$ is a strongly continuous semigroup on M_p. To do so, it suffices to show the strong continuity of the map $t \mapsto T_t$. Fix $f \in X$ and $0 \leq s \leq t$, and consider the limit

$$\lim_{t\to s} \|T_t f - T_s f\|_{\mathrm{L}^p}^p = \lim_{t\to s} \int_{-1}^{0} \|(T_t f)(\sigma) - (T_s f)(\sigma)\|^p \mathrm{d}\sigma$$

$$= \lim_{t\to s} \int_{-s}^{0} \|T(t+\sigma)\,f - T(s+\sigma)\,f\|^p \mathrm{d}\sigma$$

$$+ \lim_{t\to s} \int_{-t}^{-s} \|T(t+\sigma)\,f\|^p \mathrm{d}\sigma,$$

which is equal to 0 by Lebesgue's dominated convergence theorem (see Theorem A.23). Hence, the map $t \mapsto T_t$ is strongly right continuous, and analogously one can show that it is also strongly left continuous.

Next we show the semigroup property, and compute first the product

$$\begin{pmatrix} T(t) & 0 \\ T_t & L(t) \end{pmatrix} \begin{pmatrix} T(s) & 0 \\ T_s & L(s) \end{pmatrix} = \begin{pmatrix} T(t)T(s) & 0 \\ T_t T(s) + L(t)T_s & L(t)L(s) \end{pmatrix}$$

for $s, t \geq 0$. By the definition of the operators T_t, we obtain the identity

$$(T_t T(s)\, f)(\tau) = \begin{cases} T(t+\tau)T(s)f & \text{if } t+\tau > 0, \\ 0 & \text{if } t+\tau \leq 0. \end{cases}$$

Similarly,

$$(L(t)T_s\, f)(\tau) = \begin{cases} (T_s f)(t+\tau) & \text{if } t+\tau \leq 0, \\ 0 & \text{if } t+\tau > 0, \end{cases}$$

$$= \begin{cases} T(s+t+\tau)f & \text{if } t+\tau \leq 0 \text{ and } s+t+\tau > 0, \\ 0 & \text{if } t+\tau \leq 0 \text{ and } s+t+\tau \leq 0, \\ 0 & \text{if } t+\tau > 0. \end{cases}$$

---

[14]See Example 12.13

Combining these expressions yields

$$(T_t T(s) + L(t) T_s) f(\tau) = \begin{cases} T(s + t + \tau) f & \text{if } t + \tau > 0, \\ T(s + t + \tau) f & \text{if } t + \tau \leq 0 \text{ and } s + t + \tau > 0, \\ 0 & \text{if } s + t + \tau \leq 0, \end{cases}$$
$$= (T_{s+t} f)(\tau)$$

for each $\tau \in [-1, 0]$. Hence, $(\mathcal{T}_0(t))_{t \geq 0}$ is a strongly continuous semigroup on the product space M_p.

Let us denote its generator by A_2 and show that $A_2 = A_0$. First we compute the resolvent of A_2 using Laplace transforms and obtain

$$R(\lambda, A_2) \begin{pmatrix} f \\ \varphi \end{pmatrix} = \int_0^\infty e^{-\lambda s} \mathcal{T}_0(s) \begin{pmatrix} f \\ \varphi \end{pmatrix} ds$$

$$= \int_0^\infty e^{-\lambda s} \begin{pmatrix} T(s) f \\ T_s f + L(s) \varphi \end{pmatrix} ds$$

$$= \begin{pmatrix} \int_0^\infty e^{-\lambda s} T(s) f \, ds \\ \int_0^\infty e^{-\lambda s} T_s f \, ds + \int_0^\infty e^{-\lambda s} L(s) \varphi \, ds \end{pmatrix}$$

$$= \begin{pmatrix} R(\lambda, B) f \\ \int_0^\infty e^{-\lambda s} T_s f \, ds + R(\lambda, D_0) \varphi \end{pmatrix}.$$

Here, based on (9.11), we used the identities

$$\int_0^\infty e^{-\lambda s} T(s) f \, ds = R(\lambda, B) f$$

and

$$\int_0^\infty e^{-\lambda s} L(s) \varphi \, ds = R(\lambda, D_0) \varphi.$$

Observe that

$$\left(\int_0^\infty e^{-\lambda s} T_s f \, ds \right)(\sigma) = \int_0^\infty e^{-\lambda s} (T_s f)(\sigma) \, ds$$

$$= \int_{-\sigma}^\infty e^{-\lambda s} T(s + \sigma) f \, ds$$

$$= \int_0^\infty e^{-\lambda s + \lambda \sigma} T(s) f \, ds$$

$$= e^{\lambda \sigma} R(\lambda, B) f.$$

Hence, for $\operatorname{Re}\lambda > \max\{\omega_0(T),0\}$, we obtain

$$R(\lambda,A_2)\binom{f}{\varphi} = \begin{pmatrix} R(\lambda,B)f \\ e^{\lambda(\cdot)}\,R(\lambda,B)\,f + R(\lambda,\mathrm{D}_0)\varphi \end{pmatrix}.$$

We show now that this expression for $R(\lambda,A_2)\binom{f}{\varphi}$ equals the action of the resolvent of A_0. To do this, consider

$$\binom{g}{\psi} := \begin{pmatrix} R(\lambda,B)f \\ e^{\lambda(\cdot)}\,R(\lambda,B)\,f + R(\lambda,\mathrm{D}_0)\varphi \end{pmatrix}.$$

This implies that $\psi \in \mathrm{W}^{1,p}([-1,0],X)$. Using that $R(\lambda,\mathrm{D}_0)\varphi \in D(\mathrm{D}_0)$, we obtain $R(\lambda,\mathrm{D}_0)\varphi(0) = 0$, hence, $\psi(0) = g$. Applying $(\lambda - A_0)$ to this expression, we see that

$$(\lambda - A_0)\binom{g}{\psi} = \begin{pmatrix} (\lambda - B)R(\lambda,B)f \\ (\lambda - \mathrm{D})e^{\lambda(\cdot)}\,R(\lambda,B)\,f + (\lambda - \mathrm{D})R(\lambda,\mathrm{D}_0)\varphi \end{pmatrix} = \binom{f}{\varphi}$$

since the differentiation D is an extension of D_0 and $(\lambda - \mathrm{D})e^{\lambda(\cdot)}\,R(\lambda,B)\,f = 0$. Hence, this expression is a right inverse of $\lambda - A_0$. In an analogous way one can prove that the operator is a left inverse of $\lambda - A_0$, and is hence the inverse of $\lambda - A_0$. Thus, the operator A_0 is the generator of $(\mathcal{T}_0(t))_{t\geq0}$. $\qquad\square$

Applying the Miyadera perturbation theorem, Theorem 13.6, to the splitting (15.5) we obtain the generation result for our operator A.

Theorem 15.4. *If B generates a C_0-semigroup on X, then $A = A_0 + A_1$ generates a C_0-semigroup on M_p.*

Proof. We estimate

$$\int_0^\alpha \left\| A_1\,\mathcal{T}_0(r)\binom{f}{\varphi} \right\| \mathrm{d}r = \int_0^\alpha \left\| \begin{pmatrix} 0 & \Phi \\ 0 & 0 \end{pmatrix}\begin{pmatrix} T(r) & 0 \\ T_r & L(r) \end{pmatrix}\binom{f}{\varphi} \right\| \mathrm{d}r$$
$$= \int_0^\alpha \|\Phi(T_r f + L(r)\varphi)\|\,\mathrm{d}r.$$

Then fix a constant $c > \max\{|\eta|([-1,0]),1\}$, where $|\eta|$ is the positive Borel measure on $[-1,0]$ defined by the total variation of η. Consider the equivalent norm

$$\left\| \binom{f}{\varphi} \right\|_c := \|f\| + c\|\varphi\|_p, \quad \binom{f}{\varphi} \in M_p.$$

To apply the Miyadera theorem we have to show that there exist constants $\alpha > 0$ and $0 < \gamma < 1$ such that

$$\int_0^\alpha \|\Phi(T_r f + L(r)\varphi)\|\,\mathrm{d}r \leq \gamma \left\| \binom{f}{\varphi} \right\|_c \qquad (15.7)$$

holds for all $\left(\begin{smallmatrix} f \\ \varphi \end{smallmatrix}\right) \in D(A)$. For $0 < \alpha < 1$ we compute

$$\int_0^\alpha \|\Phi(T_r f + L(r)\varphi)\| \, dr$$

$$= \int_0^\alpha \left\| \int_{-1}^{-r} d\eta(\sigma)\, \varphi(r + \sigma) + \int_{-r}^0 d\eta(\sigma)\, T(r + \sigma)f \right\| \, dr$$

$$\leq \int_0^\alpha \int_{-1}^{-r} \|\varphi(r + \sigma)\| d|\eta|(\sigma) \, dr + \int_0^\alpha \int_{-r}^0 \|T(r + \sigma)f\| d|\eta|(\sigma) \, dr$$

$$\leq \int_{-\alpha}^0 \int_\sigma^0 \|\varphi(s)\| \, ds d|\eta|(\sigma) + \int_{-1}^{-\alpha} \int_\sigma^{\alpha + \sigma} \|\varphi(s)\| \, ds d|\eta|(\sigma)$$

$$+ \int_0^\alpha M \|f\| |\eta|([-1, 0]) \, dr$$

$$\leq c\|\varphi\|_p (c^{-1}|\eta|([-1, 0])) + \alpha M \|f\| |\eta|([-1, 0])$$

$$\leq \max\{c^{-1}|\eta|([-1, 0]), \alpha M |\eta|([-1, 0])\} \left\| \begin{pmatrix} f \\ \varphi \end{pmatrix} \right\|_c$$

where $M := \sup_{r \in [0,1]} \|T(r)\|$. Hence, if we choose $\alpha > 0$ small enough and since $c^{-1}|\eta|([-1, 0]) < 1$, we find $\gamma \in (0, 1)$ such that (15.7) holds. $\qquad\square$

Assume now that E is a Banach lattice. Applying Proposition 13.7 we also establish the positivity of the delay semigroup generated by A.

Corollary 15.5. *If B generates a positive semigroup on a Banach lattice E and $\eta \in \mathrm{BV}([-1, 0], \mathcal{L}(E)_+)$, then the delay semigroup is positive. Hence, for all $f \geq 0$ and $\varphi \geq 0$ the solutions of delay equation (15.3) remain positive.*

15.2 Gearhart's Theorem

We start with a reformulation of Datko's theorem (Theorem 12.5). To this aim we recall the convolution

$$(T * \varphi)(t) := \int_0^t T(t - s)\varphi(s)ds,$$

when $(T(t))_{t \geq 0}$ is a strongly continuous semigroup and $\varphi \in L^p(\mathbb{R}_+, X)$ for some $1 \leq p < \infty$.

Proposition 15.6. *Let A generate a strongly continuous semigroup $(T(t))_{t \geq 0}$ on X and fix $1 \leq p < \infty$. Then*

$$\omega_0(T) < 0 \iff T * \varphi \in L^p(\mathbb{R}_+, X) \text{ for all } \varphi \in L^p(\mathbb{R}_+, X).$$

Proof. Assume first that $\omega_0(T) < 0$. Take $\varphi \in L^p(\mathbb{R}_+, X)$ and define its extension to the whole line as

$$\tilde{\varphi}(t) := \begin{cases} \varphi(t) & \text{if } t \geq 0, \\ 0 & \text{if } t < 0. \end{cases} \tag{15.8}$$

Similarly, let

$$\tilde{T}(t) := \begin{cases} T(t) & \text{if } t \geq 0, \\ 0 & \text{if } t < 0. \end{cases}$$

By assumption, we have $\tilde{T}(\cdot)f \in \mathrm{L}^1(\mathbb{R}, X)$ for all $f \in X$. Hence, the Hausdorff–Young inequality (A.6) implies that $\tilde{T} * \tilde{\varphi} \in \mathrm{L}^p(\mathbb{R}, X)$. Therefore,

$$\left(\tilde{T} * \tilde{\varphi} \right) |_{\mathbb{R}_+} = T * \varphi \in \mathrm{L}^p(\mathbb{R}_+, X).$$

To prove the converse implication, take $w > \max\{0, \omega_0(T)\}$, $f \in X$, and $t \geq 0$. Then

$$\int_0^t T(t-s)\mathrm{e}^{-ws}T(s)f\mathrm{d}s = \frac{1}{w}\left(1 - \mathrm{e}^{-wt}\right)T(t)f.$$

Defining $\psi(s) := \mathrm{e}^{-ws}T(s)f$ and choosing $M \geq 1$ and $\omega_0(T) < \delta < w$ such that $\|T(t)\| \leq M\mathrm{e}^{\delta t}$, we see that

$$\|\psi(s)\| \leq M\mathrm{e}^{s(\delta - w)},$$

and hence $\psi \in \mathrm{L}^p(\mathbb{R}_+, X)$ because of the choice of the numbers δ and w. We thus have

$$T * \psi = \frac{1}{w}\left(1 - \mathrm{e}^{-w(\cdot)}\right)T(\cdot)f \in \mathrm{L}^p(\mathbb{R}_+, X),$$

which implies

$$T(\cdot)f \in \mathrm{L}^p(\mathbb{R}_+, X).$$

So, the conditions of Datko's theorem, Theorem 12.5, are satisfied and we obtain $\omega_0(T) < 0$. $\square$

For the following recall the definition of the Fourier transform $\mathcal{F}$ from Appendix A.10. We also need the notion of a bounded Fourier multiplier.

Definition 15.7. A function $m : \mathbb{R} \to \mathcal{L}(X)$ such that for all $f \in X$ the mapping $t \mapsto m(t)f$ belongs to $\mathrm{L}^\infty(\mathbb{R}, X)$[15], is called an $\mathrm{L}^p(\mathbb{R}, X)$-*Fourier multiplier*, if

$$\mathcal{F}^{-1}(m\mathcal{F}\varphi) = \mathcal{F}^{-1}(m) * \varphi$$

is well defined for all $\varphi \in \mathrm{L}^p(\mathbb{R}, X) \cap \mathrm{L}^1(\mathbb{R}, X)$ and extends to a bounded linear operator on $\mathrm{L}^p(\mathbb{R}, X)$.

We denote the set of $\mathrm{L}^p(\mathbb{R}, X)$-Fourier multipliers by $\mathcal{M}_p^{\mathcal{L}(X)}$ and the corresponding operator norm by $\|m\|_{\mathcal{M}_p^{\mathcal{L}(X)}}$.

We now characterize the growth bound of a strongly continuous semigroup in terms of Fourier multipliers.

---

[15] We denote this fact by the notation $m \in \mathrm{L}^\infty(\mathbb{R}, \mathcal{L}(X))$.

Theorem 15.8. *Let A generate the strongly continuous semigroup $(T(t))_{t\geq 0}$ on the Banach space X and fix $1 \leq p < \infty$. Then*

$$\omega_0(T) = \inf\left\{ \mu > \mathrm{s}(A) \ : \ \sup_{\alpha \geq \mu} \|R(\alpha + \mathrm{i}\cdot, A)\|_{\mathcal{M}_p^{\mathcal{L}(X)}} < \infty \right\}.$$

Proof. Denote the quantity in the right-hand side of the last equality by $r_0(A)$.

First, assume by contradiction that $\omega_0(T) < r_0(A)$ and take $\omega_0(T) < \alpha < r_0(A)$. Using the notation $T_\alpha(t) := \mathrm{e}^{-\alpha t} T(t)$ and

$$\tilde{T}_\alpha(t) := \begin{cases} T_\alpha(t) & \text{if } t \geq 0, \\ 0 & \text{if } t < 0, \end{cases}$$

we see that $\tilde{T}_\alpha(\cdot)f \in \mathrm{L}^1(\mathbb{R}, X)$ for all $f \in X$. As in the proof before, take $\varphi \in \mathrm{L}^p(\mathbb{R}_+, X)$ and define by (15.8) its extension to the whole line $\tilde{\varphi} \in \mathrm{L}^p(\mathbb{R}, X)$. The Hausdorff–Young inequality (A.6) again implies that $\tilde{T}_\alpha * \tilde{\varphi} \in \mathrm{L}^p(\mathbb{R}, X)$. Taking Fourier transforms we see that

$$\mathcal{F}(\tilde{T}_\alpha)(\cdot) = \int_0^\infty \mathrm{e}^{\mathrm{i}t\cdot} T_\alpha(t)\mathrm{d}t = R(\mathrm{i}\cdot, A - \alpha) = R(\alpha + \mathrm{i}\cdot, A) \in \mathcal{M}_p^{\mathcal{L}(X)}.$$

Take numbers $\omega_0(T) < \omega < \alpha$ and $M \geq 1$ such that $\|T(t)\| \leq M\mathrm{e}^{\omega t}$ holds. Then

$$\|R(\alpha + \mathrm{i}\cdot, A)\|_{\mathcal{M}_p^{\mathcal{L}(X)}} \leq \left\|\mathcal{F}^{-1}(R(\alpha + \mathrm{i}\cdot, A))\right\|_{\mathrm{L}^1(\mathbb{R}, \mathcal{L}(X))} = \left\|\tilde{T}_\alpha(\cdot)\right\|_{\mathrm{L}^1(\mathbb{R}, \mathcal{L}(X))}$$

$$= \int_0^\infty \|\mathrm{e}^{-\alpha t} T(t)\|\mathrm{d}t \leq M \int_0^\infty \mathrm{e}^{(\omega - \alpha)t}\mathrm{d}t = \frac{M}{\alpha - \omega},$$

which shows that $R(\alpha + \mathrm{i}\cdot, A)$ is a bounded Fourier multiplier. This contradicts the choice of α and shows that $\omega_0(T) \geq r_0(A)$.

To show equality, we use a contradiction argument again and assume that $r_0(A) < \omega_0(T)$. This means that

$$\sup_{\alpha \geq \omega_0(T)} \|R(\alpha + \mathrm{i}\cdot, A\|_{\mathcal{M}_p^{\mathcal{L}(X)}} =: N < \infty.$$

We choose numbers $\omega_0(T) < \omega_1 < \omega_2$ and see by the resolvent equality that

$$R(\omega_1 + \mathrm{i}s, A) = R(\omega_2 + \mathrm{i}s, A) + (\omega_2 - \omega_1)R(\omega_1 + \mathrm{i}s, A)R(\omega_2 + \mathrm{i}s, A)$$

for $s \in \mathbb{R}$. Taking Fourier multiplier norms, we obtain

$$\|R(\omega_1 + \mathrm{i}\cdot, A)\|_{\mathcal{M}_p^{\mathcal{L}(X)}} \leq (1 + M(\omega_2 - \omega_1))\|R(\omega_2 + \mathrm{i}\cdot, A)\|_{\mathcal{M}_p^{\mathcal{L}(X)}}.$$

Looking at the limit as $\omega_1 \to \omega_0(T)$, we see that

$$\|R(\omega_0(T) + \mathrm{i}\cdot, A)\|_{\mathcal{M}_p^{\mathcal{L}(X)}} \leq (1 + M(\omega_2 - \omega_0(T)))\|R(\omega_2 + \mathrm{i}\cdot, A)\|_{\mathcal{M}_p^{\mathcal{L}(X)}} < \infty.$$

This means that $T_{\omega_0(T)} * \varphi \in L^p(\mathbb{R}_+, X)$ for all $\varphi \in L^p(\mathbb{R}_+, X)$, where we again used the notation

$$T_{\omega_0(T)}(t) := e^{-\omega_0(A)t} T(t).$$

By Proposition 15.6, this implies $\omega_0(T_{\omega_0(T)}) < 0$, which contradicts the definition of $\omega_0(T)$. $\square$

While the previous result looks rather technical, it turns out that in Hilbert spaces uniform exponential stability of a semigroup can be characterized in terms of its generator. Plancherel's theorem, Theorem A.42, immediately implies that on a Hilbert space H one has

$$\mathcal{M}_2^{\mathcal{L}(H)} = L^\infty(\mathbb{R}, \mathcal{L}(H)).$$

Thus, Theorem 15.8 yields the following.

Theorem 15.9 (Gearhart). *Let $(T(t))_{t\geq 0}$ be a C_0-semigroup on a Hilbert space H with generator A. Then $(T(t))_{t\geq 0}$ is uniformly exponentially stable if and only if*

$$\{\lambda \in \mathbb{C} : \operatorname{Re}\lambda > 0\} \subseteq \rho(A) \quad and \quad M := \sup_{\operatorname{Re}\lambda > 0} \|R(\lambda, A)\| < \infty.$$

Remark 15.10. Consider a strongly continuous semigroup $(T(t))_{t\geq 0}$ with generator A in the Banach space X. We define the *abscissa of uniform boundedness of the resolvent* as

$$s_0(A) := \inf\left\{\mu > s(A) : \sup_{\alpha \geq \mu} \|R(\alpha + i\cdot, A)\| < \infty\right\}.$$

Clearly, $s(A) \leq s_0(A)$. Proposition 9.33 implies that $s_0(A) \leq \omega_0(T)$. By a rescaling argument, Gearhart's theorem can be reformulated as follows: For a strongly continuous semigroup $(T(t))_{t\geq 0}$ with generator A on the Hilbert space H we have

$$s_0(A) = \omega_0(T).$$

As an immediate consequence we obtain for positive semigroups the following extension of Theorem 12.17.

Corollary 15.11. *Let A be the generator of a positive C_0-semigroup $(T(t))_{t\geq 0}$ on a Hilbert lattice H. Then $\omega_0(T) = s(A)$ holds.*

Proof. Fix $\mu > s(A)$. Corollary 12.8 implies that

$$\Lambda := \{\lambda \in \mathbb{C} : \operatorname{Re}\lambda > 0\} \subseteq \rho(A - \mu)$$

and

$$\|R(\lambda, A - \mu)\| \leq \|R(\operatorname{Re}\lambda, A - \mu)\| \leq \|R(\mu, A)\|$$

for all $\lambda \in \Lambda$. So, by Theorem 15.9, we have $\omega_0(T) - \mu < 0$ and hence

$$\omega_0(T) \leq s(A).$$ $\square$

15.3 Stability of Delay Equations

We return to the delay equations as introduced in Section 15.1. Suppose that H is a Hilbert lattice and B generates a positive strongly continuous semigroup $(T(t))_{t \geq 0}$ on it. Further, let

$$\eta : [-1, 0] \longrightarrow \mathcal{L}(H)_+$$

be a positive operator-valued function of bounded variation. We refer to this fact by saying that the delay operator Φ, defined by

$$\Phi \varphi := \int_{-1}^{0} d\eta(s)\varphi(s),$$

is positive.

We consider the abstract delay equation

$$\begin{cases} \dot{u}(t) = Bu(t) + \int_{-1}^{0} d\eta(s)u(t + s), & s \geq 0, \\ u(s) = \varphi(s), & s \in [-1, 0], \end{cases} \tag{15.9}$$

where $\varphi : [-1, 0] \to H$ is some given function.

Recall from Section 15.1 that we can associate to equation (15.9) an operator matrix

$$A := \begin{pmatrix} B & \Phi \\ 0 & D \end{pmatrix}$$

with domain

$$D(A) := \left\{ \begin{pmatrix} f \\ \varphi \end{pmatrix} \in D(B) \times H^1([-1, 0], H) \ : \ \varphi(0) = f \right\},$$

where D denotes the weak operator of differentiation.

Theorem 15.4 states that A generates a *positive* strongly continuous semigroup $(\mathcal{T}(t))_{t \geq 0}$ on $M_2 = H \times L^2([-1, 0], H)$. Since M_2 can be considered as a Hilbert space, we immediately infer from Corollary 15.11 the following.

Corollary 15.12. *With the notations and assumptions introduced above in this section, the delay operator A satisfies*

$$\omega_0(\mathcal{T}) = s(A).$$

Hence, to obtain stability it suffices to look at the spectral bound of A. As a first step, we characterize the spectrum and the resolvent operator of A using the notation

$$\Phi_\lambda f := \Phi \left(e^{\lambda(\cdot)} f \right) = \int_{-1}^{0} d\eta(s) e^{\lambda s} f \, ds \tag{15.10}$$

and

$$\varepsilon_\lambda(s) := e^{\lambda s}$$

for $s \in [-1, 0]$.

Proposition 15.13. *For $\lambda \in \mathbb{C}$ and for all $1 \le p < \infty$, we have*

$$\lambda \in \rho(A) \text{ if and only if } \lambda \in \rho(B + \Phi_\lambda).$$

Moreover, for $\lambda \in \rho(A)$ the resolvent $R(\lambda, A)$ is given by

$$\begin{pmatrix} R(\lambda, B + \Phi_\lambda) & R(\lambda, B + \Phi_\lambda)\Phi R(\lambda, D_0) \\ \varepsilon_\lambda R(\lambda, B + \Phi_\lambda) & [\varepsilon_\lambda R(\lambda, B + \Phi_\lambda)\Phi + I]R(\lambda, D_0) \end{pmatrix}. \tag{15.11}$$

Proof. For $\lambda \in \rho(B + \Phi_\lambda)$ the matrix (15.11) is a bounded operator from M_2 to $D(A)$ defining the inverse of $\lambda - A$.

To see this, we show as a first step that the range of the matrix (15.11) is contained in $D(A)$. Take $\begin{pmatrix} g \\ \psi \end{pmatrix} \in M_2$ and consider

$$\begin{pmatrix} f \\ \varphi \end{pmatrix} := \begin{pmatrix} R(\lambda, B + \Phi_\lambda)g + R(\lambda, B + \Phi_\lambda)\Phi R(\lambda, D_0)\psi \\ \varepsilon_\lambda R(\lambda, B + \Phi_\lambda)g + \varepsilon_\lambda R(\lambda, B + \Phi_\lambda)\Phi R(\lambda, D_0)\psi + R(\lambda, D_0)\psi \end{pmatrix}.$$

Since $R(\lambda, B + \Phi_\lambda)$ maps into $D(B)$, we know that $f \in D(B)$. For the second component, note that the function ε_λ is smooth, and the range of $R(\lambda, D_0)$ is contained in $\mathrm{H}^1([-1, 0], H)$. Hence, $\varphi \in \mathrm{H}^1([-1, 0], H)$. Since D_0 was the first derivative with Dirichlet boundary conditions, we see that $R(\lambda, D_0)\psi(0) = 0$. From $(\varepsilon_\lambda f)(0) = f$ we obtain that $\varphi(0) = f$, so we finally have $\begin{pmatrix} f \\ \varphi \end{pmatrix} \in D(A)$.

Denoting $R_\lambda := R(\lambda, B + \Phi_\lambda)$, we compute the matrix product

$$(\lambda - A) \begin{pmatrix} R_\lambda & R_\lambda \Phi R(\lambda, D_0) \\ \varepsilon_\lambda R_\lambda & [\varepsilon_\lambda R_\lambda \Phi + I]R(\lambda, D_0) \end{pmatrix}$$

$$= \begin{pmatrix} (\lambda - B)R_\lambda - \Phi \varepsilon_\lambda R_\lambda & (\lambda - B)R_\lambda \Phi R(\lambda, D_0) - \Phi[\varepsilon_\lambda R_\lambda \Phi + I]R(\lambda, D_0) \\ (\lambda - D)\varepsilon_\lambda R_\lambda & (\lambda - D)[\varepsilon_\lambda R_\lambda \Phi + I]R(\lambda, D_0) \end{pmatrix}.$$

The identities

$$(\lambda - B)R_\lambda - \Phi(\varepsilon_\lambda R_\lambda) = (\lambda - B - \Phi_\lambda)R_\lambda = I$$

and

$$(\lambda - B)R_\lambda \Phi R(\lambda, D_0) - \Phi[\varepsilon_\lambda R_\lambda \Phi + I]R(\lambda, D_0)$$
$$= (\lambda - B - \Phi_\lambda)R_\lambda \Phi R(\lambda, D_0) - \Phi R(\lambda, D_0) = 0$$

hold. Moreover, we have $(\lambda - D)(\varepsilon_\lambda R_\lambda) = 0$. Using this identity we also obtain

$$(\lambda - D)[\varepsilon_\lambda R_\lambda \Phi + I]R(\lambda, D_0) = (\lambda - D)R(\lambda, D_0) = I.$$

So the operator in (15.11) is a right inverse of $\lambda - A$. In an analogous way, we can prove that the operator in (15.11) is also a left inverse of $\lambda - A$, hence is the inverse of $\lambda - A$.

Conversely, if $\lambda \in \rho(A)$, then for every $\binom{g}{\psi} \in M_2$ there exists a unique $\binom{f}{\varphi} \in D(A)$ such that

$$(\lambda - A)\binom{f}{\varphi} = \binom{(\lambda - B)f - \Phi\varphi}{\lambda\varphi - \varphi'} = \binom{g}{\psi}. \tag{15.12}$$

We integrate the second row of equation (15.12) and, using the variation of parameters formula, we obtain

$$\varphi = \varepsilon_\lambda \, \varphi(0) + R(\lambda, D_0) \, \psi, \tag{15.13}$$

where $R(\lambda, D_0)$ is the resolvent of D_0. Note that, since D_0 generates the nilpotent shift semigroup, its spectrum is $\sigma(D_0) = \emptyset$ and $R(\lambda, D_0)$ exists for all $\lambda \in \mathbb{C}$. Since $\varphi(0) = f$, we can rewrite equation (15.13) as

$$\varphi = \varepsilon_\lambda f + R(\lambda, D_0) \, \psi.$$

Taking into account also the first row of equation (15.12), we see that f has to satisfy the equation

$$(\lambda - B - \Phi_\lambda)f = \Phi R(\lambda, D_0)\psi + g.$$

In particular, for $\psi = 0$ and for every $g \in H$ there exists a unique $f \in D(B)$ such that

$$(\lambda - B - \Phi_\lambda)f = g.$$

This means that $\lambda - B - \Phi_\lambda$ is invertible, i.e., $\lambda \in \rho(B + \Phi_\lambda)$. $\qquad\square$

For the analysis of the spectral bound of the operator A, we have to carry out a series of preliminary calculations. First we study the operator-valued map $R : \rho \subset \mathbb{C}^2 \to \mathcal{L}(H)$ defined by

$$\begin{aligned}(\lambda, \mu) &\longmapsto R(\lambda, \mu) := R(\lambda, B + \Phi_\mu) = (\lambda - B - \Phi_\mu)^{-1} \quad \text{for}\\ (\lambda, \mu) &\in \rho := \{(r, s) \in \mathbb{C}^2 : r \in \rho(B + \Phi_s)\}.\end{aligned} \tag{15.14}$$

Lemma 15.14. *The set $\rho \subset \mathbb{C}^2$ is open and the mapping $R(\cdot, \cdot)$ is analytic.*

Proof. Let $(\lambda_0, \mu_0) \in \rho$ and $(\lambda, \mu) \in \mathbb{C}^2$. Then

$$(\lambda - B - \Phi_\mu) - (\lambda_0 - B - \Phi_{\mu_0}) = (\lambda - \lambda_0) - (\Phi_\mu - \Phi_{\mu_0}) =: \Delta_{\lambda,\mu}.$$

Note that $(\Phi_\mu - \Phi_{\mu_0})f = \Phi((\varepsilon_\mu - \varepsilon_{\mu_0})f)$. Since $\lim_{(\lambda,\mu) \to (\lambda_0,\mu_0)} \|\Delta_{\lambda,\mu}\| = 0$ and

$$\lambda - B - \Phi_\mu = (I + \Delta_{\lambda,\mu}R(\lambda_0, \mu_0))(\lambda_0 - B - \Phi_{\mu_0}), \tag{15.15}$$

we see that $\lambda - B - \Phi_\mu$ is invertible for $\|(\lambda, \mu) - (\lambda_0, \mu_0)\|$ sufficiently small, meaning that the set ρ is open. It also follows from relation (15.15) that for $\|(\lambda, \mu) - (\lambda_0, \mu_0)\|$ sufficiently small, one has

$$R(\lambda, \mu) = R(\lambda_0, \mu_0) \sum_{k=0}^{\infty} [\Delta_{\lambda,\mu} R(\lambda_0, \mu_0)]^k.$$

Since this series converges uniformly on small balls around (λ_0, μ_0) and the map $\mathbb{C}^2 \ni (\lambda, \mu) \mapsto \Delta_{\lambda,\mu} \in \mathcal{L}(H)$ is analytic, one obtains the analyticity of $R(\cdot, \cdot)$ as well. $\square$

With the help of this lemma we can now derive important properties of the *spectral bound function*, $s : \mathbb{R} \to \mathbb{R} \cup \{-\infty\}$ defined by

$$s(\lambda) := s(B + \Phi_\lambda). \tag{15.16}$$

Proposition 15.15. *Let B generate a positive semigroup on a Hilbert lattice H and assume that Φ is positive. Then the spectral bound function $s(\cdot)$ defined in (15.16) is decreasing and continuous from the left on $\mathbb{R}$. If, in addition, the point $s(\mu_0)$ is isolated in $\sigma(B + \Phi_{\mu_0}) \cap \mathbb{R}$, then $s(\cdot)$ is even continuous in $\mu_0 \in \mathbb{R}$.*

Proof. Observe first that the definition of Φ_λ in formula (15.10) implies that for $\mu_0 \leq \mu_1$ we have $\Phi_{\mu_1} \leq \Phi_{\mu_0}$. Therefore, also $s(B + \Phi_{\mu_1}) \leq s(B + \Phi_{\mu_0})$, see Proposition 12.11. This shows that the function $s(\cdot)$ is decreasing. By Corollary 12.9, we also have $s(\mu) \in \sigma(B + \Phi_\mu)$.

To show that the function $s(\cdot)$ is left-continuous, assume by contradiction that

$$s(\mu_0) < s^- := \lim_{\varepsilon \downarrow 0} s(\mu_0 - \varepsilon)$$

for some $\mu_0 \in \mathbb{R}$. Then $s^- \in \rho(B + \Phi_{\mu_0})$ and therefore $(s^-, \mu_0) \in \rho$. This contradicts the fact that $\rho \in \mathbb{C}^2$ is open, since by Corollary 12.9 we have

$$s(\mu_0 - \varepsilon) \in \sigma(B + \Phi_{\mu_0 - \varepsilon}),$$

implying that $(s(\mu_0 - \varepsilon), \mu_0 - \varepsilon) \notin \rho$ for all $\varepsilon > 0$, while $(s(\mu_0 - \varepsilon), \mu_0 - \varepsilon) \to (s^-, \mu_0)$ when $\varepsilon \to 0$.

Assume now that $s(\mu_0)$ is isolated in $\sigma(B + \Phi_{\mu_0}) \cap \mathbb{R}$. In order to show that $s(\cdot)$ is right-continuous, we proceed again a contradiction and assume that

$$s^+ := \lim_{\varepsilon \downarrow 0} s(\mu_0 + \varepsilon) < s(\mu_0).$$

Then, by assumption, there exists $\lambda \in \rho(B + \Phi_{\mu_0}) \cap \mathbb{R}$ satisfying

$$s^+ < \lambda < s(\mu_0).$$

In particular, $(\lambda, \mu_0) \in \rho$ and we conclude from Lemma 15.14 that

$$R(\lambda, \mu_0) = \lim_{\varepsilon \downarrow 0} R(\lambda, \mu_0 + \varepsilon) \geq 0.$$

This contradicts Corollary 12.10, and so the function $\lambda \mapsto s(\lambda)$ is continuous. $\quad\square$

Remark 15.16. We mention that the spectral bound function is continuous at a point $\lambda_0 \in \mathbb{R}$ if B has compact resolvent or Φ_{λ_0} is compact.

Now we justify our notation for the spectral bound function. It yields the following estimates for the spectral bound of the generator A.

Proposition 15.17. *Let B generate a positive semigroup on the Hilbert lattice H and suppose that Φ is positive. Then the following assertions hold.*

a) *If $s(B + \Phi_\lambda) < \lambda$, then $s(A) < \lambda$.*
b) *If $s(B + \Phi_\lambda) = \lambda$, then $s(A) = \lambda$.*

Proof. a) Let $\lambda > s(\lambda)$. From the monotonicity of the function $\lambda \mapsto s(\lambda)$ one obtains that

$$\mu \geq \lambda > s(\lambda) \geq s(\mu)$$

for all $\mu \geq \lambda$. This implies $\mu \in \rho(B + \Phi_\mu)$, and therefore $\mu \in \rho(A)$ for all $\mu \geq \lambda$, by Proposition 15.13. On the other hand, by Corollary 12.9, one has $s(A) \in \sigma(A)$, hence $\lambda > s(A)$, as claimed.

b) If $\lambda = s(\lambda)$, then, again by Corollary 12.9, one has $\lambda \in \sigma(B + \Phi_\lambda)$ and hence $\lambda \in \sigma(A)$. On the other hand, one can show as in a) that $\mu \in \rho(A)$ for all $\mu > \lambda$, which implies $\lambda = s(A)$. $\quad\square$

The spectral bound of the operator A defined on $M_2 = H \times L^2([-1, 0], H)$ can be completely characterized by the spectral bound of the operator $B + \Phi_\lambda$ defined on H.

Lemma 15.18. *Let B generate a positive semigroup on H and assume that Φ is positive. If $\sigma(B + \Phi_\lambda) \neq \emptyset$ for some $\lambda \in \mathbb{R}$, then*

$$s(A) = \sup\{\lambda \in \mathbb{R} \ : \ s(B + \Phi_\lambda) \geq \lambda\}. \tag{15.17}$$

In the other case, one has $s(A) = -\infty$.

Proof. If $\sigma(B + \Phi_\lambda) = \emptyset$ for all $\lambda \in \mathbb{R}$, then $s(A) = -\infty$, since by Proposition 15.13 also $\sigma(A) = \emptyset$.

We now assume that $\sigma(B + \Phi_\lambda) \neq \emptyset$ and denote

$$\mu := \sup\{\lambda \in \mathbb{R} \ : \ s(\lambda) > \lambda\}.$$

Then it follows from the left-continuity of the function $\lambda \mapsto s(\lambda)$ that $s(\mu) \geq \mu$.

Now, if $s(\mu) = \mu$, then $s(A) = \mu$ by Proposition 15.17.b) and the assertion follows. If, on the other hand, $s(\mu) > \mu$, then we proceed in two steps. First, we show that the inclusion

$$(\mu, s(B + \Phi_\mu)] \subset \sigma(B + \Phi_\mu) \tag{15.18}$$

holds. To do so, assume, by contradiction, that there exists

$$r \in (\mu, s(B + \Phi_\mu)] \cap \rho(B + \Phi_\mu).$$

Then $(r, \mu) \in \rho$, where the set ρ is defined in (15.14), and by the definition of μ one obtains

$$r + \varepsilon > \mu + \varepsilon > s(B + \Phi_{\mu+\varepsilon})$$

for all $\varepsilon > 0$. Hence

$$R(r, \mu) = R(r, B + \Phi_\mu) = \lim_{\varepsilon \downarrow 0} R(r + \varepsilon, B + \Phi_{\mu+\varepsilon}) \geq 0,$$

which contradicts the fact that $r \leq s(B+\Phi_\mu)$ by Corollary 12.10. Hence, inclusion (15.18) is proved, and from the closedness of the spectrum we deduce that $\mu \in \sigma(B + \Phi_\mu)$. Applying Proposition 15.13 we get that $\mu \in \sigma(A)$ and hence $s(A) \geq \mu$.

As a second step, we assume by contradiction that $s(A) > \mu$. Then from the definition of μ it immediately follows that

$$s(B + \Phi_{s(A)}) < s(A),$$

and hence $s(A) \in \rho(B + \Phi_{s(A)})$. The spectral characterization from Proposition 15.13 now implies $s(A) \in \rho(A)$, contradicting Corollary 12.9. $\square$

In concrete cases it could be quite difficult to solve the equation in (15.17). However, in order to show that the spectral bound $s(A)$ of A is negative, we incorporate Gearhart's theorem, Theorem 15.9, in this result.

Theorem 15.19. *Suppose that B generates a positive semigroup on a Hilbert lattice H and that Φ is a positive delay operator. Then*

$$\omega_0(\mathcal{T}) < 0 \quad \Longleftrightarrow \quad s(B + \Phi_0) < 0.$$

Proof. By Corollary 15.12, $\omega_0(\mathcal{T}) = s(A)$. We can assume that there exists $\lambda \in \mathbb{R}$ such that $\sigma(B + \Phi_\lambda) \neq \emptyset$, since otherwise $s(A) = s(B + \Phi_\lambda) = -\infty$ by Lemma 15.18.

Suppose first that $s(A) < 0$. Then $s(B + \Phi_0) < 0$, since otherwise $s(A) \geq 0$ by Lemma 15.18. This proves the implication from the left to the right. The other implication follows immediately from Proposition 15.17.a), hence the proof is complete. $\square$

Example 15.20. As an illustration we present the following equation modeling delayed heat conduction in a rod whose endpoints are kept at a constant temperature:

$$\begin{cases} \partial_t u(x,t) = \partial_{xx}^2 u(x,t) + bu(x,t-1), & x \in [0,\ell],\ t \geq 0, \\ u(0,t) = u(\ell,t) = 0, & t \geq 0, \\ u(x,t) = f(x,t), & x \in [0,\ell],\ t \in [-1,0], \end{cases}$$

where $\ell > 0$ is the length of the rod and $b > 0$ the delay coefficient. To treat this as an abstract delay equation, we introduce the following spaces and operators:

- the Hilbert space $H := \mathrm{L}^2((0,\ell))$;
- the operator $B := \Delta$ with Dirichlet boundary conditions, i.e., $D(B) := \mathrm{H}_0^1((0,\ell)) \cap \mathrm{H}^2((0,\ell))$;
- the functions $\mathbb{R}_+ \ni t \mapsto u(t) = u(\cdot,t) \in \mathrm{L}^2((0,\ell))$ and $u_t : [-1,0] \to \mathrm{L}^2((0,\ell))$, $u_t(s) := u(t+s)$;
- the operator Φ defined as $\Phi\varphi := b\delta_{-1}\varphi = b\varphi(-1)$.

In order to apply Theorem 15.19, recall that

$$\sigma(B) = \left\{ -\frac{n^2\pi^2}{\ell^2} : n \in \mathbb{N} \right\}.$$

Hence, using that $\Phi_\lambda f = be^{-\lambda} f$, we arrive at

$$\mathrm{s}(B + \Phi_0) = b - \frac{\pi^2}{\ell}.$$

So, the solutions of the delayed heat equation converge to zero exponentially if

$$b < \frac{\pi^2}{\ell}.$$

Note that this condition is independent of the size of the delay.

15.4 Notes and Remarks

Gearhart's theorem goes back to Gearhart [51], where he proved it for contraction semigroups. Later it was generalized by Prüß [114] and others. We refer to the monograph by van Neerven [103] for detailed references and for other generalizations to Banach spaces. This result has many applications to partial differential equations. As an example, we mention applications to nonlinear Schrödinger equations as surveyed in Cramer and Latushkin [27], and to dissipative systems as in Liu and Zheng [89].

The generalization of Gearhart's theorem through Fourier multipliers originates from Clark et al. [25], and our proof follows Hieber [64]. In Proposition

15.6 we follow van Neerven [103, Theorem 3.3.1]. The consequences of Gearhart's theorem for positive semigroups, stated in Theorem 15.11, were first realized by Greiner and Nagel [56].

Abstract delay equations were treated extensively in Bátkai and Piazzera [13], and the presentation here follows the exposition in Section 2.4 of that book. Closely related results appeared already in Nagel (ed.) [101, Section B-IV.3].

15.5 Exercises

1. Show for the generator A of a positive semigroup in a Banach lattice E that the equality $s(A) = s_0(A)$ holds.

2. Give an example for a positive strongly continuous semigroup $(T(t))_{t \geq 0}$ with generator A on a Banach lattice such that $s_0(A) \neq \omega_0(T)$.

3. Suppose that A generates an exponentially stable semigroup on the Hilbert space H, and that for $B \in \mathcal{L}(H)$ we have $\sup_{s \in \mathbb{R}} \|BR(is, A)\| < 1$. Show that the semigroup generated by $A + B$ is also exponentially stable.

4. Consider the scalar delay differential equation

$$\dot{u}(t) = -au(t) + bu(t-1)$$

with $a, b \in \mathbb{R}$, and analyse the stability of the solutions if

  a) $a = 0$;

  b) $a = 2$, $b = 1$;

  c) $a = 1$, $b = 1$.

5. Show the decomposition

$$\lambda - A = \begin{pmatrix} I & -\Phi R(\lambda, D_0) \\ 0 & I \end{pmatrix} \begin{pmatrix} \lambda - B - \Phi_\lambda & 0 \\ 0 & \lambda - D_0 \end{pmatrix} \begin{pmatrix} I & 0 \\ -\varepsilon_\lambda \otimes I & I \end{pmatrix}$$

  for the delay operator A.

6. Deduce from the previous exercise the spectral characterizations

$$\lambda \in \sigma_p(A) \quad \Longleftrightarrow \quad \lambda \in \sigma_p(B + \Phi_\lambda),$$
$$\lambda \in \sigma_{ess}(A) \quad \Longleftrightarrow \quad \lambda \in \sigma_{ess}(B + \Phi_\lambda).$$

Chapter 16

Koopman Semigroups

We present here a class of positive operator semigroups that arise in studying dynamical systems. The main idea is to linearize a given (nonlinear) system by considering another state space. The linear operator which acts on this new space is called the Koopman operator. It is named after B. O. Koopman, who used this in the 1930s together with G. D. Birkhoff and J. von Neumann to prove the so-called ergodic theorems.

We start with a nonlinear system of ordinary differential equations, associate a semiflow to it, and then derive the corresponding Koopman semigroup. Subsequently we present the main properties of this semigroup and its generator. At the end we show some properties of the semiflow that can be deduced from the appropriate properties of the associated Koopman semigroup or its generator.

In this chapter we assume some general knowledge of measure theory.

16.1 Ordinary Differential Equations and Semiflows

Consider the ordinary differential equation

$$\begin{cases} \dot{x}(t) = F(x(t)), & t \geq 0, \\ x(0) = x_0 \in \Omega, \end{cases} \tag{16.1}$$

where $\Omega \subset \mathbb{R}^n$ is an open set. We make the following standing assumptions.

Assumptions 16.1.

a) $F : \mathbb{R}^n \to \mathbb{R}^n$ is continuously differentiable.

b) Equation (16.1) has global solutions for all $x_0 \in \Omega$.

c) $\overline{\Omega} \subset \mathbb{R}^n$ is *positively invariant* for the solution of equation (16.1), i.e.,

$$x_0 \in \overline{\Omega} \implies x(t) \in \overline{\Omega} \text{ for } t \geq 0.$$

We comment on these assumptions based on standard theorems from ordinary differential equations. First recall that Assumption 16.1.a) implies the existence and uniqueness of local solutions to equation (16.1). Assumption 16.1.b) is satisfied whenever F grows at most linearly, i.e., if there are constants $c, d > 0$ such that

$$\|F(x)\| \leq c\|x\| + d.$$

Finally, the set $\bar{\Omega}$ is positively invariant if the subtangent condition

$$\liminf_{h\downarrow 0} \frac{1}{h} d(x + hF(x), \Omega) = \liminf_{h\downarrow 0} \frac{1}{h} \inf_{z\in\Omega} \|x + hF(x) - z\| = 0$$

holds for every $x \in \partial\Omega$. If Ω is convex, then this is equivalent to the angle condition

$$(F(x)|y) \leq 0$$

for $x \in \partial\Omega$ and y being an outer normal vector to Ω at x (see Figure 16.1).

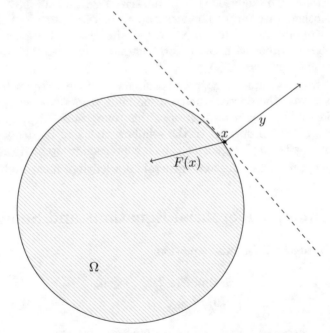

Figure 16.1: The angle condition $\sphericalangle(y, F(x)) \geq \frac{\pi}{2}$.

The assumptions imply that there exists a continuous mapping solving the differential equation in (16.1). More precisely, there exists a function $\varphi : \mathbb{R}^+ \times \overline{\Omega} \to \overline{\Omega}$, which is continuously differentiable in its first variable, satisfying

$$\begin{aligned}
\varphi(0, x) &= x && \text{for all } x \in \overline{\Omega} \text{ and} \\
\varphi(t, \varphi(s, x)) &= \varphi(t + s, x) && \text{for all } t, s \geq 0, \ x \in \overline{\Omega},
\end{aligned} \qquad (16.2)$$

such that the solutions of equation (16.1) are given by

$$x(t) = \varphi(t, x_0).$$

Such a mapping φ is called a continuous *semiflow*.

It was the fundamental observation of Koopman and von Neumann that such nonlinear dynamical systems give rise to linear ones. One motivation for this construction is that in many situations we do not see the state space Ω and the dynamics on it, but we only observe some quantity (heat, concentration, density, etc.). Hence, an *observable* is simply a (scalar) function defined on Ω. The nonlinear action of φ induces a linear action on observables.

To make these considerations precise, we suppose that Ω is a bounded open set, choose the Banach function space $E = C(\overline{\Omega})$, and define the operators

$$T(t)f(x) := f(\varphi(t, x)) \tag{16.3}$$

for $f \in E$, $t \geq 0$ and $x \in \overline{\Omega}$. Recall that the linear operator T on the Banach lattice E is called a *lattice homomorphism* if $|Tf| = T|f|$ for every $f \in E$ (see Definition 10.19).

Proposition 16.2. *The family $(T(t))_{t\geq 0}$, defined by formula (16.3), is a C_0-semigroup of positive contractions on the Banach lattice E. Its generator A is given as the closure of the operator*

$$(A_0 f)(x) = (\nabla f(x)|F(x))$$

with $f \in D(A_0) = C^1(\overline{\Omega})$ and $x \in \overline{\Omega}$. Further, $T(t)$ is a lattice homomorphism for each $t \geq 0$.

Here

$$C^1(\overline{\Omega}) := \left\{ f \in C(\overline{\Omega}) : \exists U \text{ open}, \Omega \subset U, \text{ and } g \in C^1(U) \text{ such that } f = g|_{\overline{\Omega}} \right\}.$$

The operator semigroup $(T(t))_{t\geq 0}$ from Proposition 16.2 is usually called the *Koopman semigroup* associated to the semiflow φ.

Proof. Clearly, each operator $T(t)$ is linear, positive, even a lattice homomorphism, and $T(t)\mathbb{1} = \mathbb{1}$ holds. By Lemma 10.27, the operators $T(t)$ are all contractions. For $t, s \geq 0$ and $x \in \overline{\Omega}$ we see that, by (16.2),

$$(T(t)T(s)f)(x) = (T(t)f)(\varphi(s, x)) = f(\varphi(t, \varphi(s, x)))$$
$$= f(\varphi(t + s, x)) = (T(t + s)f)(x),$$

hence the semigroup property holds. To show strong continuity of the map $T(\cdot)$, note that $\varphi : [0, 1] \times \overline{\Omega} \to \overline{\Omega}$ is uniformly Lipschitz continuous in its first variable. We will denote its Lipschitz constant by L. Take $f \in E = C(\overline{\Omega})$. Then for each $\varepsilon > 0$ there is $\delta > 0$ such that for all $x, y \in \overline{\Omega}$ with $\|x - y\| < \delta$ we have

$$\|f(x) - f(y)\| < \varepsilon.$$

The Lipschitz continuity of the semiflow implies that if $0 \leq t < \delta/L$ holds, then $\|\varphi(t,x) - x\| < \delta$. Hence, for $0 \leq t < \delta/L$ we have

$$\|T(t)f - f\| = \sup_{x \in \overline{\Omega}} \|f(\varphi(t,x)) - f(x)\| \leq \varepsilon.$$

Now we turn our attention to the characterization of the generator A of $(T(t))_{t \geq 0}$. Taking $f \in D(A_0)$, $x \in \overline{\Omega}$, we note that the function $t \mapsto (T(t)f)(x) = f(\varphi(t,x))$ is also continuously differentiable, hence $T(t)D(A_0) \subset D(A_0)$. Since $D(A_0)$ is dense in E, we see that it is a core for the generator[16]. It only remains to show that for $f \in D(A_0)$ we have $Af = A_0 f$. To verify this, take $f \in \mathrm{C}^1(\overline{\Omega})$ and consider

$$I := \left\| \frac{T(t)f - f}{t} - A_0 f \right\| = \sup_{x \in \overline{\Omega}} \left| \frac{(T(t)f)(x) - f(x)}{t} - (\nabla f(x)|F(x)) \right|$$

$$= \sup_{x \in \overline{\Omega}} \left| \frac{f(\varphi(t,x)) - f(\varphi(0,x))}{t} - (\nabla f(\varphi(0,x))|F(\varphi(0,x))) \right|.$$

By the mean value theorem, for every $t > 0$ and $x \in \overline{\Omega}$, there is $0 < \xi = \xi(t,x) < t$ such that

$$I = \sup_{x \in \overline{\Omega}} \left| (\nabla f(\varphi(\xi(t,x),x))|F(\varphi(\xi(t,x),x))) - (\nabla f(\varphi(0,x))|F(\varphi(0,x))) \right|.$$

If $t \downarrow 0$, then $\xi(t,x) \to 0$ uniformly in x. Since $f \in \mathrm{C}^1(\overline{\Omega})$, and φ and F are also continuously differentiable in their first variable, it follows that $I \to 0$ as $t \downarrow 0$. This shows that $f \in D(A)$ and $A_0 f = Af$. $\qquad\square$

Remark 16.3. If $\overline{\Omega}$ is not invariant, then we have to require boundary conditions on the domain of the generator A of the Koopman semigroup. Typical here are the so-called *Wentzell boundary conditions*. As an illustration, we present the simplest situation without a proof.

Suppose that Ω is a convex domain with C^1-boundary. In this case we have at each boundary point $x \in \partial\Omega$ a unique outer normal vector $\nu(x)$. Our main assumption is now that

$$(F(x)|\nu(x)) > 0$$

for all $x \in \partial\Omega$. We consider the operator $A_0 f := (\nabla f|F)$ with domain

$$D(A_0) = \{f \in \mathrm{C}^1(\overline{\Omega}) : A_0 f(x) = 0 \text{ for } x \in \partial\Omega\}.$$

Then the closure of this operator generates a positive semigroup of lattice homomorphisms.

---

[16] A subspace $D \subset D(A)$ is called a *core* of A if it is dense in the graph norm of A. This implies that if we know the generator on a core, the we can determine it uniquely. An important result in semigroup theory states that if a dense set $D \subset A$ is invariant under the semigroup, then it is a core.

16.2   Koopman Semigroups

In this section we collect some properties characterizing Koopman semigroups and their generators. We start with some technical lemmata.

Lemma 16.4. *Let K be a compact Hausdorff space and $f^* \in C(K)^*$ a continuous linear functional. Then the following statements are equivalent.*

(i) *f^* is a lattice homomorphism.*

(ii) *There are a $c \geq 0$ and $x \in K$ such that $f^* = c\delta_x$, i.e., $f^*(f) = \langle f, f^* \rangle = cf(x)$ holds for all $f \in C(K)$.*

Proof. The implication (ii) $\implies$ (i) is straightforward, hence we only treat the other one. Let $f^* \in C(K)^*$ be a lattice homomorphism. By the Riesz representation theorem (see Theorem A.29), there is a Borel measure μ such that

$$\langle f, f^* \rangle = \int_K f \, d\mu$$

for all $f \in C(K)$.

Assume by contradiction that $x, y \in K$ are two distinct points in the support of μ, and take two disjoint open sets $U, V \subset K$ such that $x \in U$ and $y \in V$. Applying Urysohn's lemma (see Lemma A.19) to these sets, we obtain functions $f, g \in C(K)$ such that

$$f(x) = 1 \text{ and } f(s) = 0 \text{ for } s \in K \setminus U,$$
$$g(y) = 1 \text{ and } g(s) = 0 \text{ for } s \in K \setminus V.$$

Then $f \wedge g = 0$ holds, but $\langle f, f^* \rangle \wedge \langle g, f^* \rangle > 0$, a contradiction. $\qquad\square$

This allows us to characterize lattice homomorphisms between $C(K)$ spaces.

Lemma 16.5. *Let K, L be compact Hausdorff spaces and consider $T : C(K) \to C(L)$. Then the following are equivalent.*

(i) *T is a lattice homomorphism.*

(ii) *There exist a unique positive function $g \in C(L)$ and a unique function $\psi : L \to K$ which is continuous on the set $\{y \in L : g(y) > 0\}$, such that*

$$(Tf)(s) = g(s)f(\psi(s))$$

holds for all $f \in C(K)$, $s \in L$.

Proof. Again, as in the proof of the previous lemma, we only have to prove the implication (i) $\Longrightarrow$ (ii). Suppose that T is a lattice homomorphism, and let $y \in L$. Since

$$(Tf)(y) = \delta_y(Tf) = (\delta_y \circ T)(f),$$

we see that $\delta_y \circ T \in C(L)^*$ is a lattice homomorphism. By Lemma 16.4, there exist a scalar $g(y) \geq 0$ and a point $\psi(y) \in K$ satisfying

$$(Tf)(y) = g(y)f(\psi(y)).$$

Clearly, $g = T\mathbb{1}_K \in C(L)$. It remains to show that ψ is continuous. To this end, take a net $(y_\alpha) \subset L$ such that $y_\alpha \to y$ in L, $g(y) > 0$ and $g(y_\alpha) > 0$ for all α. Then $g(y_\alpha) \to g(y)$, and for all $f \in C(K)$ we have

$$g(y_\alpha)f(\psi(y_\alpha)) = (Tf)(y_\alpha) \longrightarrow (Tf)(y) = g(y)f(\psi(y)).$$

Since $g(y_\alpha), g(y) \neq 0$, this implies that

$$f(\psi(y_\alpha)) \longrightarrow f(\psi(y))$$

for all $f \in C(K)$. This establishes the continuity of ψ. $\square$

Recall that a positive linear operator $T : C(K) \to C(L)$ is called a *Markov operator*, if the identity $T\mathbb{1}_K = \mathbb{1}_L$ holds, see Definition 10.28. Further, a semiflow on K is defined analogously to the previous section: it is a continuous mapping $\varphi : \mathbb{R}^+ \times K \to K$ satisfying

$$\begin{aligned}
\varphi(0, x) &= x && \text{for all } x \in K, \text{ and} \\
\varphi(t, \varphi(s, x)) &= \varphi(t + s, x) && \text{for all } t, s \geq 0, \ x \in K.
\end{aligned} \tag{16.4}$$

This leads to the following characterization of Koopman semigroups on $C(K)$ spaces.

Theorem 16.6. *Let K be a compact Hausdorff space, $E = C(K)$, and $(T(t))_{t \geq 0}$ a strongly continuous semigroup on E. Then the following are equivalent.*

(i) *Each $T(t)$, $t \geq 0$, is a Markov lattice homomorphism.*

(ii) *$(T(t))_{t \geq 0}$ is induced by a continuous semiflow $\varphi : \mathbb{R}^+ \times K \to K$, i.e.,*

$$(T(t)f)(x) = f(\varphi(t, x)), \quad f \in E, \ x \in K, \ t \geq 0.$$

Proof. (ii) $\Longrightarrow$ (i): This implication was the content of Proposition 16.2.

(i) $\Longrightarrow$ (ii): From the proof of Lemma 16.5 we see that if the operator T is Markov, then we have $g(s) = 1$ for each $s \in L$. Hence, Lemma 16.5 implies the existence of a function $\varphi : \mathbb{R}^+ \times K \to K$ such that $(T(t)f)(x) = f(\varphi(t, x))$. The semiflow property defined in (16.4) follows directly from the semigroup property of the operators $T(t)$, since

$$f(\varphi(t+s,x)) = (T(t+s)f)(x) = (T(t)T(s)f)(x)$$
$$= (T(t)(f(\varphi(s,x)))) = f(\varphi(t,\varphi(s,x)))$$

holds for each $f \in C(K)$, implying that $\varphi(t+s,x) = \varphi(t,\varphi(s,x))$. It thus remains to show that φ is continuous in t. Take a sequence $(t_n) \subset \mathbb{R}^+$ such that $t_n \to t$. Then, by the strong continuity of the semigroup,

$$f(\varphi(t_n,x)) = (T(t_n)f)(x) \longrightarrow (T(t)f)(x) = f(\varphi(t,x))$$

for all $f \in C(K)$ and $x \in K$. Hence, $\varphi(t_n,x) \to \varphi(t,x)$, implying continuity of φ in t. $\qquad\square$

Remark 16.7. Note that $C(K)$ is not only a Banach lattice, but also a C^*-*algebra*. If the conditions of Lemma 16.5 are satisfied, then T is also a C^*-algebra homomorphism. This leads us to the following additional equivalences in Theorem 16.6.

(iii) Each operator $T(t)$ is a C^*-algebra homomorphism.

(iv) The generator A is a *derivation*, i.e., the domain $D(A)$ is a *-subalgebra and the identities $A(f \cdot g) = f \cdot Ag + Af \cdot g$ and $\overline{Af} = A\overline{f}$ hold for all $f, g \in D(A)$.

It seems desirable to have not only an algebraic, but also an order theoretic characterization of generators of Koopman semigroups. Before deriving one such characterization, let us make the following informal comment. If $(T(t))_{t \geq 0}$ is a semigroup of lattice homomorphisms, then

$$T(t)|f| = |T(t)f|$$

for each $f \in E$. Accepting that the real function $\mathrm{abs}(s) = |s|$ has derivative $\mathrm{abs}'(s) = \mathrm{sgn}\, s$, we obtain formally that

$$A|f| = \frac{\mathrm{d}}{\mathrm{d}t}T(t)|f|\Big|_{t=0} = \frac{\mathrm{d}}{\mathrm{d}t}|T(t)f|\Big|_{t=0} = (\mathrm{sgn}\, f)Af.$$

To give this calculation a meaning and to see how to interpret it correctly, we need the following preparations.

Definition 16.8. Let X be a Banach space, $\eta : X \to X$ a mapping, $f, u \in X$. The mapping η is called *right Gâteaux differentiable in f in direction u*, if

$$\partial_u \eta(f) := \lim_{t \downarrow 0} \frac{\eta(f+tu) - \eta(f)}{t} \tag{16.5}$$

exists. The function η is called *right Gâteaux differentiable in f* if $\partial_u \eta(f)$ exists for all directions $u \in X$.

Example 16.9. Let $X = \mathbb{C}$ and consider the function $\eta(z) = |z|$. Using the notation $\operatorname{sgn} z = \frac{z}{|z|}$ for $z \neq 0$ and $\operatorname{sgn} 0 = 0$, we can show that η is right Gâteaux differentiable and

$$\partial_u \eta(z) = \begin{cases} \operatorname{Re}((\operatorname{sgn} \bar{z})u) & \text{if } z \neq 0, \\ |u| & \text{if } z = 0. \end{cases}$$

This can be obtained by direct calculations which are left as Exercise 3.

We shall use the following chain rule for the right Gâteaux derivative.

Lemma 16.10. *Let* $\psi : \mathbb{R} \to X$ *be right differentiable at* $a \in \mathbb{R}$ *with the right derivative* $\psi'(a)$, *and suppose that* $\eta : E \to E$ *is Lipschitz continuous. If* η *is right Gâteaux differentiable at the point* $\psi(a)$ *in the direction* $\psi'(a)$, *then* $\eta \circ \psi$ *is right differentiable at* a *and its right derivative in* a *equals*

$$(\eta \circ \psi)'(a) = \partial_{\psi'(a)} \eta(\psi(a)).$$

Proof. Take $L > 0$ such that $\|\eta(f) - \eta(g)\| \leq L\|f - g\|$ holds for all $f, g \in X$. Then

$$\lim_{t \downarrow 0} \left\| \frac{1}{t} \big(\eta(\psi(a+t)) - \eta(\psi(a)) \big) - \partial_{\psi'(a)} \eta(\psi(a)) \right\|$$

$$\leq \limsup_{t \downarrow 0} \left\| \frac{1}{t} \big(\eta(\psi(a+t)) - \eta(\psi(a) + t\psi'(a)) \big) \right\|$$

$$+ \limsup_{t \downarrow 0} \left\| \frac{1}{t} \big(\eta(\psi(a) + t\psi'(a)) - \eta(\psi(a)) \big) - \partial_{\psi'(a)} \eta(\psi(a)) \right\|$$

$$\leq L \limsup_{t \downarrow 0} \left\| \frac{1}{t} (\psi(a+t) - \psi(a)) - \psi'(a) \right\| + 0 = 0,$$

by the right Gâteaux differentiability of η at $\psi(a)$ in the direction of $\psi'(a)$ and by the right differentiability of ψ at $a \in \mathbb{R}$. $\qquad \square$

Let us introduce some further notation. For $f, g \in \mathrm{C}(K)$ we write[17]

$$((\operatorname{sgn} f)(g))(x) := (\operatorname{sgn} f(x)) \cdot g(x)$$

and

$$((\widehat{\operatorname{sgn}} f)(g))(x) := \begin{cases} (\operatorname{sgn} f)(g)(x) & \text{if } f(x) \neq 0, \\ |g(x)| & \text{if } f(x) = 0. \end{cases}$$

Note that, though this may not be a continuous map, we can extend the duality map to it by the following straightforward construction. For $\mu \in \mathrm{C}(K)^*$ we define

$$\langle (\widehat{\operatorname{sgn}} f)(g), \mu \rangle = \int_K (\widehat{\operatorname{sgn}} f)(g) \mathrm{d}\mu.$$

---

[17]Compare with (14.17).

Lemma 16.11. *Let $(T(t))_{t \geq 0}$ be a strongly continuous semigroup on $E = C(K)$ with generator A. Then for every $f \in D(A)$ and $\mu \in E^*$ we have*

$$\frac{d}{dt} \langle |T(t)f|, \mu \rangle \Big|_{t=0} = \langle \text{Re}((\widehat{\text{sgn}}\bar{f})(Af)), \mu \rangle.$$

Proof. For $f \in D(A)$ and $x \in K$, define the function

$$\eta(t) = (T(t)f)(x).$$

It is right differentiable in 0 with $\eta'(0) = (Af)(x)$. The chain rule, Lemma 16.10, and Example 16.9 imply that

$$|\eta(0)|' = \text{Re}((\widehat{\text{sgn}}\bar{f})(Af))(x).$$

Moreover,

$$\frac{1}{t}\big||T(t)f| - |f|\big| \leq \frac{1}{t}|T(t)f - f|$$

implies that

$$\sup_{0 < t \leq 1} \frac{1}{t}\big\||T(t)f| - |f|\big\| < \infty.$$

Hence, the functions

$$k_t(x) = \frac{1}{t}(|T(t)f(x)| - |f(x)|)$$

are uniformly bounded on K. Lebesgue's dominated convergence theorem (see Theorem A.23) then implies that

$$\frac{d}{dt} \langle |T(t)f|, \mu \rangle \Big|_{t=0} = \lim_{t \downarrow 0} \langle k_t, \mu \rangle = \langle \text{Re}((\widehat{\text{sgn}}\bar{f})(Af)), \mu \rangle. \qquad \square$$

We are now able to characterize a Koopman semigroup by its generator in the following way.

Theorem 16.12. *A strongly continuous semigroup $(T(t))_{t \geq 0}$ with the generator A on the Banach lattice $E = C(K)$ is a semigroup of lattice homomorphisms if and only if the Kato equality*

$$\langle \text{Re}\left((\widehat{\text{sgn}}\bar{f})(Af)\right), \mu \rangle = \langle |f|, A^*\mu \rangle \tag{16.6}$$

holds for all $f \in D(A)$ and $\mu \in D(A^*)$.

Proof. Suppose that $(T(t))_{t \geq 0}$ is a semigroup of lattice homomorphisms and let $f \in D(A)$ and $\mu \in D(A^*)$. Lemma 16.11 implies that

$$\langle \text{Re}((\widehat{\text{sgn}}\bar{f})(Af)), \mu \rangle = \frac{d}{dt} \langle |T(t)f|, \mu \rangle \Big|_{t=0} = \frac{d}{dt} \langle T(t)|f|, \mu \rangle \Big|_{t=0} = \langle |f|, A^*\mu \rangle.$$

Conversely, suppose that (16.6) holds. We have to show that

$$T(t)|f| = |T(t)f|$$

holds for all $t > 0$ and $f \in E$. Since $D(A)$ is dense, it is sufficient to show this equality for $f \in D(A)$. Moreover, it suffices to show

$$\langle |T(t)f|, \mu \rangle = \langle T(t)|f|, \mu \rangle \tag{16.7}$$

for all $\mu \in D(A^*)$.

Let $\mu \in D(A^*)$, $t > 0$, and define the function

$$\xi(s) = \langle T(t-s)|T(s)f|, \mu \rangle$$

for $s \in [0, t]$. If we show that ξ is constant, then $\xi(0) = \xi(t)$, which is exactly relation (16.7).

Since $\mu \in D(A^*)$,

$$\lim_{h \downarrow 0} \frac{1}{h} \langle g, (T(t-(s+h)) - T(t-s))^* \mu \rangle = -\langle g, A^* T(t-s)^* \mu \rangle$$

holds for all $g \in E$. Consequently, by the Uniform Boundedness Principle (see Theorem A.15), we see that

$$\limsup_{h \downarrow 0} \frac{1}{h} \|(T(t-(s+h)) - T(t-s))^* \mu\| < \infty.$$

Hence,

$$\lim_{h \downarrow 0} \frac{1}{h} \langle |T(s+h)f|, (T(t-(s+h)) - T(t-s))^* \mu \rangle = -\langle |T(s)f|, A^* T(t-s)^* \mu \rangle.$$

Using this equality, we obtain

$$\lim_{h \downarrow 0} \frac{1}{h} (\xi(s+h) - \xi(s))$$

$$= \lim_{h \downarrow 0} \frac{1}{h} (\langle T(t-(s+h))|T(s+h)f|, \mu \rangle - \langle T(t-s)|T(s+h)f|, \mu \rangle$$

$$+ \langle T(t-s)|T(s+h)f| - T(t-s)|T(s)f|, \mu \rangle)$$

$$= -\langle |T(s)f|, A^* T(t-s)^* \mu \rangle + \lim_{h \downarrow 0} \frac{1}{h} \langle T(t-s)|T(s+h)f| - T(t-s)|T(s)f|, \mu \rangle.$$

By Lemma 16.11,

$$\lim_{h \downarrow 0} \frac{1}{h} (\xi(s+h) - \xi(s))$$

$$= -\langle |T(s)f|, A^* T(t-s)^* \mu \rangle + \langle \mathrm{Re}((\widehat{\mathrm{sgn}\, T(s)f})(AT(s)f)), T(t-s)^* \mu \rangle.$$

By the Kato Equality (16.6), this last expression equals zero, proving that ξ is constant. $\square$

16.3   Applications of Koopman Semigroups

We show by some simple examples how one can translate properties of the semiflow into appropriate properties of the Koopman semigroup. As before, let K be a compact Hausdorff space, $\varphi : \mathbb{R}^+ \times K \to K$ a continuous semiflow on it, and $(T(t))_{t \geq 0}$ the associated Koopman semigroup.

Lemma 16.13. *A closed subset $L \subset K$ is invariant under φ if and only if the ideal J_L generated by L is invariant under the Koopman semigroup $(T(t))_{t \geq 0}$.*

Proof. Recall from Proposition 10.13 the characterization of closed ideals in $C(K)$. Suppose now that L is invariant under φ, that is, $\varphi(t, L) \subset L$. Then, by definition, taking a function $f \in J_L$, we see that $f \circ \varphi(t, \cdot) \in J_L$, showing that

$$T(t)J_L \subset J_L.$$

For the other direction, let us assume that $T(t)J_L \subset J_L$ and $x \in K \setminus L$. By Urysohn's lemma (see Lemma A.19), there is $f \in J_L$ such that $f(x) = 1$. On the other hand, since $T(t)J_L \subset J_L$, we see that

$$f(\varphi(t, y)) = 0 \quad \text{for all } y \in L.$$

This implies that $x \notin \varphi(t, L)$. Since $x \in K \setminus L$ was arbitrary, the invariance $\varphi(t, L) \subset L$ follows. $\qquad\square$

We need some standard notions from topological dynamical systems.

A semiflow is called *minimal* if it has no nontrivial closed invariant sets. In view of the characterization of ideals in $C(K)$ in Proposition 10.13, the following is a straightforward consequence of Lemma 16.13.

Corollary 16.14. *The semiflow φ is minimal if and only if the Koopman semigroup $(T(t))_{t \geq 0}$ is irreducible.*

The semiflow is called *topologically (forward) transitive* if there is a point $x \in K$ such that its orbit $\operatorname{orb}(x) := \{\varphi(t, x) : t \geq 0\}$ is dense in K.

Proposition 16.15. *If the semiflow is topologically transitive, then the generator A of the Koopman semigroup satisfies*

$$\dim \ker(A) = 1.$$

Proof. If $f \in C(K)$ is a constant function, then $T(t)f = f$, hence $f \in \ker(A)$, meaning that $\dim \ker(A) \geq 1$ is true for every Koopman semigroup.

Further, if $T(t)f = f$ holds for $f \in C(K)$, then $f(\varphi(t, x)) = T(t)f(x) = f(x)$ implies that f is constant along orbits. Hence, if there is a point $x \in K$ with dense orbit, then f has to be constant, implying that $\ker(A)$ consists of the constant functions. $\qquad\square$

Definition 16.16. We say that a positive measure $\mu \in C(K)^*$ is an *invariant Borel measure* for the semiflow φ if

$$\mu(\varphi^{-1}(t, H)) = \mu(H)$$

holds for each $t \geq 0$ and for each Borel measurable set $H \subset K$.

Interesting and important is the connection between invariant measures and the adjoint of the Koopman semigroup.

Lemma 16.17. *A measure $\mu \in C(K)^*$ is an invariant probability measure for the continuous semiflow φ if and only if it is an eigenvector associated to the eigenvalue 1 for the adjoint of the corresponding Koopman semigroup, i.e., if and only if*

$$T^*(t)\mu = \mu$$

holds for all $t \geq 0$.

Proof. Note that, by the definition of the Koopman semigroup, a measure μ is invariant if and only if

$$\langle f, T^*(t)\mu \rangle = \langle T(t)f, \mu \rangle = \int_K f(\varphi(t, x)) \mathrm{d}\mu(x)$$

$$= \int_K f(x) \mathrm{d}\mu(\varphi^{-1}(t, x)) = \int_K f(x) \mathrm{d}\mu(x) = \langle f, \mu \rangle$$

holds for all $f \in C(K)$ and $t \geq 0$, and hence

$$T^*(t)\mu = \mu$$

for all $t \geq 0$. $\qquad\square$

It turns out that continuous flows always have an invariant measure.

Theorem 16.18 (Krylov–Bogoliubov). *Let K be a compact Hausdorff space and $\varphi : \mathbb{R}^+ \times K \to K$ a continuous semiflow. Then there is at least one invariant probability measure for the semiflow φ.*

Proof. Fix $y \in K$ and let $\mu_0 := \delta_y$ be the Dirac measure supported at y. Define the probability measures μ_t for which

$$\int_K f(x) \mathrm{d}\mu_t(x) = \frac{1}{t} \int_0^t \int_K T(t)f(x) \mathrm{d}\mu_0(x) \mathrm{d}t$$

holds for all $f \in C(K)$. This means that

$$\mu_t = \frac{1}{t} \int_0^t T^*(t)\mu_0 \mathrm{d}t,$$

where the integral is defined in the weak*-topology. By the Banach–Alaoglu theorem, Theorem A.31, there is a weak*-accumulation point μ of μ_t as $t \to \infty$. Since μ_t are positive probability measures, so is μ. By Exercise 5, this measure satisfies

$$T^*(t)\mu = \mu,$$

hence it is an invariant measure. □

We can apply the theory of irreducible semigroups here.

Proposition 16.19. *If the semiflow φ is minimal, then all invariant probability measures are strictly positive.*

Proof. This is a straightforward consequence of Proposition 14.12. □

Suppose again that K is a compact Hausdorff space and $\varphi : \mathbb{R}^+ \times K \to K$ a continuous semiflow, and let μ be an invariant measure for φ.

For each $t > 0$ and $f \in \mathrm{L}^1(K, \mu)$, we define the measure

$$\nu_t(H) := \int_{\varphi^{-1}(t,H)} f \mathrm{d}\mu.$$

Notice that if for $B \subset K$ we have $\mu(B) = 0$, then $\nu_t(B) = 0$. Hence, by the Radon–Nikodým theorem (see Theorem A.25), there is a unique $P(t)f \in \mathrm{L}^1(K, \mu)$ such that

$$\int_H P(t)f \mathrm{d}\mu = \int_{\varphi^{-1}(t,H)} f \mathrm{d}\mu.$$

Definition 16.20. The operator family $(P(t))_{t \geq 0}$ defined above is called the *Perron–Frobenius semigroup* associated with the semiflow φ.

We summarize the main properties of the Perron–Frobenius semigroup.

Proposition 16.21. *The Perron–Frobenius semigroup is a strongly continuous positive contraction semigroup on $\mathrm{L}^1(K, \mu)$. It can be identified with the restriction of the adjoint of the Koopman semigroup when $\mathrm{L}^1(K, \mu)$ is identified with the set of absolutely continuous measures with respect to μ.*

Note that, since the Perron–Frobenius semigroup is a restriction of the adjoint of the Koopman semigroup to an invariant subspace, the only thing to prove is its strong continuity. Then one shows directly that the Perron–Frobenius semigroup is weakly continuous. The proof is finished by applying the fact that weakly continuous semigroups are already strongly continuous.

16.4   Notes and Remarks

A standard reference for ordinary differential equations is Amann [3]. The generator A is sometimes referred to as the "Lie generator" of the semiflow φ, see for example Neuberger [104]. The contents of Proposition 16.2 and the boundary conditions appearing in Remark 16.3 are due to Ulmet [144]. For the fact that a dense and invariant subset of the domain of the generator is a core in the proof of Proposition 16.2, see Engel and Nagel [43, Proposition II.1.7].

The operator theoretic characterization of Koopman semigroups as Markov lattice homomorphism and the Kato equality are due to Nagel, Arendt, and the research group in Tübingen, and is documented in Nagel (ed.) [101, Sections B-II.2,3]. Relation (16.6) has its origins in Kato's investigations on the positivity of Schrödinger semigroups in L^2, see [72].

For the proof of Proposition 16.21 the main technical tool is the fact that weakly continuous semigroups are already strongly continuous, see Engel and Nagel, [43, Theorem I.5.8] and Exercise 6.

Basic properties of Perron-Frobenius and Koopman semigroups can be also found in the book by Lasota and Mackey [83]. A characterization of generators of Koopman semigroups in L^p spaces can be found in Edeko and Kühner [38].

Applications of Koopman semigroups to ergodic theory are numerous. The connection of nonlinear dynamical systems and their "linearization" using the Koopman operator has a long history, and in these notes here we only scratched the surface. For a comprehensive treatment of the time discrete case we refer to the monograph by Eisner, Farkas, Haase and Nagel [40].

16.5   Exercises

1. Let $F : \mathbb{R} \to \mathbb{R}$ be continuously differentiable with $\sup_{x \in \mathbb{R}} |F'(x)| < \infty$. Define the flow $\varphi : \mathbb{R} \times \mathbb{R} \to \mathbb{R}$ as the solution of the nonlinear ODE

$$\begin{cases} \dot{y}(t) = F(y(t)), \\ y(0) = s, \end{cases}$$

i.e., $\varphi(t, s) := y(t)$. Take $E := C_0(\mathbb{R})$ and define

$$\big(T(t)f\big)(s) := f\big(\varphi(t, s)\big)$$

for $t \geq 0$, $s \in \mathbb{R}$.

   a) Show that $(T(t))_{t \geq 0}$ is a positive contraction semigroup (i.e., of type $(1, 0)$) and identify its generator.

   b) What is the corresponding abstract Cauchy problem? Which partial differential equation can we associate with it? Relate the semigroup $(T(t))_{t \geq 0}$ to the method of characteristics.

2. Consider the ordinary differential equation

$$\begin{cases} \dot{y}(t) = y(t)(y(t) - 1), \\ y(0) = s \in (0, 1), \end{cases}$$

Write down the explicit formula for the corresponding Koopman semigroup and its generator. Identify all invariant ideals of this semigroup.

3. Show that $z \mapsto |z|$ is right Gâteaux differentiable as stated in Example 16.9.

4. Let $\Omega \subset \mathbb{R}^n$ be an open set, φ a continuous semiflow on it, and define the Koopman semigroup on $\mathrm{L}^2(\Omega)$ by the rule (16.3).

   a) Show that it consists of unitary operators.

   b) Show that it is strongly continuous. [Hint: Use that it is continuous for $f \in \mathrm{C}_0(\Omega)$.]

5. Consider the adjoint $(T^*(t))_{t\geq 0}$ of the Koopman semigroup $(T(t))_{t\geq 0}$ and show that

$$(I - T^*(t))\frac{1}{r}\int_0^r T^*(s)\nu ds = (I - T^*(r))\frac{1}{r}\int_0^t T^*(s)\nu ds$$

for $r > t$. Use this identity to show that each weak*-accumulation point of the Cesàro-means in the proof of Theorem 16.18 is a fixed point of the adjoint semigroup.

6. Show that the Perron–Frobenius semigroup is weakly continuous.

Chapter 17

Linear Boltzmann Transport Equations with Scattering

In this chapter we give an application of positive semigroup theory to linear transport equations. This is a wonderful piece of mathematics modeling neutron transport in a reactor which uses much of the theory we developed in this text.

17.1 The Reactor Problem

We want to model the time evolution of the motion of neutrons in an absorbing and scattering homogeneous medium. This problem is known as the *reactor problem*. We use the linearized Boltzmann transport equation which was originally developed in 1872 by L. Boltzmann when studying the kinetic theory of gases and is given by the integro-differential equation

$$\partial_t u(t,x,v) = -\sum_{j=1}^{3} v_j \partial_{x_j} u(t,x,v) - \sigma(x,v)u(t,x,v) + \int_V \zeta(x,v,v')u(t,x,v')\mathrm{d}v'.$$

(17.1)

Here, $u(t,x,v)$ represents the density distribution of the neutrons at time t depending on the space variable $x \in D \subseteq \mathbb{R}^3$ and on the velocity v. By D we denote the interior of the vessel in which neutron transport takes place and which is filled by some background material surrounded by a total absorber. The neutrons migrate in D and are scattered by the inner or absorbed by the outer material. We suppose that neutrons do not interact with each other.

The term $-\sum_{j=1}^{3} v_j \partial_{x_j} u$ in equation (17.1) is called the *free streaming term* and is responsible for the motion of particles between collisions. The second term on the right in equation (17.1) corresponds to collisions causing *absorption*, while the third term describes scattering of neutrons: particles at the position x with the incoming velocity v' go into particles at x with the outgoing velocity v. This transition is governed by a *scattering kernel* $\zeta(x,v,v')$.

The fact that $u(t, \cdot, \cdot)$ should describe a density suggests to require that $u(t, \cdot, \cdot)$ is an element of $L^1(D \times V)$ for all $t \geq 0$. Following this line and introducing the vector-valued function $u(t) := u(t, \cdot, \cdot)$, equation (17.1) is equivalent to the abstract equation

$$\dot{u}(t) = (A + K_\zeta)u(t) := (A_0 - M_\sigma)u(t) + K_\zeta u(t), \quad t \geq 0,$$

where $u(t) \in L^1(D \times V)$, $A_0 := -\sum_{j=1}^{3} v_j D_j$ denotes the *free streaming operator* on a suitable domain, M_σ is the multiplication operator induced by σ and called the *absorption operator*, while the *scattering operator* K_ζ is defined by

$$(K_\zeta f)(x, v) := \int_V \zeta(x, v, v')f(x, v')\mathrm{d}v', \quad (x, v) \in D \times V,\ f \in L^1(D \times V).$$

We take an appropriate domain $D(A + K_\zeta)$ to show that the operator $A + K_\zeta$ generates a positive C_0-semigroup on $L^1(D \times V)$ and to describe the qualitative properties of the solution.

17.2   The One-dimensional Reactor Problem

In order to simplify our exposition we study here only the one-dimensional case. The *one-dimensional reactor problem* is given by the following transport equation

$$\begin{cases} \partial_t u(t, x, v) = -v\partial_x u(t, x, v) - \sigma(x, v)u(t, x, v) + \int_V \zeta(x, v, v')u(t, x, v')\mathrm{d}v', \\ \qquad\qquad\qquad\qquad\qquad t \geq 0,\ (x, v) \in D \times V, \\ u(t, 0, v) = 0 \text{ if } v > 0 \quad \text{and} \quad u(t, 1, v) = 0 \text{ if } v < 0, \quad t \geq 0, \\ u(0, x, v) = f(x, v), \qquad (x, v) \in D \times V, \end{cases}$$

(17.2)

where $0 \leq \sigma \in L^\infty(D \times V)$, $0 \leq \zeta \in L^\infty(D \times V \times V)$, $D := [0, 1]$, and $V := \{v \in \mathbb{R} : v_{\min} \leq |v| \leq v_{\max}\}$ for given constants $0 < v_{\min} < v_{\max} < \infty$. Here ∂_x denotes the partial derivative with respect to x in the sense of distributions, see Appendix A.11 for the definition.

We rewrite problem (17.2) as an abstract Cauchy problem on the Banach lattice $L^1(D \times V)$. To do so we define the *free streaming operator* A_0 by

$$(A_0 f)(x, v) := -v\partial_x f(x, v), \quad \text{with}$$

$$D(A_0) := \left\{ f \in L^1(D \times V) : v\partial_x f \in L^1(D \times V), \quad \begin{array}{l} f(0, v) = 0 \text{ if } v > 0 \\ f(1, v) = 0 \text{ if } v < 0 \end{array} \right\},$$

the *absorption operator* M_σ by

$$(M_\sigma f)(x, v) := \sigma(x, v)f(x, v), \quad (x, v) \in D \times V,\ f \in L^1(D \times V),$$

and the *scattering operator* K_ζ as above by

$$(K_\zeta f)(x, v) := \int_V \zeta(x, v, v') f(x, v') dv', \quad (x, v) \in D \times V, \, f \in L^1(D \times V).$$

We note that M_σ and K_ζ are both in $\mathcal{L}(L^1(D \times V))$, since $\sigma \in L^\infty(D \times V)$ and $\zeta \in L^\infty(D \times V \times V)$. Moreover, if $f \in D(A_0)$, then for a.e. $v \in V$, the function $f(\cdot, v)$ belongs to $W^{1,1}(D) = AC(D)$, the set of absolutely continuous functions on D, see Exercise 1. Hence, it can be extended to a continuous function on $\overline{D}$, and so $f(0, v)$ and $f(1, v)$ make sense for a.e. fixed $v \in V$.

We first study the free streaming operator A_0. By a direct computation one can see that $(0, \infty) \subseteq \rho(A_0)$ and

$$(R(\lambda, A_0)f)(x, v) = \begin{cases} \dfrac{1}{v} \displaystyle\int_0^x e^{-\frac{\lambda}{v}(x-x')} f(x', v) \, dx' & \text{if } v > 0, \\ -\dfrac{1}{v} \displaystyle\int_x^1 e^{-\frac{\lambda}{v}(x-x')} f(x', v) \, dx' & \text{if } v < 0, \end{cases} \qquad (17.3)$$

for $(x, v) \in D \times V$ and $f \in L^1(D \times V)$. Hence,

$$\|R(\lambda, A_0)\| \leq \frac{1}{\lambda} \qquad \text{for all } \lambda > 0.$$

Therefore, by Theorem 11.1, A_0 with domain $D(A_0)$ generates a C_0-semigroup $(T_0(t))_{t>0}$ of contractions on $L^1(D \times V)$. Moreover, $(T_0(t))_{t \geq 0}$ is positive since $R(\lambda, A_0) \geq 0$ for all $\lambda > 0$, see Corollary 11.4. On the other hand, formula (17.3) implies that

$$(R(\lambda, A_0)f)(x, v) = \int_0^\infty e^{-\lambda t} \chi_D(x - vt) f(x - vt, v) \, dt$$

for $(x, v) \in D \times V$, $f \in L^1(D \times V)$. So, by the uniqueness of the Laplace transform, we obtain

$$(T_0(t)f)(x, v) = \chi_D(x - tv) f(x - tv, v), \quad (x, v) \in D \times V, \, f \in L^1(D \times V). \quad (17.4)$$

Moreover, since the absorption operator M_σ is bounded, it follows that

$$A := A_0 - M_\sigma \quad \text{with } D(A) = D(A_0)$$

generates a C_0-semigroup. It is not difficult to see that this semigroup is given by the formula

$$(T(t)f)(x, v) = e^{-\int_0^t \sigma(x - \tau v, v) d\tau} (T_0(t)f)(x, v) \qquad (17.5)$$

for $(x, v) \in D \times V$, $f \in L^1(D \times V)$ and is also positive. It is called the *streaming semigroup*.

We now suppose that the scattering kernel ζ satisfies

$$\zeta(x, v, v') > 0 \text{ for all } (x, v, v') \in D \times V \times V. \tag{17.6}$$

Then the scattering operator K_ζ is positive. Therefore, by Corollary 11.7 and Proposition 11.6, the *transport operator* $A + K_\zeta$ with domain $D(A_0)$ generates the positive C_0-semigroup $(S(t))_{t\geq0}$ given by the Dyson–Phillips expansion

$$S(t) = \sum_{k=0}^{\infty} U_k(t), \quad \text{where } U_0(t) = T(t) \text{ and}$$

$$U_{k+1}(t)f = \int_0^t U_k(t-s)K_\zeta T(s)f \, ds, \quad f \in \mathrm{L}^1(D \times V), \, t \geq 0, \, k \in \mathbb{N}. \tag{17.7}$$

This semigroup is called the *transport semigroup* and enjoys the following properties.

Proposition 17.1. *For the streaming semigroup $(T(t))_{t\geq0}$ and the transport semigroup $(S(t))_{t\geq0}$ we have*

$$0 \leq T(t) \leq S(t) \text{ for all } t \geq 0 \text{ and } \omega_0(S) = s(A + K_\zeta). \tag{17.8}$$

Proof. The first assertion follows from the positivity of K_ζ and the Dyson–Phillips expansion (17.7). The second is a consequence of Theorem 12.17. □

We shall see that under our assumptions the transport semigroup $(S(t))_{t\geq0}$ is even irreducible. First we show the following.

Lemma 17.2. *If a closed ideal I in $\mathrm{L}^1(D \times V)$ is $S(\cdot)$-invariant, then it is invariant under both $T_0(\cdot)$ and K_ζ.*

Proof. Assume that I is $S(\cdot)$-invariant. Since $0 \leq T(t) \leq S(t)$ for all $t \geq 0$, we deduce that I is $T(\cdot)$-invariant. Thus, formula (17.5) implies that I is $T_0(\cdot)$-invariant. By Proposition 11.6, we have

$$\lim_{t\downarrow0} \frac{1}{t}(S(t)f - T(t)f) = \lim_{t\downarrow0} \frac{1}{t}\int_0^t T(t-s)K_\zeta S(s)f \, ds = K_\zeta f$$

for $f \in \mathrm{L}^1(D \times V)$. Since I is closed and invariant under both $S(t)$ and $T(t)$, we conclude that I is also K_ζ-invariant. □

Lemma 17.3. *For the transport semigroup $(S(t))_{t\geq0}$ defined in (17.7) the following holds.*

a) *The remainder $R_2(t) := \sum_{k=2}^{\infty} U_k(t)$, $t \geq 0$, of the Dyson–Phillips expansion (17.7) is a weakly compact operator on $\mathrm{L}^1(D \times V)$.*

b) *If the scattering kernel ζ satisfies condition (17.6), then the transport semigroup $(S(t))_{t\geq0}$ is irreducible.*

Proof. a) By equation (17.5), $T(t) \leq T_0(t)$ for $t \geq 0$. For $0 \leq f \in L^1(D \times V)$ and the positive operator K_ζ we have

$$(K_\zeta T(t) K_\zeta f)(x, v) \leq (K_\zeta T_0(t) K_\zeta f)(x, v)$$

$$\leq \|\zeta\|_\infty^2 \int_V \int_D \chi_D(x - tv'') f(x - tv'', v') dv'' dv'$$

$$\leq t^{-1} \|\zeta\|_\infty^2 \int_V \int_D f(x', v') \, dx' dv' \quad \text{for } t > 0.$$

Hence

$$K_\zeta T(t) K_\zeta \leq \frac{\|\zeta\|_\infty^2}{t} \Xi, \tag{17.9}$$

where Ξ is the bounded linear operator on $L^1(D \times V)$ defined by

$$\Xi f := \left(\int_V \int_D f(x, v) \, dx \, dv \right) \chi_{D \times V}, \qquad f \in L^1(D \times V).$$

By the definition of the terms $U_k(t)$ in the Dyson–Phillips series in (17.7) one can see that

$$R_{m+1}(t) := \sum_{k=m+1}^{\infty} U_k(t) = \int_0^t T(t - s) K_\zeta R_m(s) \, ds, \quad t \geq 0, \, m \in \mathbb{N}. \tag{17.10}$$

In particular,

$$R_2(t) = \int_0^t \int_0^{t-s_2} T(s_1) K_\zeta T(s_2) K_\zeta S(t - s_1 - s_2) \, ds_1 \, ds_2 \quad \text{for } t \geq 0.$$

Taking $t > \varepsilon > 0$ and considering

$$R_{2,\varepsilon}(t) := \int_\varepsilon^t \int_0^{t-s_2} T(s_1) K_\zeta T(s_2) K_\zeta S(t - s_1 - s_2) \, ds_1 \, ds_2,$$

it can be verified that

$$\lim_{\varepsilon \to 0} \|R_{2,\varepsilon}(t) - R_2(t)\| = 0 \quad \text{for all } t > 0.$$

On the other hand, inequality (17.9) implies that

$$R_{2,\varepsilon}(t) \leq \|\zeta\|_\infty^2 \int_\varepsilon^t \int_0^{t-s_2} \frac{1}{s_2} T(s_1) \Xi S(t - s_1 - s_2) \, ds_1 \, ds_2.$$

From the definition of $(T_0(t))_{t \geq 0}$ and since $0 \leq T(t) < T_0(t)$, we see that

$$T(t) \Xi \leq \Xi$$

for the order in $\mathcal{L}(\mathrm{L}^1(D \times V))$. Now, for $0 \leq f \in \mathrm{L}^1(D \times V)$ and $s_1 + s_2 \leq t$, we obtain

$$\Xi S(t - s_1 - s_2) f = \left(\int_V \int_D (S(t - s_1 - s_2) f)(x, v) \,\mathrm{d}x \,\mathrm{d}v \right) \chi_{D \times V}$$

$$\leq M e^{\omega(t - s_1 - s_2)} \left(\int_V \int_D f(x, v) \,\mathrm{d}x \,\mathrm{d}v \right) \chi_{D \times V} = M e^{\omega(t - s_1 - s_2)} \Xi f,$$

where $M \geq 1$ and $\omega \in \mathbb{R}$ are such that $\|S(t)\| \leq M e^{\omega t}$ for all $t \geq 0$. Consequently,

$$R_{2,\varepsilon}(t) \leq M \|\zeta\|_\infty^2 \left(\int_\varepsilon^t \frac{1}{s_2} \int_0^{t - s_2} e^{\omega(t - s_1 - s_2)} \,\mathrm{d}s_1 \,\mathrm{d}s_2 \right) \Xi$$

$$= \frac{M \|\zeta\|_\infty^2}{\omega} \left(\int_\varepsilon^t \frac{e^{\omega(t - s_2)} - 1}{s_2} \,\mathrm{d}s_2 \right) \Xi$$

holds for all $t \geq 0$. This implies that $R_{2,\varepsilon}(t)$ is dominated by a one-dimensional operator. So, by Proposition A.37.b), we obtain that $R_{2,\varepsilon}(t)$ is weakly compact. Since $R_{2,\varepsilon}(t)$ converges in the operator norm to $R_2(t)$, Proposition A.37.a) implies that $R_2(t)$ is weakly compact for all $t \geq 0$. This proves the first assertion.

b) We recall from Proposition 10.15 that every closed ideal in $\mathrm{L}^1(D \times V)$ has the form
$$I = \{f \in \mathrm{L}^1(D \times V) : f \text{ vanishes a.e. on } \Omega\}$$
for some measurable subset $\Omega \subseteq D \times V$. We assume that I is $S(t)$-invariant for all $t \geq 0$. Then, by Lemma 17.2, I is K_ζ-invariant. Assume that $\Omega \neq \emptyset$. Since $\chi_{D \times V \setminus \Omega} \in I$, we obtain

$$(K_\zeta \chi_{D \times V \setminus \Omega})(x, v) = \int_V \zeta(x, v, v') \chi_{D \times V \setminus \Omega}(x, v') \mathrm{d}v'$$

$$= \int_{V \setminus \Omega_x} \zeta(x, v, v') \mathrm{d}v' = 0$$

for $(x, v) \in \Omega$ and $\Omega_x := \{v \in V : (x, v) \in \Omega\}$. Since ζ is strictly positive, we infer that $\lambda(V \setminus \Omega_x) = 0$, where λ is the Lebesgue measure on V. Hence $\chi_\Omega = \chi_{Y \times V}$ for some measurable subset Y of D.

On the other hand, again by Lemma 17.2, the set I is $T_0(t)$-invariant for all $t \geq 0$. Thus, it is also $R(\lambda, A_0)$-invariant for all $\lambda > 0$. Hence,

$$(R(\lambda, A_0) \chi_{D \times V \setminus \Omega})(x, v) = 0$$

for a.e. $(x, v) \in \Omega$. So, by using (17.3), one can see that

$$\int_0^x \chi_{D \setminus Y}(s) \,\mathrm{d}s = 0 \quad \text{and} \quad \int_x^1 \chi_{D \setminus Y}(s) \,\mathrm{d}s = 0.$$

Therefore, $\int_0^1 \chi_{D \setminus Y}(s) \,\mathrm{d}s = 0$, and this implies that $\chi_{D \setminus Y} = 0$. Consequently, $I = \{0\}$ or $I = \mathrm{L}^1(D \times V)$, and b) is proved. $\square$

We can now apply Theorem 14.18 to describe the asymptotic behaviour of the transport semigroup.

Theorem 17.4. *Suppose that ζ satisfies the positivity assumption* (17.6). *Then the transport semigroup* $(S(t))_{t \geq 0}$ *has balanced exponential growth. More precisely, there are strictly positive functions* $\varphi \in L^1(D \times V)$ *and* $\psi \in L^\infty(D \times V)$ *satisfying*

$$\iint_{D \times V} \varphi(x, v) \psi(x, v) \mathrm{dv} \ \mathrm{d}x = 1$$

such that

$$\|e^{-s(A + K_\zeta)t} S(t) - \psi \otimes \varphi\| \leq M e^{-\varepsilon t}$$

for all $t \geq 0$ and some constants $M \geq 0$ and $\varepsilon > 0$.

Proof. Since $v_{\min} > 0$, it follows that $(T(t))_{t \geq 0}$ is a nilpotent semigroup, i.e., there is $t_0 > 0$ such that

$$T(t) = 0 \quad \text{for all } t \geq t_0. \tag{17.11}$$

Here one can take $t_0 = \frac{1}{v_{\min}}$. Hence, $r(T(t)) = r_{\mathrm{ess}}(T(t)) = 0$ for all $t > 0$. So, by Lemma 17.3.a) and Theorem A.35, we have

$$\omega_{\mathrm{ess}}(S) = -\infty.$$

On the other hand, equation (17.11) implies that

$$U_1(t) = \int_0^t T(s) K_\zeta T(t - s) \ \mathrm{d}s = 0$$

for all $t \geq 2t_0$, and therefore

$$R_2(t) = S(t)$$

for all $t \geq 2t_0$. So, by Lemma 17.3.b) we see that $R_2(t)$ is irreducible for all $t \geq 2t_0$. Now, one can apply Lemma 17.3.a), Proposition A.36 and Theorem A.38, to conclude that $r(S(t)) = r(R_2(t)) > 0$ for all $t \geq 2t_0$. Therefore,

$$-\infty = \omega_{\mathrm{ess}}(S) < \omega_0(S).$$

Applying Theorem 14.18 to the transport semigroup $(S(t))_{t \geq 0}$, we finally obtain the assertions. $\qquad \square$

17.3 Notes and Remarks

The transport equation has been studied by many authors. Our presentation is based on the approach used in the papers by Greiner [55], Vidav [147], and Voigt [150, 151]. For more information see the monographs by Belleni-Morante [16], Belleni-Morante and McBride [17], Mokhtar–Kharroubi [97], [98], Engel and Nagel [43, Section IV.2], and Banasiak and Arlotti [11].

17.4   Exercises

1. Let $I := (a, b)$ be an open interval in $\mathbb{R}$. A function f is called *absolutely continuous* on I if for any $\varepsilon > 0$ there is $\delta > 0$ such that for every finite sequence of disjoint intervals $(a_k, b_k) \subset I$ with $\sum_k (b_k - a_k) < \delta$ we have $\sum_k |f(b_k) - f(a_k)| < \varepsilon$. Prove that the Sobolev space $W^{1,1}(I)$ is equal to $AC(I)$, the set of absolutely continuous functions on I. For the definition of Sobolev spaces, see Appendix A.11.

2. Let $(T(t))_{t \geq 0}$ be a C_0-semigroup with generator A such that the mapping $t \mapsto T(t)$ is norm continuous for $t \geq t_0$ and some $t_0 \geq 0$ and that $R(\lambda, A)T(t_0)$ is compact for some (and hence all) $\lambda \in \rho(A)$. Prove that the operators $T(t)$ are compact for all $t \geq t_0$.

3. For the problem studied in Section 17.2 verify the following:

   a) the expressions for the resolvent $R(\lambda, A_0)$ in (17.3),

   b) the formula for the semigroup $(T(t))_{t \geq 0}$ in (17.5),

   c) the formula for the remainder $R_{m+1}(t) := \sum_{k=m+1}^{\infty} U_k(t)$ in (17.10).

4. Consider the operator defined on $L^1(\mathbb{R}_+)$ by

$$Af := -f' - \mu f,$$

$$D(A) := \left\{ f \in L^1(\mathbb{R}_+) : f' \in L^1(\mathbb{R}_+) \text{ and } f(0) = \int_0^\infty \beta(a) f(a) \mathrm{d}a \right\},$$

   where $\mu, \beta \in L^\infty(\mathbb{R}_+)$ are two nonnegative functions.

   a) Prove that A generates a positive C_0-semigroup $(T(t))_{t \geq 0}$.

   b) Show that $(T(t))_{t \geq 0}$ is irreducible if and only if there is no $\tau \geq 0$ such that $\beta|_{(\tau, \infty)} = 0$ almost everywhere.

5. On $E := L^1[\alpha/2, 1]$ consider the operators defined by

$$A_0 f = -f' - (\mu + b) f,$$

with

$$D(A_0) = \{ f \in E : f' \in E, f(\alpha/2) = 0 \},$$

$$B f(s) = \begin{cases} 4b(2s) f(2s) & \text{if } \alpha/2 \leq s \leq 1/2, \\ 0 & \text{if } 1/2 \leq s \leq 1, \end{cases}$$

   for $f \in E$, where $0 \leq \mu \in C[\alpha/2, 1]$ and b is continuous and such that $b(s) > 0$ for $s \in (\alpha, 1)$ and $b(s) = 0$ otherwise.

   a) Prove that A_0 generates a positive semigroup $(T(t))_{t \geq 0}$, given by

$$T(t) f(s) = \begin{cases} e^{-\int_{s-t}^{s} (\mu(r) + b(r))\, \mathrm{d}r} f(s - t) & \text{if } s - t > \alpha/2, \\ 0 & \text{otherwise.} \end{cases}$$

b) Deduce that $A := A_0 + B$ with domain $D(A) = D(A_0)$ generates a positive C_0-semigroup $(S(t))_{t \geq 0}$ on E.

c) Prove that $S(t)$ is compact for all $t > 1 - \alpha/2$ (use Exercise 2).

d) Prove that $(S(t))_{t \geq 0}$ is irreducible on E.

e) Deduce that $(S(t))_{t \geq 0}$ possesses asynchronous exponential growth. (Use Theorem A.38 and Theorem 14.18.)

Chapter 18

Transport Problems in Networks

Consider a closed network of pipes or wires in which some material (water, electrons, information packets, goods, etc.) is flowing at constant speed on each edge, with no friction or loss. In the nodes of the network the material is redistributed into the pipes according to Kirchhoff's laws. Simplifying the physical laws and concentrating on the structure of the network, this situation can be described by a system of linear transport equations on the edges of a graph.

Such problems can be solved by positive semigroups and the theory we developed in the previous chapters. We are able to describe the asymptotic behavior of the solutions in terms of the properties of the underlying graph. It is a nice interplay of finite- and infinite-dimensional results on positive matrices and positive semigroups.

18.1 The Model and the Associated Abstract Cauchy Problem

The network is given by a simple, directed graph $G = (V, E)$ with vertices $V = \{v_1, \ldots, v_n\}$ and directed edges $E = \{e_1, \ldots, e_m\}$. We assume that G is connected, not necessarily strongly connected (recall the definitions in Section 6.1), but without *sinks*, that is, every vertex has an outgoing edge. We describe the transport process by the mass distribution on the edges and use the following basic assumptions.

- On every edge e_i the particles are flowing in only one direction with constant velocity $c_i > 0$.

- No mass is gained or lost during the process. In particular, no absorption takes place along the edges, and in every node a *Kirchhoff law* holds, i.e.,

$$\sum \text{incoming material} = \sum \text{outgoing material}.$$

- In every vertex v_i the incoming material is distributed into the outgoing edges e_j according to weights $w_{ij} \geq 0$ satisfying

$$\sum_{j:v_i \to e_j} w_{ij} = 1 \text{ for each } i = 1, \ldots, n. \tag{18.1}$$

The graph structure is described by the following $n \times m$ matrices:

- the *outgoing incidence matrix* $\Phi^- = \left(\Phi_{ij}^-\right)$, with

$$\Phi_{ij}^- := \begin{cases} 1 & \text{if } v_i \xrightarrow{e_j}, \\ 0 & \text{otherwise,} \end{cases}$$

- *incoming incidence matrix* $\Phi^+ = \left(\Phi_{ij}^+\right)$, with

$$\Phi_{ij}^+ := \begin{cases} 1 & \text{if } \xrightarrow{e_j} v_i, \\ 0 & \text{otherwise,} \end{cases}$$

- the *weighted outgoing incidence matrix* $\Phi_w^- = \left(\Phi_{w,ij}\right)$, with

$$\Phi_{w,ij}^- := \begin{cases} w_{ij} & \text{if } v_i \xrightarrow{e_j}, \\ 0 & \text{otherwise.} \end{cases}$$

The matrix $\Phi = \Phi^+ - \Phi^-$ is the *incidence matrix* of the directed graph G.

Incidence matrices describe the structure of the network completely and we obtain the $n \times n$ transposed adjacency matrix of the weighted graph G defined in Chapter 1 as

$$\mathbb{A} := \Phi^+ \left(\Phi_w^-\right)^\top. \tag{18.2}$$

This means that the nonzero entries of $\mathbb{A}$ correspond exactly to the edges of the graph, keeping track also of the appropriate weights:

$$\mathbb{A}_{ij} = \begin{cases} w_{jk} & \text{if } v_j \xrightarrow{e_k} v_i, \\ 0 & \text{otherwise.} \end{cases}$$

Analogously, the $m \times m$ matrix

$$\mathbb{B} := \left(\Phi_w^-\right)^\top \Phi^+ \tag{18.3}$$

is the *transposed adjacency matrix of the line graph* of G, which is roughly the graph obtained from G by exchanging the roles of the vertices and edges (maintaining the directions and the weights). Calculating its entries, one obtains

$$\mathbb{B}_{ij} = \begin{cases} w_{ki} & \text{if } \xrightarrow{e_j} v_k \xrightarrow{e_i}, \\ 0 & \text{otherwise.} \end{cases}$$

The following relations hold:

$$A\Phi^+ = \Phi^+\mathbb{B}, \quad \left(\Phi_w^-\right)^{\mathsf{T}} A = \mathbb{B}\left(\Phi_w^-\right)^{\mathsf{T}} \quad \text{and} \quad \Phi^-\left(\Phi_w^-\right)^{\mathsf{T}} = I_{\mathbb{C}^n}. \tag{18.4}$$

We shall also need the matrix

$$\mathbb{B}_C := C^{-1}\mathbb{B}C \quad \text{for} \quad C = \operatorname{diag}\left(c_j\right). \tag{18.5}$$

By condition (18.1), the matrices A and $\mathbb{B}$ as well as their powers are column stochastic, and by Gelfand's formula for the spectral radius we have

$$r(A) = \lim_{k\to\infty} \|A^k\|^{1/k} = 1 = r(\mathbb{B}). \tag{18.6}$$

In particular, 1 is an eigenvalue of both matrices, with $A^{\mathsf{T}}\mathbb{1} = \mathbb{1}$ and $\mathbb{B}^{\mathsf{T}}\mathbb{1} = \mathbb{1}$. Moreover, from definition of these matrices in (18.2), (18.3), and (18.5) we infer that

$$\sigma(A) \setminus \{0\} = \sigma(\mathbb{B}) \setminus \{0\} = \sigma(\mathbb{B}_C) \setminus \{0\}. \tag{18.7}$$

Example 18.1. Let us write the corresponding graph matrices for the graph presented in Figure 18.1. We obtain the entries of the adjacency matrix directly from the left picture:

$$A = \begin{pmatrix} 0 & 0 & 0 & 1 \\ 1-w & 0 & 0 & 0 \\ w & 0 & 0 & 0 \\ 0 & 1 & 1 & 0 \end{pmatrix}.$$

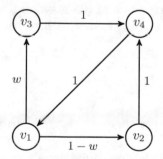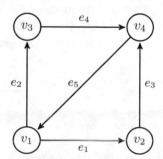

Figure 18.1: A weighted directed graph, left with weights, right with edge labels.

The entries in Φ_w^-, Φ^+, and $\mathbb{B}$ depend on the labelling of edges, and for the labelling indicated in Figure 18.1 we have

$$\Phi_w^- = \begin{pmatrix} 1-w & w & 0 & 0 & 0 \\ 0 & 0 & 1 & 0 & 0 \\ 0 & 0 & 0 & 1 & 0 \\ 0 & 0 & 0 & 0 & 1 \end{pmatrix}, \quad \Phi^+ = \begin{pmatrix} 0 & 0 & 0 & 0 & 1 \\ 1 & 0 & 0 & 0 & 0 \\ 0 & 1 & 0 & 0 & 0 \\ 0 & 0 & 1 & 1 & 0 \end{pmatrix},$$

$$\mathbb{B} = \begin{pmatrix} 0 & 0 & 0 & 0 & 1-w \\ 0 & 0 & 0 & 0 & w \\ 1 & 0 & 0 & 0 & 0 \\ 0 & 1 & 0 & 0 & 0 \\ 0 & 0 & 1 & 1 & 0 \end{pmatrix}.$$

We can also compute the spectra of the adjacency matrices

$$\sigma(\mathbb{A}) = \left\{0, e^{2\pi i/3}, e^{4\pi i/3}, 1\right\} = \sigma(\mathbb{B}).$$

We now identify the edges as $e_j \equiv [0,1]$, where we choose the parametrization contrary to the direction of the flow, i.e., the material flows from 1 to 0. The distribution of the material along an edge e_j at time $t \geq 0$ is described by the functions $x_j(s,t)$ for $s \in (0,1)$.

The transport in the network can now be described by the following system of equations:

$$\partial_t x_j(s,t) = c_j \partial_s x_j(s,t), \qquad s \in (0,1), t \geq 0, \qquad (18.8)$$

$$x_j(s,0) = f_j(s), \qquad s \in (0,1), \qquad (18.9)$$

$$\Phi_{ij}^- c_j x_j(1,t) = w_{ij} \sum_{k=1}^{m} \Phi_{ik}^+ c_k x_k(0,t), \qquad t \geq 0, \qquad (18.10)$$

for $i = 1, \ldots, n$, and $j = 1, \ldots, m$.

Equation (18.8) describes transport on every edge and (18.9) is the initial condition, where we denote by $f_j(s)$ the distribution of the mass on the edge e_j at time $t = 0$. The boundary conditions in (18.10) together with relations (18.1) imply the Kirchhoff law in the vertices:

$$\sum_{j=1}^{m} \Phi_{ij}^- c_j x_j(1,t) = \sum_{j=1}^{m} \Phi_{ij}^+ c_j x_j(0,t), \qquad i = 1, \ldots, n, \qquad (18.11)$$

i.e., in each vertex the total outgoing flow per time unit is equal to the total incoming flow per time unit.

Our next step is to transform the problem into an equivalent abstract Cauchy problem. To this aim we take the Banach lattice

$$X := L^1\left([0,1], \mathbb{C}^m\right),$$

on which we define the operator

$$A = \begin{pmatrix} c_1 \frac{d}{ds} & & 0 \\ & \ddots & \\ 0 & & c_m \frac{d}{ds} \end{pmatrix} \qquad (18.12)$$

with domain

$$D(A) = \left\{ g \in W^{1,1}\left([0,1], \mathbb{C}^m\right) : g(1) = \mathbb{B}_C g(0) \right\}. \tag{18.13}$$

If we can show that this abstract Cauchy problem corresponds to our original transport problem, we can study it by means of semigroup theory. We use the notation $x(t, \cdot) \equiv x(t) \in X$.

Proposition 18.2. *The abstract Cauchy problem on X*

$$\begin{cases} \dot{x}(t) = Ax(t), & t \geq 0, \\ x(0) = (f_j), \end{cases} \tag{18.14}$$

is equivalent to the transport problem in (18.8)–(18.10).

Proof. We only need to show that the condition (18.13) in the domain is equivalent to the boundary conditions in (18.10). First choose $g \in D(A)$. Taking its jth component we obtain

$$c_j g_j(1) = \mathbb{B}_j C g(0) = \left(\Phi_w^-\right)_j^\top \Phi^+ C g(0).$$

Note that the jth row of $\left(\Phi_w^-\right)^\top$ corresponds to the edge e_j and has exactly one nonzero element, say w_{ij}, corresponding to the starting vertex v_i for this edge (i.e., $\Phi_{ij}^- = 1$). Therefore

$$\Phi_{ij}^- c_j g_j(1) = w_{ij} \sum_{k=1}^m \Phi_{ik}^+ c_k g_k(0).$$

To prove the converse, we compute the jth component

$$\begin{aligned}
(\mathbb{B}_C g(0))_j &= c_j^{-1} \left(\Phi_w^-\right)_j^\top \Phi^+ C g(0) \\
&= c_j^{-1} w_{ij} \sum_{k=1}^m \Phi_{ik}^+ c_k g_k(0) \\
&= c_j^{-1} \Phi_{ij}^- c_j g_j(1) = g_j(1),
\end{aligned}$$

where in the last row we applied the boundary condition from (18.10) and used again that $\Phi_{ij}^- = 1$. $\qquad\square$

18.2   The Simple Case $c_j = 1$

Let us study first the simpler case when the velocities of the flow on all edges coincide. We may assume in this case that $c_j = 1$ for all $j = 1, \ldots, m$.

The operator A is

$$A := \operatorname{diag}\left(\frac{\mathrm{d}}{\mathrm{d}s}\right), \quad \text{with domain}$$

$$D(A) := \left\{g \in \mathrm{W}^{1,1}\left([0,1], \mathbb{C}^m\right) : g(1) = \mathbb{B}g(0)\right\}.$$

(18.15)

We start by computing the resolvent of this operator.

Proposition 18.3. *For* $\operatorname{Re}\lambda > 0$ *the resolvent* $R(\lambda, A)$ *of the operator* A *is given by*

$$\left(R(\lambda, A)f\right)(s) = \sum_{k=0}^{\infty} \mathrm{e}^{-\lambda k}\int_0^1 \mathrm{e}^{-\lambda(t+1-s)}\mathbb{B}^{k+1}f(t)\,\mathrm{d}t + \int_s^1 \mathrm{e}^{\lambda(s-t)}f(t)\,\mathrm{d}t,$$

for any $f \in X$ *and* $s \in [0,1]$

Proof. Let $f \in X$ and $g \in D(A)$ such that

$$(\lambda - A)g = f, \quad \text{i.e.,} \quad \lambda g - g' = f.$$

Using the variation of constants formula we obtain

$$g(s) = \mathrm{e}^{\lambda s}\cdot d + \int_s^1 \mathrm{e}^{\lambda(s-t)}f(t)\,\mathrm{d}t, \quad s \in [0,1]$$

(18.16)

for some $d \in \mathbb{C}^m$. Since $g \in D(A)$, we have $g(1) = \mathbb{B}g(0)$, that is,

$$d \cdot \mathrm{e}^{\lambda} = \mathbb{B}d + \mathbb{B}\int_0^1 \mathrm{e}^{-\lambda t}f(t)\,\mathrm{d}t,$$

which is further equivalent to

$$\left(I - \mathrm{e}^{-\lambda}\mathbb{B}\right)d = \mathbb{B}\int_0^1 \mathrm{e}^{-\lambda(t+1)}f(t)\,\mathrm{d}t.$$

(18.17)

By (18.6) we see that $\mathrm{r}\left(\mathrm{e}^{-\lambda}\mathbb{B}\right) < 1$ for $\operatorname{Re}\lambda > 0$, therefore we may use the Neumann series for the resolvent $R\left(1, \mathrm{e}^{-\lambda}\mathbb{B}\right)$ (see Corollary 2.15) and obtain d as

$$d = \sum_{k=1}^{\infty}\left(\mathrm{e}^{-\lambda}\mathbb{B}\right)^k \mathbb{B}\int_0^1 \mathrm{e}^{-\lambda(t+1)}f(t)\,\mathrm{d}t.$$

Plugging this into (18.16) yields the desired formula. $\square$

As a corollary we also obtain a so-called *characteristic equation* for the spectrum of A in terms of the eigenvalues of the adjacency matrices.

Corollary 18.4. *For $\lambda \in \mathbb{C}$ the following equivalences hold.*

$$\lambda \in \sigma(A) \iff e^\lambda \in \sigma(\mathbb{B}) \iff e^\lambda \in \sigma(\mathbb{A}).$$

Proof. The first equivalence follows from relation (18.17), while for the second recall that the nonzero spectra of $\mathbb{B}$ and $\mathbb{A}$ coincide, see (18.7). $\qquad\square$

Example 18.5. For the graph presented in Example 18.1 we obtain

$$\lambda \in \sigma(A) \iff \lambda = \frac{2k\pi i}{3}, \quad k \in \mathbb{Z}.$$

Formula (18.6) yields the spectral bound of the generator A.

Corollary 18.6. *The spectral bound of the operator A given by (18.15) satisfies*

$$s(A) = 0 \in \sigma(A)$$

and is a pole of the resolvent $R(\lambda, A)$.

As we will see in the following proposition, the operator A is the generator of the strongly continuous semigroup given explicitly as

$$T(t)f(s) = \mathbb{B}^k f(t + s - k), \tag{18.18}$$

for $k \in \mathbb{N}_0$, $k \le t + s < k + 1$, and $f \in X$. It acts as a left shift with a "jump" caused by the matrix $\mathbb{B}$.

Proposition 18.7. *The operators $(T(t))_{t \ge 0}$ defined by (18.18) form a positive C_0-semigroup of contractions on X with generator A given by (18.15).*

Proof. The semigroup law, the strong continuity, and the contraction property are straightforward, while the positivity follows from the fact that $\mathbb{B}$ is positive. Hence we only need to prove that A defined in (18.15) is indeed the generator of this semigroup.

Denote by B the generator of $(T(t))_{t \ge 0}$. We can compute its resolvent for any $\operatorname{Re} \lambda > 0$ using the integral representation formula in (9.11) and obtain, for $f \in D(A)$ and $s \in [0, 1]$,

$$\left(R(\lambda, B)f\right)(s) = \int_0^\infty e^{-\lambda t} \left(T(t)f\right)(s)\, dt$$

$$= \int_0^{1-s} e^{-\lambda t} f(t + s)\, dt + \sum_{k=1}^\infty \int_{k-s}^{k-s+1} e^{-\lambda t} \mathbb{B}^k f(t + s - k)\, dt.$$

Here we note that, by the Rellich–Sobolev embedding theorem (see Theorem A.45), every $f \in D(A)$ is continuous, and hence the computations above make sense for any $s \in [0, 1]$.

Substituting $\tau := s + t$ in the first and $\tau := s + t - k$ in other terms and taking $k' := k - 1$, we obtain $\left(R(\lambda, A)f\right)(s)$ given in Proposition 18.3. Since $R(\lambda, A)$ and $R(\lambda, B)$ coincide on the dense set $D(A)$, we conclude that $A = B$. $\qquad\square$

As a consequence we obtained the well-posedness of our problem.

Corollary 18.8. *The abstract Cauchy problem* (18.14), *with* $c_j = 1$ *for all* $j = 1, \ldots, m$, *is well posed.*

We now proceed to study the qualitative behavior of the solutions $x(t) = T(t)f$ to the problem (18.8)–(18.10).

We have seen already in Chapter 6 that strong connectedness of a graph yields the irreducibility of its adjacency matrix. Amazingly, the same holds for the semigroup $(T(t))_{t \geq 0}$ (recall equivalent characterizations of irreducibility of a semigroup from Proposition 14.10).

Proposition 18.9. *The following properties of a graph G, its adjacency matrices $\mathbb{A}$ and $\mathbb{B}$, and the corresponding semigroup $(T(t))_{t \geq 0}$ defined in* (18.18) *are equivalent.*

(i) *G is strongly connected.*

(ii) *$\mathbb{A}$ and/or $\mathbb{B}$ are irreducible.*

(iii) *$(T(t))_{t \geq 0}$ is irreducible.*

Proof. (i) $\Longleftrightarrow$ (ii): Since a graph is strongly connected if and only if its line graph is strongly connected, this equivalence follows by Proposition 6.1.

(ii) $\Longrightarrow$ (iii): We use the representation of the resolvent given in Proposition 18.3 and the fact that, by Corollary 6.2, the powers of $\mathbb{B}$ are eventually strictly positive. Therefore, we obtain that $R(\lambda, A)f$ is a.e. strictly positive for $\lambda > 0$ and $f \gneq 0$. Hence the semigroup $(T(t))_{t \geq 0}$ is irreducible (see also Example 14.11.a)).

(iii) $\Longrightarrow$ (ii): Let the semigroup $(T(t))_{t \geq 0}$ be irreducible and take $f(s) = x_j$ ($s \in [0, 1]$), the jth unit basis vector, for any $j \in \{1, \ldots, m\}$. Then $R(\lambda, A)f$ must be strictly positive, which means that the vector $\sum_{k=0}^{\infty} e^{-\lambda k} \mathbb{B}^{k+1} x_j \gg 0$. Hence, for each $i \in \{1, \ldots, m\}$ there must exist an N such that $\left(\mathbb{B}^N\right)_{ij} > 0$. Since j was arbitrary, $\mathbb{B}$ is irreducible by Corollary 6.2. $\qquad \square$

Having shown these properties of the semigroup $(T(t))_{t \geq 0}$, we can now describe the asymptotic behavior of the solutions to our transport problem.

Theorem 18.10. *Let G be a strongly connected graph. Then the semigroup $(T(t))_{t \geq 0}$ is asymptotically periodic, that is, the space X can be decomposed as*

$$X = X_{\mathrm{r}} \oplus X_{\mathrm{s}},$$

where X_r and X_s are closed, $T(\cdot)$-invariant subspaces such that the following properties are fulfilled.

a) *$T_{\mathrm{s}}(\cdot) := T(\cdot)|_{X_{\mathrm{s}}}$ is uniformly exponentially stable on the space X_{s}.*

b) *$T_{\mathrm{r}}(\cdot) := T(\cdot)|_{X_{\mathrm{r}}}$ is a periodic group on the space X_r with the period equal to the greatest common divisor of the lengths of all cycles in the graph G.*

Proof. Using formula (18.18), we only need to examine the behavior of $(\mathbb{B}^k)$. By Corollary 5.23, the sequence $(\mathbb{B}^k)$ is asymptotically periodic with period h equal to the index of imprimitivity of the matrix $\mathbb{B}$. By Definition 5.22, this means that we can decompose

$$\mathbb{C}^m = Y_r \oplus Y_s$$

so that $\mathbb{B}$ is strongly stable on Y_s and periodic with period h on Y_r. Moreover, Y_r consists of the spectral subspaces of $\mathbb{B}$ that correspond to the eigenvalues in the boundary spectrum, and Y_s of the spectral spaces corresponding to the inner part of the spectrum of $\mathbb{B}$ (see the proof of Corollary 5.23).

Let $\mathbb{P}$ be the projection that belongs to this decomposition, i.e., $Y_s = \ker \mathbb{P}$ and $Y_r = \operatorname{im} \mathbb{P}$. Further, let P be the projection on X induced by $\mathbb{P}$,

$$(Pf)(s) := \mathbb{P}f(s) \text{ for a.e. } s \in [0,1].$$

This projection commutes with every $T(t)$ and yields a decomposition

$$X = \operatorname{im} P \oplus \ker P := X_r \oplus X_s$$

and

$$T(\cdot) = T_r(\cdot) \oplus T_s(\cdot),$$

where $T_r(t) := T(t)|_{X_r}$ and $T_s(t) := T(t)|_{X_s}$. Note that $f \in X_r$ ($f \in X_s$) iff $f(s) \in Y_s$ ($f(s) \in Y_r$) for a.e. $s \in [0,1]$. Therefore, we have a spectral decomposition for $T(\cdot)$.

By the irreducibility of $(T(t))_{t \geq 0}$, the boundary spectrum of its generator consist of eigenvalues (see Lemma 14.16). By Corollary 18.4 and the spectral mapping theorem for the point spectrum (Theorem A.33), we obtain

$$e^\lambda \in \sigma_b(\mathbb{B}) \iff e^{t\lambda} \in \sigma_b(T_r(t)), \quad t \geq 0.$$

Hence periodicity of $\mathbb{B}$ on Y_r implies that the semigroup $(T_r(t))_{t \geq 0}$ is periodic on X_r with the same period h. By Proposition 6.4, the number h is the greatest common divisor of all cycle lengths in graph G.

Note that also the spectra of $T(1)$ and $\mathbb{B}$ coincide, hence $r(T_s(1)) < 1$, and by Proposition 12.4, the semigroup $(T_s(t))_{t \geq 0}$ is uniformly exponentially stable on X_s. □

Observe that the period does not depend on the weights on the edges.

Example 18.11. For the graph in Figure 18.1, the period of the periodic group is 3, while for the one in Figure 18.2 it is 1.

18.3   The General Case

In the previous section we have explicitly constructed the semigroup generated by A by using the adjacency matrix $\mathbb{B}$. Many properties of this semigroup were caused by the positive irreducible column stochastic matrix $\mathbb{B}$. Since, in the general case, we are not able to write down the semigroup explicitly, we shall use the theory of positive semigroups in infinite dimensions to obtain results on the qualitative behavior. Again, the graph structure and appropriate positive matrices play an important role.

Let us first compute the resolvent of A. The formula in this case is more complicated and we need some notation. Define matrices

$$E_\lambda(s) := \operatorname{diag}\left(e^{(\lambda/c_k)s}\right), \quad s \in [0,1], \tag{18.19}$$

and yet another (transposed) weighted adjacency matrix of the graph G and its line graph,

$$\mathbb{A}_\lambda := \Phi^+ E_\lambda(-1)\left(\Phi_w^-\right)^\top \quad \text{and} \quad \mathbb{B}_{C,\lambda} := E_\lambda(-1)\mathbb{B}_C.$$

For the adjacency matrices notice that $\mathbb{A}_0 = \mathbb{A}$ and $\mathbb{B}_{C,0} = \mathbb{B}_C$, while

$$\mathrm{r}(\mathbb{A}_\lambda) \le \|\mathbb{A}_\lambda\| < 1 \quad \text{and} \quad \mathrm{r}(\mathbb{B}_{C,\lambda}) \le \|\mathbb{B}_{C,\lambda}\| < 1 \quad \text{for } \operatorname{Re}\lambda > 0. \tag{18.20}$$

Furthermore, the symbol δ_0 denotes the point evaluation at 0, and

$$(R_\lambda f)(s) := \int_s^1 E_\lambda(s-t)\, C^{-1} f(t)\, \mathrm{d}t, \;\; s \in [0,1], \; f \in X.$$

Proposition 18.12. *For* $\operatorname{Re}\lambda > 0$ *the resolvent* $R(\lambda, A)$ *of* A *is given by*

$$R(\lambda, A) = \left(I_X + E_\lambda(\cdot - 1)C^{-1}\left(\Phi_w^-\right)^\top (1 - \mathbb{A}_\lambda)^{-1}\Phi^+ C \otimes \delta_0\right) R_\lambda \tag{18.21}$$

$$= \left(I_X + E_\lambda(\cdot)\,(1 - \mathbb{B}_{C,\lambda})^{-1}\mathbb{B}_{C,\lambda} \otimes \delta_0\right) R_\lambda. \tag{18.22}$$

Proof. We proceed as in the proof of Proposition 18.3. Let $f \in X$ and $g \in D(A)$ be such that

$$(\lambda - A)g = f.$$

The variation of constants formula yields

$$g(s) = E_\lambda(s)d + \int_s^1 E_\lambda(s-t)C^{-1}f(t)\, \mathrm{d}t, \quad s \in [0,1], \tag{18.23}$$

for some $d \in \mathbb{C}^m$. Since $g \in D(A)$, we have $g(1) = \mathbb{B}_C g(0)$, that is,

$$E_\lambda(1)d = \mathbb{B}_C d + \mathbb{B}_C \int_0^1 E_\lambda(-t)C^{-1}f(t)\, \mathrm{d}t.$$

Multiplying this relation by $E_\lambda(-1)$ and reorganizing it, we obtain

$$(I - \mathbb{B}_{C,\lambda}) \, d = \mathbb{B}_{C,\lambda} \, (R_\lambda f) \, (0).$$

By inequalities (18.20), the resolvent $(I - \mathbb{B}_{C,\lambda})^{-1}$ exists for $\mathrm{Re}\,\lambda > 0$, and

$$d = (I - \mathbb{B}_{C,\lambda})^{-1} \, \mathbb{B}_{C,\lambda} \, (R_\lambda f) \, (0).$$

Plugging this into relation (18.23) yields formula (18.22).

To obtain relation (18.21) first observe that by (18.20) we may use the Neumann series for both resolvents $R\,(1, \mathbb{B}_{C,\lambda})$ and $R\,(1, \mathbb{A}_\lambda)$. Taking also the relations between the graph matrices in (18.2), (18.3), (18.4), and (18.5) into account, we finally obtain

$$(I - \mathbb{B}_{C,\lambda})^{-1} = \sum_{k=0}^{\infty} (\mathbb{B}_{C,\lambda})^k$$

$$= E_\lambda(-1) C^{-1} \left(\Phi_w\right)^\top \sum_{k=0}^{\infty} (\mathbb{A}_\lambda)^k \, \Phi^+ C$$

$$= E_\lambda(-1) C^{-1} \left(\Phi_w^-\right)^\top (1 - \mathbb{A}_\lambda)^{-1} \Phi^+ C. \qquad \square$$

Notice that $R(\lambda, A)$ is a compact and positive operator. Moreover, we have the following equation characterizing the spectrum of A in terms of the adjacency matrices.

Corollary 18.13. *For every $\lambda \in \mathbb{C}$ we have*

$$\lambda \in \sigma\,(A) \iff \det\,(1 - \mathbb{A}_\lambda) = 0 \iff \det\,(1 - \mathbb{B}_{C,\lambda}) = 0.$$

In particular, $\lambda \in \rho(A)$ for $\mathrm{Re}\,\lambda > 0$, and the spectral bound satisfies

$$s\,(A) = 0 \in \sigma\,(A)$$

and is a pole of the resolvent $R(\lambda, A)$.

Now we prove that A generates a positive contractive C_0-semigroup.

Lemma 18.14. *The operator A defined in (18.12)–(18.13) is dispersive on the Banach lattice X.*

Proof. Let $g \in D\,(A)$. Define $\chi := (\chi_{\{g>0\}})$ and observe that $\chi \in \mathrm{L}^\infty\,([0,1], \mathbb{C}^m)$ satisfies $\chi \in \mathcal{I}^+(g)$. So, by Proposition 11.12, it suffices to prove that $\langle Ag, \chi \rangle \leq 0$.

From the definition of A and χ we have

$$\langle Ag, \chi \rangle = \sum_{k=1}^{m} \int_0^1 c_k g_k'(s) \chi_{\{g_k > 0\}} \, ds$$

$$= \left\langle [Cg(1)]^+ - [Cg(0)]^+ , \mathbb{1} \right\rangle_{\mathbb{R}^m}$$

$$\leq \left\langle [\mathbb{B}_C g(0)]^+ - [g(0)]^+ , C\mathbb{1} \right\rangle_{\mathbb{R}^m},$$

where $\mathbb{1}$ denotes the constant 1 vector in $\mathbb{R}^m$ and we use the positivity and symmetry of C as well as the condition $g(1) = \mathbb{B}_C g(0)$ for $g \in D(A)$. Continuing the above estimate and using the positivity of $\mathbb{B}_C$ and the equality $\mathbb{B}_C^\top C\mathbb{1} = C\mathbb{1}$, we finally obtain

$$\langle Ag, \chi \rangle \leq \left\langle \mathbb{B}_C \left[g(0)\right]^+ - \left[g(0)\right]^+ , C\mathbb{1} \right\rangle_{\mathbb{R}^m}$$
$$= \left\langle \left[g(0)\right]^+ , (\mathbb{B}_C^\top - I)C\mathbb{1} \right\rangle_{\mathbb{R}^m} = 0. \qquad \square$$

Since $D(A)$ is dense in X, the dispersivity shown above, Proposition 18.12, and Theorem 11.10 yield the generation result.

Corollary 18.15. *The operator A defined in (18.12)–(18.13) generates a positive contractive C_0-semigroup $(T(t))_{t\geq 0}$. Hence, the abstract Cauchy problem (18.14) is well posed.*

By using the Jacobs–de Leeuw–Glicksberg splitting theorem, see Theorem A.39, and Proposition A.40, we already obtain a projection that decomposes the space X into a pair of $T(\cdot)$-invariant subspaces such that the semigroup splits into a reversible part and a stable part. More can be deduced for irreducible semigroups. As before, the irreducibility of the semigroup is equivalent to the strong connectedness of the underlying graph (compare with Proposition 18.9).

Proposition 18.16. *The following properties for a graph G, its adjacency matrices $\mathbb{A}_\lambda$ and $\mathbb{B}_{C,\lambda}$, and the semigroup $(T(t))_{t\geq 0}$ corresponding to the operator A defined in (18.12)–(18.13) are equivalent.*

  (i) *G is strongly connected.*
 (ii) *$\mathbb{A}$ and/or $\mathbb{B}$ are irreducible.*
(iii) *$\mathbb{A}_\lambda$ and/or $\mathbb{B}_{C,\lambda}$ are irreducible.*
 (iv) *$(T(t))_{t\geq 0}$ is irreducible.*

Proof. The equivalence (i) $\Longleftrightarrow$ (ii) holds by Proposition 18.9 and (ii) $\Longleftrightarrow$ (iii) is clear from the definition of $\mathbb{A}_\lambda$ and $\mathbb{B}_{C,\lambda}$.

(iii) $\Longrightarrow$ (iv): It suffices to show that for $\lambda > 0$ and $f \gneqq 0$ the function $R(\lambda, A)f$ is strictly positive a.e. We use formula (18.22) and first observe that $R_\lambda f \in X$ is strictly positive everywhere except on the largest interval $(1-\varepsilon, 1]$ for which $f|_{(1-\varepsilon,1]} = 0$. Applying $\mathbb{B}_{C,\lambda} \otimes \delta_0$ to $R_\lambda f$, we obtain a strictly positive vector. Since the matrix $\mathbb{B}_{C,\lambda}$ is positive, irreducible, and (18.20) holds for $\operatorname{Re}\lambda > 0$, its resolvent $(1 - \mathbb{B}_{C,\lambda})^{-1}$ is strictly positive (see Exercise 5.4). Observe also that applying E_λ yields a strictly positive vector, and we obtain $R(\lambda, A)f \gg 0$ a.e.

(iv) $\Longrightarrow$ (iii): If $R(\lambda, A)f \gg 0$ a.e. for $\lambda > 0$ and $f > 0$, we use the Neumman series for $(1 - \mathbb{B}_{C,\lambda})^{-1} = \sum_{k=0}^\infty (\mathbb{B}_{C,\lambda})^k$ and proceed as in the proof of Proposition 18.9 concluding that $\mathbb{B}_{C,\lambda}$ is irreducible. $\qquad \square$

Combining all the properties obtained so far and using Theorem 14.19, we shall be able to describe the asymptotic behavior as soon as we determine the boundary spectrum of the generator. By Theorem 14.17, there is $\alpha_0 \geq 0$ such that

$$\sigma_b(A) = i\alpha_0 \mathbb{Z}$$

and $\sigma_b(A)$ contains only simple poles. Moreover, by Corollary 18.13, $\sigma_b(A)$ can be characterized as

$$i\alpha \in \sigma_b(A) \iff 1 \in \sigma(\mathbb{B}_{C,i\alpha})$$

for some $\alpha \in \mathbb{R}$. The following lemma relates this property of the adjacency matrix to a property of the corresponding weighted directed graph.

Lemma 18.17. *Let G be a strongly connected graph. For any $\alpha \in \mathbb{R}$, the following properties are equivalent.*

(i) $1 \in \sigma(\mathbb{B}_{C,i\alpha})$,
(ii) $e^{-i\alpha\tau(j_1,\ldots,j_k)} = 1$ *for all cycles* $e_{j_1}, \ldots, e_{j_k}$ *in G, where*

$$\tau(j_1, \ldots, j_k) := \frac{1}{c_{j_1}} + \cdots + \frac{1}{c_{j_k}}. \tag{18.24}$$

Proof. Since $\mathbb{B}_{C,i\alpha} = C^{-1}\mathbb{B}_{I,i\alpha}C$, there is no loss of generality to study only the simpler matrix

$$\mathbb{B}_{I,i\alpha} = \operatorname{diag}\left(e^{-\frac{i\alpha}{c_k}}\right)\mathbb{B}. \tag{18.25}$$

(i) $\implies$ (ii): Assume first that $1 \in \sigma(\mathbb{B}_{I,i\alpha})$ and $\mathbb{B}_{I,i\alpha}y = y$. Since $\mathbb{B}$ is a positive irreducible matrix with $r(\mathbb{B}) = 1$ and $|\mathbb{B}_{I,i\alpha}| \leq \mathbb{B}$, by Wielandt's lemma (see Lemma 5.18) we have

$$\mathbb{B}_{I,i\alpha} = D\mathbb{B}D^{-1}, \tag{18.26}$$

where D is a diagonal matrix with $|D| = I$. From the proof of this lemma it is also clear that $|y| \gg 0$ and we can take for the diagonal entries D_{kk} of D the coordinates of y: $D_{kk} = y_k$, $k = 1, \ldots, m$. For every two connected edges e_s and e_r such that the head of e_s coincides with the tail of e_r, we have $\mathbb{B}_{rs} \neq 0$, hence (18.26) implies

$$e^{-\frac{i\alpha}{c_r}} = \frac{y_r}{y_s}.$$

If a cycle in G consists of the edges $e_{j_1}, \ldots, e_{j_k}$, then

$$e^{-i\alpha\tau(j_1,\ldots,j_k)} = \prod_{l=1}^{k} e^{-\frac{i\alpha}{c_{j_l}}} = \prod_{l=1}^{k} \frac{y_{j_{l+1}}}{y_{j_l}} - \frac{y_{j_{k+1}}}{y_{j_1}}.$$

Since the eigenvector y is determined only up to a scalar multiple, we may take $y_{j_{k+1}} = y_{j_1}$, and we are done.

(ii) $\Longleftarrow$ (i): Recall the definition of the determinant of a $m \times m$ matrix B as

$$\det B = \sum_{\pi \in \mathcal{S}_m} (-1)^{\text{sign}(\pi)} b_{1,\pi(1)} \cdots b_{m,\pi(m)},$$

where $\mathcal{S}_m$ is the set of all permutations of order m. Remember also that any permutation can be expressed as a product of cycles. Now plug $B = I - \mathbb{B}_{I,\mathrm{i}\alpha}$ in the above expression and note that the nonzero products of the entries correspond to closed walks in G. Assuming (ii) and using (18.25) we thus have

$$\det\left(I - \mathbb{B}_{I,\mathrm{i}\alpha}\right) = \det\left(I - \mathbb{B}\right) = 0, \quad \text{i.e., } 1 \in \sigma\left(\mathbb{B}_{I,\mathrm{i}\alpha}\right). \qquad \square$$

As a consequence, a kind of linear dependency condition for the velocities over $\mathbb{Q}$ plays a crucial role:

$$\text{There is } 0 < d \in \mathbb{R} \text{ such that } d \cdot \tau(j_1, \ldots, j_k) \in \mathbb{N} \tag{18.27}$$
$$\text{for all cycles } e_{j_1}, \ldots, e_{j_k} \text{ in } G.$$

Note that this condition was trivially fulfilled in the simple case of Section 18.2.

Proposition 18.18. *Let G be a strongly connected graph. The boundary spectrum of the generator A defined in (18.12)–(18.13) equals*

$$\sigma_b(A) = \begin{cases} \frac{2\pi\mathrm{i}}{\tau}\mathbb{Z} & \text{if (18.27) is satisfied,} \\ \{0\} & \text{otherwise,} \end{cases}$$

where

$$\tau := \frac{1}{d} \gcd\left\{d \cdot \tau(j_1, \ldots, j_k) \; : \; e_{j_1}, \ldots, e_{j_k} \text{ is a cycle in } G\right\}, \tag{18.28}$$

with $\tau(j_1, \ldots, j_k)$ defined in (18.24).

Proof. By Theorem 14.17 and Corollary 18.13, $0 \in \sigma_b(A)$. Moreover, any nonzero element in $\sigma_b(A)$ is of the form $\mathrm{i}\alpha$ for some $\alpha \neq 0$ and in this case $1 \in \sigma\left(\mathbb{B}_{C,\mathrm{i}\alpha}\right)$. We now use Lemma 18.17. If condition (18.27) holds, there exists $d > 0$ such that for every cycle $e_{j_1}, \ldots, e_{j_k}$ in G one has

$$\alpha\tau(j_1, \ldots, j_k) \in 2\pi\mathbb{Z} \quad \Longleftrightarrow \quad \frac{\alpha}{d} d\tau(j_1, \ldots, j_k) \in 2\pi\mathbb{Z}$$

$$\Longleftrightarrow \quad \frac{\alpha}{d} d\tau \in 2\pi\mathbb{Z} \Longleftrightarrow \mathrm{i}\alpha \in \frac{2\pi\mathrm{i}}{\tau}\mathbb{Z},$$

with τ given in (18.28). Further, if there is a nonzero element $\mathrm{i}\alpha \in \sigma_b(A)$, Lemma 18.17 guarantees that $\frac{\alpha}{2\pi}\tau(j_1, \ldots, j_k) \in \mathbb{Z}$ for every cycle $e_{j_1}, \ldots, e_{j_k}$ in G, which imposes condition (18.27). $\qquad \square$

This two different forms of the boundary spectrum are reflected by the two different asymptotical behaviors of the solution semigroup, which is either asymptotically periodic or converges towards an equilibrium.

Theorem 18.19. *Suppose that the graph G is strongly connected. Then the space X and the semigroup $T(\cdot)$ can be decomposed as*

$$X = X_r \oplus X_s \quad and \quad T(\cdot) = T_r(\cdot) \oplus T_s(\cdot)$$

such that the following holds.

a) *$T_s(\cdot)$ is strongly stable on X_s.*

b) *In the case when condition (18.27) is fulfilled, $T_r(\cdot)$ is a periodic irreducible group on*

$$X_r = \overline{\operatorname{span}} \left\{ f \in D(A) \; : \; \exists k \in \mathbb{Z} \text{ such that } Af = \frac{2k\pi i}{\tau} \right\}$$

with period τ given in (18.28).

c) *If condition (18.27) is not fulfilled, then $T_r(\cdot)$ converges strongly towards a projection onto the one-dimensional subspace*

$$X_r = \operatorname{fix} T(\cdot) = \ker A,$$

which is spanned by a strictly positive eigenvector of A.

Proof. By Proposition 18.18 and Corollary 18.13, we may apply Theorem 14.19 and obtain the desired decomposition with $X_r = \operatorname{im} P$ and $X_s = \ker P$. The restricted semigroup $(T_s(t))_{t \geq 0}$ on X_s is strongly stable. Moreover, X_r is a closed sublattice of X and $(T_r(t))_{t \geq 0}$ is an irreducible periodic semigroup on X_r.

Now, if condition (18.27) holds, the spectral inclusion property for eigenvalues, see Corollary 9.32, implies $T_r(\tau) = I$ for τ defined in (18.28), and it also follows that τ is the period of the periodic semigroup $(T_r(t))_{t \geq 0}$ on X_r.

If, on the other hand, condition (18.27) is not fulfilled, $X_r = \ker A$ is one-dimensional and by Proposition 14.12, the corresponding projection P has the form $P = f^* \otimes f$, where f and f^* are a strictly positive eigenvectors of A and A^*, respectively. $\qquad\square$

Remark 18.20. In the case when condition (18.27) holds one can prove an even stronger result. Using a variant of the spectral mapping theorem, one obtains that $T_s(\cdot)$ is uniformly exponentially stable on X_s. Furthermore, according to Remark 14.20, it can be shown that the subspace X_r is isomorphic to $L^1(\Gamma)$, where Γ is the unit circle, and the group $T_r(\cdot)$ is isomorphic to the rotation group on $L^1(\Gamma)$ with period τ.

18.4   Vertex Control in Networks

Here we return to the simple case of equal velocities $(c_j = 1)$ studied in Section 18.2. We would like to control the flow in our network by adding or subtracting material in a chosen vertex. We can surely not reach all desired states in such a way, as the following simple example shows.

Example 18.21. Consider the network in Figure 18.2, where the weights w and $1 - w$ represent the proportions of the mass leaving vertex v_1 into edges leading to v_2 and v_3, respectively. Observe that the mass distributions on the outgoing edges from v_1 will always satisfy the ratio $\frac{w}{1-w}$. Therefore, not every mass distribution on the edges can be attained. Taking this observation into account, we shall see that all other distributions can be achieved if we control in the vertices v_2 or v_3, but not by controlling in v_1 or v_4 (see Example 18.30).

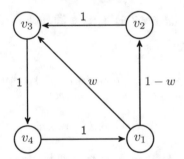

Figure 18.2: Weighted directed graph from Example 18.21.

Based on the observations in the previous example, we ask: Which states (i.e., mass distributions) in the network can be reached by controlling the flow in a single vertex?

To tackle this question, we shall use control theory. Recall the notions from the theory of positive control systems in finite dimensions introduced in Chapter 8. However, the *state space* $X := \mathrm{L}^1([0,1], \mathbb{C}^m)$ is now an infinite-dimensional Banach space. We consider systems without observation (i.e., $Y = X$ and $C = 1$), and assume the control takes place only on a small subset of X. So, besides the *control space* $U := \mathbb{C}$, we also use a *boundary space* $\partial X := \mathbb{C}^n$, both finite-dimensional.

We "split" the system operator A defined in Chapter 8 into two operators: a closed, densely defined *system operator* $A_m : D(A_m) \subseteq X \to X$ and a *boundary operator* $Q \in \mathcal{L}([D(A_m)], \partial X)$. The *control operator* B now goes from the control space U into the boundary space ∂X.

We define these operators for transport in networks. Observe that, using the relations between graph matrices in (18.3) and (18.4), the domain of A defined by (18.15) can be written as

$$D(A) = \left\{ g \in \mathrm{W}^{1,1}([0,1], \mathbb{C}^m) \ : \ g(1) \in \mathrm{im}(\Phi_w^-)^\top \text{ and } \Phi^- g(1) = \Phi^+ g(0) \right\}.$$

Define the system operator as

$$A_m := \text{diag}\left(\frac{d}{ds}\right), \quad \text{with}$$

$$D(A_m) := \{g \in W^{1,1}([0,1], \mathbb{C}^m) : g(1) \in \text{im}(\Phi_w^-)^\top\}$$

(18.29)

and the boundary operator as

$$Q := \Phi^- \delta_1 - \Phi^+ \delta_0 \in \mathcal{L}([D(A_m)], \partial X),$$

(18.30)

where δ_1 and δ_0 are the point evaluations at 1 and 0, respectively. Then the abstract Cauchy problem (18.14) takes on the form

$$\begin{cases} \dot{x}(t) = A_m x(t), & t \geq 0, \\ Q x(t) = 0, & t \geq 0, \\ x(0) = f. \end{cases}$$

We now impose control in the vertex $v = v_i \in V$ for some fixed $i \in \{1, \ldots, n\}$. In the following we identify this vertex with a vector of the canonical basis of $\mathbb{C}^n$. Let B be any (bounded) linear operator acting on the control space $U = \mathbb{C}$ as

$$B : \mathbb{C} \longrightarrow \text{span}\{v\} \subset \partial X = \mathbb{C}^n.$$

With these notations we arrive at an *abstract Cauchy problem with boundary control* written as an *abstract boundary control system* (compare with (8.1)):

$$\begin{cases} \dot{x}(t) = A_m x(t), & t \geq 0, \\ Q x(t) = B u(t), & t \geq 0, \\ x(0) = f. \end{cases}$$

(18.31)

A function $x(\cdot) = x(\cdot, f, u) \in C^1(\mathbb{R}_+, X)$ with $x(t) \in D(A_m)$ for all $t \geq 0$ solving problem (18.31) is called a *classical solution* of the boundary control problem. In order to describe the states a given system can possibly attain one defines the following space.

Definition 18.22. The *approximate reachability space* associated to problem (18.31) is

$$\mathcal{R}^{BC} := \text{cl}\{y \in X : \exists t > 0 \text{ and } u(\cdot) \in L^1([0,t], U) \text{ such that } y = x(t, 0, u)\}.$$

(18.32)

The boundary control system (18.31) is called *approximately boundary controllable* if $\mathcal{R}^{BC} = X$.

To describe this space in terms of the operators A_m, Q, and B, we define for $\lambda \in \rho(A)$

$$B_\lambda := \left(Q|_{\ker(\lambda - A_m)}\right)^{-1} B \in \mathcal{L}(U, \ker(\lambda - A_m)).$$

(18.33)

This operator is well defined under our assumptions and yields the following.

Theorem 18.23. *The approximate reachability space* $\mathcal{R}^{\mathrm{BC}}$ *of system* (18.31) *coincides with*

a) *the smallest closed, $T(\cdot)$-invariant subspace of X containing* $\mathrm{im}(B_\mu)$ *for all μ sufficiently large;*

b) *the smallest closed, $R(\mu, A)$-invariant, subspace of X containing* $\mathrm{im}(B_\mu)$ *for all $\mu > \omega_0(T)$ sufficiently large;*

c) $\overline{\mathrm{span}}\big(\bigcup_{\lambda > \omega} \mathrm{im}(B_\lambda) \big)$ *for some $\omega > \omega_0(T)$.*

On the other hand we define another type of reachability space, which is independent of the boundary and control operators.

Definition 18.24. The *maximal reachability space* associated to system (18.31) is

$$\mathcal{R}^{\mathrm{BC}}_{\max} := \overline{\mathrm{span}}\Big(\bigcup_{\lambda > \omega_0(T)} \ker(\lambda - A_m) \Big). \tag{18.34}$$

The system (18.31) is called *maximally controllable* if $\mathcal{R}^{\mathrm{BC}} = \mathcal{R}^{\mathrm{BC}}_{\max}$.

To justify the name of this space, note that the operators B_λ map onto $\ker(\lambda - A_m)$. So, by Theorem 18.23.c), $\mathcal{R}^{\mathrm{BC}} \subset \bigcup_{\lambda > \omega} \mathrm{im}(B_\lambda)$ for ω large enough. Hence, $\mathcal{R}^{\mathrm{BC}}_{\max} \supseteq \mathcal{R}^{\mathrm{BC}}$ is indeed the largest possible space of states that can be approximately reached by applying some boundary control B.

Since the eigenvectors of A_m have to satisfy the boundary conditions in the vertices (see the domain in (18.29)), we see that the space $\mathcal{R}^{\mathrm{BC}}_{\max}$ can be a proper subspace of the state space $X = \mathrm{L}^1([0,1], \mathbb{C}^m)$. This phenomena was already noticed in Example 18.21. The relevant question for the controllability of our network problem is whether $\mathcal{R}^{\mathrm{BC}} = \mathcal{R}^{\mathrm{BC}}_{\max}$ can be achieved, i.e., whether the system is maximally controllable. For this purpose we describe explicitly both reachability spaces in terms of the graph matrices.

Lemma 18.25. *The maximal reachability space* $\mathcal{R}^{\mathrm{BC}}_{\max}$ *is*

$$\mathcal{R}^{\mathrm{BC}}_{\max} = \mathrm{L}^1([0,1], \mathbb{C}) \otimes \mathrm{im}(\Phi_w^-)^\top \tag{18.35}$$
$$= \mathrm{span}\big\{ (\alpha_1 g, \ldots, \alpha_m g) \;:\; g \in \mathrm{L}^1([0,1], \mathbb{C}) \text{ and } (\alpha_1, \ldots, \alpha_m) \in \mathrm{im}(\Phi_w^-)^\top \big\}.$$

Proof. Observe that for every $\lambda \in \sigma_{\mathrm{p}}(A_m)$, the eigenspace $\ker(\lambda - A_m)$ is spanned by the functions

$$g(s) = \mathrm{e}^{\lambda(s-1)}(\alpha_1, \ldots, \alpha_m)$$

satisfying

$$g(1) = (\alpha_1, \ldots, \alpha_m) = (\Phi_w^-)^\top d \quad \text{for some } d \in \mathbb{C}^n,$$

that is by functions of the form $g(s) = \mathrm{e}^{\lambda(s-1)}(\Phi_w^-)^\top d$. By the Stone–Weierstrass theorem (see Theorem A.20), the span of $\{\mathrm{e}^{\lambda(\cdot-1)} : \lambda > \omega_0(T)\}$ is dense in the space $\mathrm{C}([0,1], \mathbb{C})$ for the sup-norm and thus also in $\mathrm{L}^1([0,1], \mathbb{C})$, hence we are done.   □

In order to compute $\mathcal{R}^{\mathrm{BC}}$ using Theorem 18.23, we need the form of the operators B_λ for λ large enough. We start by computing the inverse of the boundary operator restricted to the eigenspace $\ker(\lambda - A_m)$.

Lemma 18.26. *For* $\lambda > 0 = \omega_0(T)$ *we have*

$$\left(Q|_{\ker(\lambda - A_m)}\right)^{-1} = e^{\lambda(\cdot)}(\Phi_w^-)^\top (e^\lambda - \mathbb{A})^{-1}.$$

Proof. Remember first that $Q = \Phi^-\delta_1 - \Phi^+\delta_0$. Then compute

$$Q e^{\lambda(\cdot)}(\Phi_w^-)^\top \left(e^\lambda - \mathbb{A}\right)^{-1} = \left(e^\lambda \Phi^-(\Phi_w^-)^\top - \Phi^+(\Phi_w^-)^\top\right)\left(e^\lambda - \mathbb{A}\right)^{-1}$$

$$= \left(e^\lambda - \mathbb{A}\right)\left(e^\lambda - \mathbb{A}\right)^{-1} = I_{\mathbb{C}^n},$$

where we used the relations between the graph matrices in (18.4).

As in the proof of Lemma 18.25, we have for any $g \in \ker(\lambda - A_m)$ that

$$g(s) = e^{\lambda(s-1)}(\Phi_w^-)^\top d \quad \text{for some } d \in \mathbb{C}^n.$$

Applying Q to this function we obtain, using again the relations in (18.4),

$$Q g = \Phi^- g(1) - \Phi^+ g(0) = \Phi^-(\Phi_w^-)^\top d - e^{-\lambda}\Phi^+(\Phi_w^-)^\top d = e^{-\lambda}(e^\lambda - \mathbb{A})d,$$

and therefore

$$e^{\lambda(\cdot)}(\Phi_w^-)^\top \left(e^\lambda - \mathbb{A}\right)^{-1} Q g = e^{\lambda(\cdot-1)}(\Phi_w^-)^\top d = g.$$

This concludes the proof. $\square$

Theorem 18.23.c) now yields the following characterization of the approximate reachability space.

Corollary 18.27. *Let the standard basis vector* $v_i \in \mathbb{C}^n$ *correspond to the vertex in* G *in which the control takes place. There exists* $\omega > 0$ *such that*

$$\mathcal{R}^{\mathrm{BC}} = \overline{\mathrm{span}} \bigcup_{\lambda > \omega} \left\{ e^{\lambda(\cdot)}(\Phi_w^-)^\top \left(e^\lambda - \mathbb{A}\right)^{-1} v_i \right\} \tag{18.36}$$

$$= L^1\left([0,1], \mathbb{C}\right) \otimes (\Phi_w^-)^\top \left(\mathrm{span}\{v_i, \mathbb{A}v_i, \ldots, \mathbb{A}^{n-1}v_i\}\right). \tag{18.37}$$

Proof. The first equality is a consequence of Theorem 18.23.c) and Lemma 18.26. We only have to prove the second equality. Using formula (18.18) and relations (18.4), we have

$$T(1)\left(e^{\lambda(\cdot)}(\Phi_w^-)^\top \left(e^\lambda - \mathbb{A}\right)^{-1} v_i\right) = e^{\lambda(\cdot)}\mathbb{B}(\Phi_w^-)^\top \left(e^\lambda - \mathbb{A}\right)^{-1} v_i$$

$$= e^{\lambda(\cdot)}(\Phi_w^-)^\top \left(e^\lambda - \mathbb{A}\right)^{-1} \mathbb{A}v_i \in \mathcal{R}^{\mathrm{BC}},$$

where the last inclusion follows by the invariance of $\mathcal{R}^{\mathrm{BC}}$ under the semigroup, see Theorem 18.23.a). Applying $T(1)$ to this vector again yields

$$T(1)\left(\mathrm{e}^{\lambda(\cdot)}(\Phi_w^-)^\top\left(\mathrm{e}^\lambda - \mathbb{A}\right)^{-1}\mathbb{A}v_i\right) = \mathrm{e}^{\lambda(\cdot)}(\Phi_w^-)^\top\left(\mathrm{e}^\lambda - \mathbb{A}\right)^{-1}\mathbb{A}^2 v_i \in \mathcal{R}^{\mathrm{BC}}.$$

Continuing this procedure, we obtain that

$$\mathrm{e}^{\lambda(\cdot)}(\Phi_w^-)^\top\left(\mathrm{e}^\lambda - \mathbb{A}\right)^{-1}\mathbb{A}^k v_i \in \mathcal{R}^{\mathrm{BC}}, \quad k \in \mathbb{N}_0.$$

Since $\mathcal{R}^{\mathrm{BC}}$ is a linear subspace, we also have that

$$\mathrm{e}^\lambda \cdot \mathrm{e}^{\lambda(\cdot)}(\Phi_w^-)^\top\left(\mathrm{e}^\lambda - \mathbb{A}\right)^{-1}\mathbb{A}^k v_i - \mathrm{e}^{\lambda(\cdot)}(\Phi_w^-)^\top\left(\mathrm{e}^\lambda - \mathbb{A}\right)^{-1}\mathbb{A}^{k+1}v_i$$
$$= \mathrm{e}^{\lambda(\cdot)}(\Phi_w^-)^\top\mathbb{A}^k v_i \in \mathcal{R}^{\mathrm{BC}}, \quad k = 0, 1, \ldots, n-1.$$

Using the Stone–Weierstrass theorem, the Neumann series expansion of the resolvent $\left(\mathrm{e}^\lambda - \mathbb{A}\right)^{-1}$, and the Cayley–Hamilton theorem (see Corollary 2.12), we finally obtain the result. $\qquad\square$

We are now able to characterize the vertices of our graph in which the control operator can achieve the maximal control.

Theorem 18.28. *The following assertions are equivalent for a vertex v_i of the graph G.*

(i) $\mathcal{R}^{\mathrm{BC}} = \mathcal{R}^{\mathrm{BC}}_{\mathrm{max}}$, *i.e., the flow is* maximally controllable *in the vertex v_i.*
(ii) $\mathrm{span}\left\{v_i, \mathbb{A}v_i, \ldots, \mathbb{A}^{n-1}v_i\right\} = \mathbb{C}^n$.

Proof. Using equations (18.35) and (18.37), (i) is equivalent to

$$\begin{aligned}
\mathcal{R}^{\mathrm{BC}}_{\mathrm{max}} &= L^1\left([0,1], \mathbb{C}\right) \otimes (\Phi_w^-)^\top \mathbb{C}^n \\
&= L^1\left([0,1], \mathbb{C}\right) \otimes (\Phi_w^-)^\top\left(\mathrm{span}\left\{v_i, \mathbb{A}v_i, \ldots, \mathbb{A}^{n-1}v_i\right\}\right) = \mathcal{R}^{\mathrm{BC}}.
\end{aligned}$$

This holds if and only if $\mathrm{span}\left\{v_i, \mathbb{A}v_i, \ldots, \mathbb{A}^{n-1}v_i\right\} = \mathbb{C}^n$ since $(\Phi_w^-)^\top$ is injective, hence left invertible. $\qquad\square$

Remark 18.29.

a) The assertion (ii) in Theorem 18.28 is a *Kàlmàn-type condition*, similar to the one met in Chapter 8 (see, e.g., Corollary 8.15). In our situation, it guarantees that by controlling in the vertex v_i the largest possible space of mass distributions in the network can be (approximately) reached.

b) In Section 8.2 we have introduced *positive controllability* for finite-dimensional systems. In our network example it is also reasonable to consider the positive control problem. However, the treatment of such systems is much more complicated. Using different techniques one can prove that the flow is *maximally positive controllable* in the vertex v_i if and only if

$$\overline{\mathrm{co}}\{v_i, \mathbb{A}v_i, \mathbb{A}^2 v_i, \ldots\} = \mathbb{R}^n_+. \tag{18.38}$$

Example 18.30. We again consider the network in Figure 18.2. Using appropriate graph matrices and Lemma 18.25, we see that the maximal reachability space $\mathcal{R}_{\max}^{BC}$ is

$$\text{span}\left\{(\alpha_1 g, \ldots, \alpha_5 g) \ : \ g \in L^1\left([0,1], \mathbb{C}\right), \ (\alpha_1, \ldots, \alpha_5) \in \mathbb{C}^5, \ \alpha_3 = \frac{1-w}{w}\alpha_1\right\}.$$

Moreover, verifying the Kàlmàn condition from Theorem 18.28 one obtains that

$$\mathcal{R}_{\max}^{BC} = \mathcal{R}^{BC} \iff v_i = v_2 \text{ or } v_i = v_3.$$

18.5 Notes and Remarks

Dynamical processes taking place in networks are of enormous interest in recent years, modeling various real life phenomena. Methods from the theory of operator semigroups to treat such processes were first used by Kramar and Sikolya [76] for the transport equation on a finite network. These methods were further applied to generalizations of this problem in finite networks by Sikolya [130], Mátrai and Sikolya [91], Kunszenti–Kovács [79], and Banasiak and Namayanja [12]. The linear Boltzmann equation with scattering, as introduced in Chapter 17, was considered on a network by Radl [116]. Engel et al. [42, 41], and Boulite et al. [20] studied some related control problems, Bayazit, Dorn, and Rhandi [15] a delay problem, and Bayazit, Dorn, and Kramar Fijavž [14] a non-autonomous problem for flows in networks. A transport problem in infinite networks was considered by Dorn [33], Dorn, Keicher, and Sikolya [34], and Kunszenti–Kovács [78]. See also Dorn et al. [35] for a survey of the semigroup theory approach to transport processes in networks.

The context of the first and third sections relies on the above-cited works [76, 91] with some adaptations to the physically correct Kirchhoff laws mentioned in [12]. The second section uses results from [33] adapted to finite networks while the results of the last section originate in [42, 41]. Specifically, for the proof of Theorem 18.23 see Engel et al. [41, Theorem 2.12]. Positive controllability was studied in Boulite at al. [20], where condition (18.38) appears.

For the semigroup approach to many other dynamical processes in networks we refer to the recent monograph by Mugnolo [100].

18.6 Exercises

1. In Example 18.1 the appropriate graph matrices are given for the graph in Figure 18.1.

   a) Verify the relations (18.2), (18.3), and (18.4) for these matrices.

    b) Choose a labelling of the edges different from the one on the right-hand picture in Figure 18.1. Verify that your specific choice of labels does not affect the relations above.

2. Take the graph in Figure 18.2 and the associated operator A defined by (18.12)–(18.13). Compute the spectrum of A and describe the asymptotic behavior of the solutions for the following cases:

    a) $c_j = 1$ for all $j = 1, \ldots, m$;

    b) c_j are not all equal, but condition (18.27) is fulfilled;

    c) c_j are not all equal and condition (18.27) is not fulfilled.

In each of the last two cases choose appropriate values of c_j and use them in your computations.

3. Show that the following holds for the control flow problem considered on the graph in Figure 18.3.

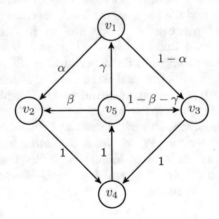

Figure 18.3: Weighted directed graph from Exercise 3.

    a) The flow is maximally controllable in any of the vertices v_2 and v_3, independently of the particular choice of the weights.

    b) The flow is not maximally controllable in any of the vertices v_4 and v_5, independently of the particular choice of the weights.

    c) The problem is maximally controllable in v_1 if and only if

$$\alpha - \beta - \alpha \cdot \gamma \neq 0.$$

4. Suppose that we control the flow in network G in the vertices $v_{i_1}, \ldots, v_{i_k}$. Show that in this case one can characterize maximal controllability by the following modification of the Kàlmàn condition given in Theorem 18.28:

$$\operatorname{span}\left\{ v_{i_1}, \mathbb{A}v_{i_1}, \ldots, \mathbb{A}^{n-1}v_{i_1}, \ldots, v_{i_k}, \mathbb{A}v_{i_k}, \ldots, \mathbb{A}^{n-1}v_{i_k} \right\} = \mathbb{C}^n.$$

5. Find examples of weighted directed graphs in which the flow is
   a) maximally controllable in every vertex;
   b) not controllable in any of its vertices.
6. Consider a generalized transport model from Section 1 in which we allow some absorption along the edges, so equation (18.8) now takes the form

$$\partial_t x_j\,(s,t) = c_j \partial_s x_j(s,t) + q_j x_j(s,t), \quad s \in (0,1), t \geq 0,$$

   for some absorption coefficients q_j, $j = 1, \ldots, m$.
   a) Write down the appropriate operator A and show well-posedness using a bounded perturbation.
   b) What can you say about the asymptotic behavior of the solutions in this case?

Chapter 19

Population Equations with Diffusion

Many applications of positive semigroups occur in mathematical biology or chemistry. In the finite-dimensional part of our text we have already discussed a very simple discrete-time population model, called the Leslie model (see Section 6.3). In this chapter we present a time-continuous age-structured population model with spatial diffusion. We present a rather advanced model in order to show the reader some generalizations and applications.

After presenting our model, we describe the appropriate abstract framework. In Sections 19.2 and 19.3 we introduce Hille–Yosida operators, extrapolation spaces and extrapolated semigroups. Since we are dealing with a non-autonomous problem in Section 19.4, we also introduce evolution families. The results are then applied to our population model in Section 19.5.

Although the presented theory is quite demanding, one can still recognize the thread we are following throughout our text. Since this chapter is meant as an outlook, its style differs from the rest of the book by not giving full proofs.

19.1 The Mathematical Model

Let us start by describing our McKendrick type model with age- and space-dependent spatial diffusion. Consider the equations

$$
\begin{cases}
\partial_t u(t,a,x) + \partial_a u(t,a,x) = \kappa(a)\Delta u(t,a,x) - \mu(a,x)u(t,a,x), \\
\qquad\qquad\qquad\qquad\qquad\qquad t \geq 0,\, a \in I,\, x \in \Omega, \\
\partial_\nu u(t,a,x) = 0, \quad t \geq 0,\, a \in I,\, x \in \Gamma_1, \\
u(t,a,x) = 0, \quad t \geq 0,\, a \subset I,\, x \in \Gamma_2, \\
u(t,0,x) = \displaystyle\int_I \beta(a,x)u(t,a,x)\,da, \quad t \geq 0,\, x \in \Omega, \\
u(0,a,x) = f(a,x), \quad a \in I,\, x \in \Omega.
\end{cases}
\tag{19.1}
$$

Here, $u(t, a, x)$ is the population density at time t, age $a \in I$, and position $x \in \Omega$, a bounded domain in ess with smooth boundary $\partial\Omega = \Gamma_1 \cup \Gamma_2$. We take $I := [0, a_{\mathrm{m}}]$ if $a_{\mathrm{m}} < \infty$ and $I = \mathbb{R}_+$ if $a_{\mathrm{m}} = \infty$, where a_{m} is the maximal life expectancy of the species. By $\partial_\nu u$ we denote the normal derivative of u. The coefficient $\kappa(a)$ is an age-dependent diffusion coefficient. Here, for simplicity, we assume that the mortality rate $0 \le \mu \in \mathrm{L}^\infty(I \times \Omega)$. One can allow μ to be singular with respect to x and a, see Remark 19.27. Nonintegrability of μ at $a = a_{\mathrm{m}}$ ensures that no individual reaches maximal age, and a singularity of $x \mapsto \mu(a, x)$ may represent a very hostile part of the domain. Finally, the birth or fertility rate $\beta \ge 0$ is supposed to be uniformly continuous with respect to a and bounded with respect to x.

We shall show existence and uniqueness of positive (generalized) solution of problem (19.1) and discuss spectral and asymptotic properties of the solution operator in $E := \mathrm{L}^1(I \times \Omega)$. This is the natural state space for such equations, because $\|u(t)\|$ gives the size of the population at time t.

Let us briefly describe our approach. Consider the realization $A_1(a)$ in $X := \mathrm{L}^1(\Omega)$ of the diffusion operator $A(a, x, D) := \kappa(a)\Delta$ subject to the mixed boundary conditions given in the second and third equation of problem (19.1). The operator $Lf = -f' + A(\cdot)f(\cdot)$ defined on a suitable domain of $E \cong \mathrm{L}^1(I, X)$ has a closure G in E. It is important for our analysis that the restriction G_0 of G to functions with $f(0) = 0$ generates the so-called *evolution semigroup*

$$(T_0(t)f)(a) := \chi_I(a - t) T_{\Delta_1}\left(\int_{a-t}^a \kappa(s)\, \mathrm{d}s\right) f(a - t), \quad t \ge 0,\, a \in I,\, f \in \mathrm{L}^1(I, X),$$

where $T_{\Delta_1}(\cdot)$ is the semigroup generated by the Laplace operator subject the boundary conditions given in the second and third identity of problem (19.1) on $\mathrm{L}^1(\Omega)$. Using the perturbation theory of Miyadera type developed in Chapter 13, we show that $D(G)$ is contained in the domain of the multiplication operator V induced by μ on E and that the operator $\mathcal{G}_V$ on $X \times E$ defined by

$$(0, f) \longmapsto (-f(0), (G - V)f) \text{ for } f \in D(G)$$

is a Hille–Yosida operator, see Section 19.2. The birth law in (19.1), given by

$$Bf := \int_I \beta(a, \cdot) f(a, \cdot)\, \mathrm{d}a,$$

can be expressed as a bounded perturbation of $\mathcal{G}_V$ in $\mathrm{L}^1(\Omega) \times \mathrm{L}^1\left(I, \mathrm{L}^1(\Omega)\right)$ of the form $(0, f) \mapsto (Bf, 0)$. In this way we obtain the existence of a positive C_0-semigroup $(S_0(t))_{t\ge 0}$ on E solving problem (19.1). For the spectral and asymptotic properties of $(S_0(t))_{t\ge 0}$ we use a perturbation theorem for the essential spectral radius, see Theorem 19.5. In particular, if β is strictly positive, then (after rescaling) the solution semigroup $(S_0(t))_{t\ge 0}$ converges exponentially to the projection on the unique positive stationary solution.

19.2 Hille–Yosida Operators and Extrapolated Semigroups

For the following we need the notion of Hille–Yosida operators and their corresponding extrapolated semigroups. We collect all important definitions and relevant properties.

A linear operator A on a Banach space X is called a *Hille–Yosida operator* if it satisfies the resolvent estimate in the Hille–Yosida theorem (see Theorem 11.1), i.e., there exist $M \geq 1$ and $\omega \in \mathbb{R}$ such that $(\omega, +\infty) \subset \rho(A)$ and

$$\sup\{(\lambda - \omega)^k \|(\lambda - A)^{-k}\| : k \in \mathbb{N},\ \lambda > \omega\} \leq M. \tag{19.2}$$

In this section we assume that A with domain $D(A)$ is a Hille–Yosida operator on a Banach space X. Let A_0 be the *part of A in* $X_0 := \overline{D(A)}$, defined as

$$\begin{aligned} D(A_0) &= \{x \in D(A) : Ax \in \overline{D(A)}\}, \\ A_0 x &= Ax \quad \text{for } x \in D(A_0). \end{aligned} \tag{19.3}$$

We quote the following result from Engel and Nagel [43, Corollary II.3.21 and Lemma IV.1.15].

Lemma 19.1. *The operator A_0 defined by the rule (19.3) generates a C_0-semigroup $(T_0(t))_{t \geq 0}$ on X_0 with $\|T_0(t)\| \leq M e^{\omega t}$ for $t \geq 0$. Moreover, $\rho(A) \subset \rho(A_0)$ and*

$$R(\lambda, A_0) = R(\lambda, A)\big|_{X_0}$$

for $\lambda \in \rho(A)$, where $R(\lambda, A)\big|_{X_0}$ is the restriction of $R(\lambda, A)$ to X_0.

For a fixed $\lambda_0 \in \rho(A)$, we introduce the norm

$$\|x\|_{-1} := \|R(\lambda_0, A_0)x\| \quad \text{for } x \in X_0.$$

The completion X_{-1} of $(X_0, \|\cdot\|_{-1})$ is called the *extrapolation space* of X associated with the operator A. The notation $\|\cdot\|_{-1}$ is independent of the choice of $\lambda_0 \in \rho(A)$, since the norms $\|R(\lambda_0, A_0)x\|_{-1}$ and $\|R(\lambda, A_0)x\|$ are equivalent for any $\lambda, \lambda_0 \in \rho(A)$.

The operator $T_0(t)$ has a unique bounded linear extension $T_{-1}(t)$ to the Banach space X_{-1} and $(T_{-1}(t))_{t \geq 0}$ is a C_0-semigroup on X_{-1}, which is called the *extrapolated semigroup* of $(T_0(t))_{t \geq 0}$. We denote by $(A_{-1}, D(A_{-1}))$ the generator of $(T_{-1}(t))_{t \geq 0}$ on the space X_{-1}.

We summarize the most important facts about this semigroup from Engel and Nagel [43, Section II.5.a].

Lemma 19.2. *The following properties hold.*

a) $\|T_{-1}(t)\|_{\mathcal{L}(X_{-1})} = \|T_0(t)\|_{\mathcal{L}(X_0)}$.

b) $D(A_{-1}) = X_0$.

c) $A_{-1} : X_0 \to X_{-1}$ is the unique continuous extension of the operator $A_0 :$ $D(A_0) \subseteq (X_0, \| \cdot \|) \to (X_{-1}, \| \cdot \|_{-1})$ and $\lambda_0 - A_{-1}$ is an isometry from $(X_0, \| \cdot \|)$ to $(X_{-1}, \| \cdot \|_{-1})$.

d) $\rho(A) \subseteq \rho(A_{-1})$. In particular, $R(\lambda, A_{-1})\big|_{X_0} = R(\lambda, A_0)$ for any $\lambda \in \rho(A)$.

e) The space $X_0 = \overline{D(A)}$ is dense in $(X_{-1}, \| \cdot \|_{-1})$. Hence, the extrapolation space X_{-1} is also the completion of $(X, \| \cdot \|_{-1})$ and we have $X \hookrightarrow X_{-1}$.

f) The operator A_{-1} is an extension of the operator A. In particular, if $\lambda \in \rho(A)$, then $R(\lambda, A_{-1})\big|_X = R(\lambda, A)$.

We state here another fundamental lemma, which is useful for perturbations of Hille–Yosida operators, see Nagel and Sinestrari [102].

Lemma 19.3. *For $f \in L^1_{\mathrm{loc}}(\mathbb{R}_+, X)$, we define*

$$(T_{-1} * f)(t) := \int_0^t T_{-1}(t - s)f(s)ds \quad \text{for } t \geq 0.$$

Then

a) $(T_{-1} * f)(t) \in X_0$, for all $t \geq 0$;

b) $\|(T_{-1} * f)(t)\| \leq M e^{\omega t} \int_0^t e^{-\omega s}\|f(s)\|\, ds$, where M is a constant independent of f;

c) $\lim_{t \downarrow 0} \|(T_{-1} * f)(t)\| = 0$.

The following perturbation theorem can be proved using Lemma 19.3 and the same arguments as in the proof of Theorem 11.5 and Proposition 11.6.

Theorem 19.4. *Let A with domain $D(A)$ be a Hille–Yosida operator on a Banach space X. Let $(T_0(t))_{t \geq 0}$ be the C_0-semigroup generated by the part A_0 of A in $X_0 = \overline{D(A)}$ and let $B \in \mathcal{L}(X_0, X)$. Then the part of $A + B$ in X_0 generates a C_0-semigroup on X_0 given by the Dyson–Phillips expansion*

$$S_0(t) = \sum_{k=0}^{\infty} U_k(t), \quad t \geq 0,$$

where

$$U_0(t) = T_0(t) \quad \text{and} \quad U_{k+1}(t) = \int_0^t T_{-1}(t - s)BU_k(s)\, ds, \quad t \geq 0, \ k \in \mathbb{N}.$$

Moreover, the following variation of constant formula holds:

$$S_0(t)x = T_0(t)x + \int_0^t T_{-1}(t - s)BS(s)x\, ds, \quad x \in X, \ t \geq 0.$$

Denote by

$$R_m(t) := \sum_{k=m}^{\infty} U_k(t)$$

the mth *remainder* of the above expansion. The following spectral result is a generalization of Voigt's theorem, see Theorem A.35. Recall that a bounded linear operator B is called *strictly power compact* if there is $k \in \mathbb{N}$ such that $(BT)^k$ is compact for every bounded linear operator T.

Theorem 19.5. *Let A be a Hille–Yosida operator on a Banach space X and B a bounded linear operator from $X_0 := \overline{D(A)}$ to X. Let $(T_0(t))_{t \geq 0}$ and $(S_0(t))_{t \geq 0}$ be the C_0-semigroups on X_0 generated by the parts of A and $A+B$ in X_0, respectively. Assume that there exist $m \in \mathbb{N}$ and a sequence $(t_k) \subset \mathbb{R}_+$, $t_k \to \infty$, such that the remainders $R_m(t_k)$ at t_k are strictly power compact for all $k \in \mathbb{N}$. Then,*

$$\omega_{\mathrm{ess}}(S_0) \leq \omega_0(T_0).$$

Proof. Assume that $\omega_0(T_0) > -\infty$. Set $\omega_\varepsilon = \omega_0(T_0) + \varepsilon$ and $\omega_\varepsilon' = \omega_0(T_0) + \frac{\varepsilon}{2}$. Then, by Lemma 19.3, we have

$$\|U_j(t)\| \leq M_\varepsilon^{j+1} \|B\|_{\mathcal{L}(E_0,E)} \frac{t^j}{j!} e^{\omega_\varepsilon' t}, \quad t \geq 0, j \in \mathbb{N}.$$

Hence, in particular, there is $\tau_\varepsilon > 0$ such that

$$\left\| \sum_{j=0}^{m-1} U_j(t) \right\| \leq e^{\omega_\varepsilon t} \quad \text{for } t \geq \tau_\varepsilon. \tag{19.4}$$

Since $R_m(t_k)$ is strictly power compact, Theorem A.35 implies that

$$r_{\mathrm{ess}}(S_0(t_k)) = r_{\mathrm{ess}}\left(\sum_{j=0}^{m-1} U_j(t_k) \right) \leq e^{\omega_\varepsilon t} \quad \text{for } t_k \geq \tau_\varepsilon.$$

Thus, $\omega_{\mathrm{ess}}(S_0) \leq \omega_\varepsilon$. This proves the result by letting $\varepsilon \to 0$ if $\omega_0(T_0) > -\infty$. In the case where $\omega_0(T_0) = -\infty$, ω_ε in inequality (19.4) is taken so that $\lim_{\varepsilon \to 0} \omega_\varepsilon = -\infty$. Hence, $\omega_{\mathrm{ess}}(S_0) = -\infty$. $\square$

19.3 Spectral Properties of Perturbed Hille–Yosida Operators

We assume that E is a Banach lattice with order continuous norm and A a Hille–Yosida operator on E. As before, we denote by $(T_0(t))_{t \geq 0}$ the C_0-semigroup generated by the part A_0 of A in $E_0 := \overline{D(A)}$, see Lemma 19.1.

Assumptions 19.6. We make the following hypotheses.

a) A is a resolvent positive operator on E, and the perturbation $0 \leq B \in \mathcal{L}(E_0, E)$ is such that the mapping $t \mapsto BT_0(t) \in \mathcal{L}(E_0, E)$ is continuous for $t > 0$ (for the operator norm).

b) For all $\varepsilon > 0$, there is a positive and compact operator $K_\varepsilon : E_0 \to E$ such that $0 \leq BT_0(t) \leq K_\varepsilon$ for $t \geq 0$ and the mapping $t \mapsto K_\varepsilon T_0(t) \in \mathcal{L}(E_0, E)$ is continuous for $t > 0$.

Remark 19.7. Let A be a resolvent positive operator on E. By Exercise 12.5.5, $E_0 = \overline{D(A)}$ is an ideal and hence a sublattice of E. So, A_0 is resolvent positive and Corollary 11.4 implies that $(T_0(t))_{t \geq 0}$ is positive.

We need the following auxiliary results of Arendt [5, Theorem 5.7].

Lemma 19.8. *Let A with domain $D(A)$ be a resolvent positive operator on a Banach lattice E with order continuous norm. Then there is a unique strongly continuous family $(S(t))_{t \geq 0}$ of operators on E satisfying*

$$0 = S(0) \leq S(r) \leq S(t), \quad 0 \leq r \leq t, \text{ and}$$

$$R(\lambda, A) = \int_0^\infty \lambda e^{-\lambda t} S(t) \mathrm{d}t, \quad \lambda > \max\{0, \mathrm{s}(A)\}.$$

Connections between domination and compactness were characterized by Aliprantis and Burkinshaw [2, Theorem 5.15].

Lemma 19.9. *Let E, F and G be Banach lattices and $T_1 \in \mathcal{L}(E, F)$, $T_2 \in \mathcal{L}(F, G)$. If G has order continuous norm and each T_i is dominated by a compact operator, then $T_2 T_1$ is a compact operator.*

We also quote the following technical result from Schaefer [126, Theorem IV.1.5].

Lemma 19.10. *If E is an L^1- Banach lattice, then for every $T \in \mathcal{L}(E)$ there exist positive operators $T_1, T_2 \in \mathcal{L}(E)$ such that $T = T_1 - T_2$.*

The following lemma will be useful in the proof of the main results. We denote by $(T_{-1}(t))_{t \geq 0}$ the *extrapolated semigroup* of $(T_0(t))_{t \geq 0}$ defined in Section 19.2.

Lemma 19.11. *Let A be a Hille–Yosida and resolvent positive operator on a Banach lattice E with order continuous norm. Then*

$$0 \leq \int_0^t T_{-1}(t - s)\varphi(s) \, \mathrm{d}s \quad \text{for } 0 \leq \varphi \in L^1_{\mathrm{loc}}(\mathbb{R}_+, E), \ t \geq 0.$$

Proof. We recall that (see Lemma 19.2)

$$R(\lambda, A)f = R(\lambda, A_{-1})f = \int_0^\infty e^{-\lambda t} T_{-1}(t)f \, \mathrm{d}t$$

$$= \int_0^\infty \lambda e^{-\lambda t} \left(\int_0^t T_{-1}(s)f \, \mathrm{d}s \right) \mathrm{d}t \tag{19.5}$$

for large λ and $f \in E$. On the other hand, by Lemma 19.8, there is a unique increasing strongly continuous family $(S(t))_{t \geq 0}$ of bounded linear operators on E such that $S(0) = 0$ and

$$R(\lambda, A)f = \int_0^\infty \lambda e^{-\lambda t} S(t)f \, dt$$

for large λ and $f \in E$. So, by relation (19.5) and the uniqueness of the Laplace transform, we have

$$S(t)f = \int_0^t T_{-1}(s)f \, ds, \quad f \in E, \, t \geq 0.$$

Thus

$$\int_0^t T_{-1}(t - s)\varphi(s) \, ds = \int_0^t \varphi(t - s) \, dS(s) \geq 0, \quad t \geq 0. \qquad \square$$

Throughout the remainder of this section, we enforce a) from Assumptions 19.6 and denote by $(S_0(t))_{t \geq 0}$ the C_0-semigroup generated by the part of $A + B$ in $E_0 := \overline{D(A)}$. This semigroup is given by the Dyson–Phillips expansion

$$S_0(t) = \sum_{k=0}^\infty U_k(t), \quad t \geq 0,$$

where $U_0(t) := T_0(t)$ and $U_{k+1}(t) := \int_0^t T_{-1}(t - s)BU_k(s) \, ds$ for $t \geq 0$ and $k \in \mathbb{N}$, see Theorem 19.4. Let $R_m(t) := \sum_{k=m}^\infty U_k(t)$ be the mth *remainder* of the expansion for $S_0(t)$. The formula

$$R_{m+1}(t)f = \int_0^t T_{-1}(t - s)BR_n(m)f \, ds, \quad f \in E_0, \, t \geq 0,$$

can be derived the same way as the one in (17.10). In the following lemma we use the approximation $R_{2,\varepsilon}(t)$ of $R_2(t)$ defined by

$$R_{2,\varepsilon}(t) := \int_\varepsilon^{t-\varepsilon} T_{-1}(t - s)B \int_0^{s-\varepsilon} T_{-1}(s - \tau)BS_0(\tau)d\tau \, ds$$

for $t \geq 2\varepsilon$.

Lemma 19.12. *Assumption 19.6.a) implies that $BR_{2,\varepsilon}(t) \to BR_2(t)$ in $\mathcal{L}(E_0, E)$ as $\varepsilon \downarrow 0$, uniformly for t in compact subsets of $\mathbb{R}_+$. Moreover, the mapping $t \mapsto BR_2(t) \in \mathcal{L}(E_0, E)$ is continuous for $t \geq 0$.*

Proof. We have

$$R_2(t)f - R_{2,\varepsilon}(t)f = \int_{t-\varepsilon}^{t} T_{-1}(t-s)B \int_0^s T_{-1}(s-\tau)BS_0(\tau)f\mathrm{d}\tau \, \mathrm{d}s$$

$$+ \, T_0(t-\varepsilon) \int_0^\varepsilon T_{-1}(\varepsilon-s)B \int_0^s T_{-1}(s-\tau)BS_0(\tau)f\mathrm{d}\tau \, \mathrm{d}s$$

$$+ \, T_0(\varepsilon) \int_\varepsilon^{t-\varepsilon} T_{-1}(t-\varepsilon-s)B \int_{s-\varepsilon}^s T_{-1}(s-\tau)BS_0(\tau)f\mathrm{d}\tau \, \mathrm{d}s$$

$$=: I_1 + I_2 + I_3$$

for $2\varepsilon \le t \le L$, $f \in E_0$ and any $L > 0$. Now Lemma 19.3 implies that

$$\|I_1\| \le M_L^1 \int_{t-\varepsilon}^{t} \left\| \int_0^s T_{-1}(s-\tau)BS_0(\tau)f\mathrm{d}\tau \right\| \mathrm{d}s$$

$$\le M_L^1 \|f\| \int_{t-\varepsilon}^{t} s \, \mathrm{d}s,$$

$$\|I_2\| \le M_L^2 \int_0^\varepsilon \left\| \int_0^s T_{-1}(s-\tau)BS_0(\tau)f\mathrm{d}\tau \right\| \mathrm{d}s$$

$$\le M_L^2 \|f\| \int_0^\varepsilon s \, \mathrm{d}s,$$

$$\|I_3\| \le M_L^3 \int_\varepsilon^{t-\varepsilon} \left\| \int_{s-\varepsilon}^s T_{-1}(s-\tau)BS_0(\tau)f\mathrm{d}\tau \right\| \mathrm{d}s$$

$$\le M_L^3 \|f\| t\varepsilon,$$

where M_L^i, $i = 1, 2, 3$, are positive constants depending on L. This proves the first claim. The continuity property can be shown by similar arguments. $\qquad\square$

As a consequence we obtain the following result which has an impact on the longterm behavior of the perturbed semigroup.

Proposition 19.13. *Assumptions* 19.6 *imply that*

$$\omega_{\mathrm{ess}}(S_0) \le \omega_0(T_0).$$

Proof. We apply Theorem 19.5. For this purpose we need to prove that $R_3(t)$ is compact for all $t \ge 0$.

Let us first show that $BR_{2,\varepsilon}(t)$ is compact. Fix $\varepsilon > 0$ and $t \ge 2\varepsilon$. Then

$$BR_{2,\varepsilon}(t)f = BT_0(\varepsilon) \int_0^{t-2\varepsilon} T_{-1}(s)BT_0(\varepsilon)$$

$$\times \int_0^{t-s-2\varepsilon} T_{-1}(\tau)BS_0(t-s-\tau-2\varepsilon)f\mathrm{d}\tau \, \mathrm{d}s \tag{19.6}$$

$$=: BT_0(\varepsilon)L_1(\varepsilon,t)\,(BT_0(\varepsilon)L_2(\varepsilon,t)f)$$

for $f \in E_0$, where $L_k(\varepsilon, t)$, $k = 1, 2$, are given by

$$L_1(\varepsilon, t) : \mathrm{L}^1\left([0, t], E\right) \longrightarrow E_0, \quad \varphi \longmapsto \int_0^{t-2\varepsilon} T_{-1}(s)\varphi(t - s - 2\varepsilon)\, \mathrm{d}s$$

$$L_2(\varepsilon, t) : E_0 \longrightarrow \mathrm{L}^1\left([0, t], E_0\right), \quad (L_2(\varepsilon, t)f)(s) := \int_0^s T_{-1}(\tau)BS_0(s - \tau)f\mathrm{d}\tau.$$

Using Lemma 19.11 one sees that $L_1(\varepsilon, t)$ and $L_2(\varepsilon, t)$ are bounded and positive operators, and b) from Assumptions 19.6 implies

$$0 \le BT_0(\varepsilon)L_1(\varepsilon, t)\varphi \le K_\varepsilon L_1(\varepsilon, t)\varphi \quad \text{and}$$
$$0 \le BT_0(\varepsilon)\left(L_2(\varepsilon, t)f\right)(s) \le K_\varepsilon\left(L_2(\varepsilon, t)f\right)(s) \tag{19.7}$$

for $0 \le \varphi \in \mathrm{L}^1([0, t], E)$, $0 \le f \in E_0$, and $s \in [0, t]$. Since K_ε is a compact operator, we infer that $K_\varepsilon L_1(\varepsilon, t) : \mathrm{L}^1([0, t], E) \to E$ is compact as well.

Let us prove now that the operator $K_\varepsilon L_2(\varepsilon, t) : E_0 \to \mathrm{L}^1([0, t], E)$ is also compact. Take $(f_k) \subseteq E_0$ with $\|f_k\| \le 1$ and set $\varphi_k := L_2(\varepsilon, t)f_k$. Then, by Lemma 19.3,

$$\|\varphi_k(s)\| = \left\|\int_0^s T_{-1}(s - \tau)BS_0(\tau)\mathrm{d}\tau\right\| \le M_t s. \tag{19.8}$$

Since $K_\varepsilon \in \mathcal{L}(E_0, E)$ is compact, one can choose for each rational $s \in [0, t]$ a subsequence $\ell_s(k)$ such that $\lim_{k\to\infty} K_\varepsilon\varphi_{\ell_s(k)} = \varphi(s) \in E$. By taking the diagonal sequence $\ell(k)$ we have

$$K_\varepsilon\varphi_{\ell(k)}(s) = K_\varepsilon \int_0^s T_{-1}(s - \tau)BS_0(\tau)f_{\ell(k)}\mathrm{d}\tau \longrightarrow \varphi(s) \quad \text{as } k \longrightarrow \infty \tag{19.9}$$

for $s \in \mathbb{Q} \cap [0, t]$. Moreover, for $t \ge s \ge s' \ge 2\varepsilon$, one obtains

$$\left\|K_\varepsilon \int_0^{s-2\varepsilon} T_{-1}(s - \tau)BS_0(\tau)f_k\mathrm{d}\tau - K_\varepsilon \int_0^{s'-2\varepsilon} T_{-1}(s' - \tau)BS_0(\tau)f_k\mathrm{d}\tau\right\|$$

$$\le \|K_\varepsilon\|\left\|\int_{s'-2\varepsilon}^{s-2\varepsilon} T_{-1}((s - \tau)BS_0(\tau)f_k\mathrm{d}\tau\right\|$$

$$+ \|K_\varepsilon(T_0(s - s' + 2\varepsilon) - T_0(\varepsilon))\|\left\|\int_0^{s'-2\varepsilon} T_{-1}(s' - 2\varepsilon - \tau)BS_0(\tau)f_k\mathrm{d}\tau\right\|$$

$$\le M_t\left(\|K_\varepsilon\|(s - s') + \|K_\varepsilon(T_0(s - s' + 2\varepsilon) - T_0(\varepsilon))\|\right).$$

Thus, the mapping $[0, t] \ni s \mapsto K_\varepsilon\varphi_k(s)$ is continuous uniformly in k by Assumptions 19.6.b) and (19.8). So, by (19.9), one can find $\varphi(s) \in E$ such that

$$K_\varepsilon\varphi_{\ell(k)}(s) \longrightarrow \varphi(s), \quad \text{as } k \longrightarrow \infty, \quad \text{for all } s \in [0, t].$$

Hence, by (19.8) and the Lebesgue dominated convergence theorem (see Theorem A.23), we have that

$$\lim_{k \to \infty} K_\varepsilon \varphi_{\ell(k)} = \varphi \quad \text{in } L^1([0,t], E).$$

That is, $K_\varepsilon L_2(\varepsilon, t)$ is compact. Thus, due to (19.6), (19.7), and the order continuity of E, one can apply Lemma 19.9 to deduce that $BR_{2,\varepsilon}(t)$ is compact. Finally, Lemma 19.12 implies that $BR_2(t)$ is compact for $t \geq 0$.

Recall that

$$R_3(t)f = \int_0^t T_{-1}(t-s)BR_2(s)f \, ds, \quad f \in E_0,\ t \geq 0.$$

Take $(f_k) \subseteq E_0$ with $\|f_k\| \leq 1$ and set $\varphi_k(s) := BR_2(s)f_k$ for $s \in [0,t]$. As above, one can find a subsequence so that $\varphi_{k_i}(s)$ converges in E for rational s. From Lemma 19.12 we know that the mapping $[0,t] \ni s \mapsto \varphi_k(s)$ is continuous uniformly in k. So, $\varphi_{k_i}(s)$ converges for all $s \in [0,t]$. Therefore, by Lebesgue's theorem, (φ_{k_i}) converges in $L^1([0,t], E)$. Finally, applying Lemma 19.3, one obtains that $R_3(t)f_{k_i}$ converges in E_0. This means that $R_3(t)$ is compact.   □

19.4   Evolution Equations with Boundary Perturbations

In this section we study a Cauchy problem with age-dependent perturbation as an abstract version of problem (19.1). Due to the presence of the operators $A(a) := \kappa(a)\Delta - \mu(a, \cdot)$, $a \in I$, in (19.1), we need the notion of evolution families. A family $(U(a,b))_{(a,b)\in D}$ of bounded linear operators on a Banach space X is called a *strongly continuous evolution family* if

  a) $U(a,c)U(c,b) = U(a,b)$ and $U(a,a) = I$ for $a,b,c \in I$ with $a \geq c \geq b$, and

  b) $D \ni (a,b) \mapsto U(a,b)$ is strongly continuous, where $D := \{(a,b) \in I^2 : a \geq b\}$.

An evolution family $(U(a,b))_{(a,b)\in D}$ is called *exponentially bounded* if its exponential *growth bound*

$$\omega(U) := \inf\{\omega \in \mathbb{R} : \exists M_\omega \geq 1 \text{ with } \|U(a,b)\| \leq M_\omega e^{\omega(a-b)} \text{ for } (a,b) \in D\}$$

satisfies $\omega(U) < \infty$. A simple example of an exponentially bounded evolution family is the following.

Example 19.14. If A generates a C_0-semigroup $(T(t))_{t\geq 0}$ on a Banach space X and $0 \leq \xi \in C_b(I)$, then

$$U(a,b) := T\left(\int_b^a \xi(\tau)d\tau\right), \quad (a,b) \in D,$$

defines an exponentially bounded evolution family on X.

Example 19.15. Consider the diffusion operator

$$A(a, x, D) := \kappa(a)\Delta$$

on $X := L^1(\Omega)$, with the boundary conditions given by the second and third relations in (19.1), i.e.,

$$D(A_1(a)) = \Big\{\varphi \in L^1(\Omega) : \varphi \in W^{1,q}(\Omega) \text{ for } 1 \leq q < \tfrac{d}{d-1};\ A(a, x, D)\varphi \in L^1(\Omega);$$

$$\langle A(a, x, D)\varphi, \psi \rangle = \langle \varphi, A(a, x, D)\psi \rangle \text{ for all } \psi \in D(\Delta_p),\ 1 < p < \infty \Big\}$$

$$= D(\Delta_1),$$

$$A_1(a)\varphi = A(a, x, D)\varphi, \quad \varphi \in D(A_1(a)),$$

where

$$D(\Delta_p) = \Big\{\varphi \in W^{2,p}(\Omega) : \varphi = 0 \text{ on } \Gamma_2,\ \frac{\partial \varphi}{\partial \nu} = 0 \text{ on } \Gamma_1 \Big\},$$

see also Tanabe [135, Section 5.4]. Then,

$$U(a, b) := T_{\Delta_1}\left(\int_b^a \kappa(\tau)\mathrm{d}\tau\right), \quad (a, b) \in D,$$

defines an exponentially bounded evolution family on X, where $T_{\Delta_1}(\cdot)$ is the positive C_0-semigroup generated by Δ_1, the L^1-realization of the Laplacian subject to boundary conditions given by the second and third relations in (19.1). It follows directly that for $\varphi \in D(\Delta_1)$, the function $U(\cdot, b)\varphi \in C^1(I \cap [b, \infty), X)$ is the unique solution to the Cauchy problem

$$u'(a) = A_1(a)u(a),\ a \geq b, \quad u(b) = \varphi.$$

Let $(U(a, b))_{(a,b) \in D}$ be an exponentially bounded evolution family on X. Then for $\omega > \omega(U)$ there exists $M_\omega \geq 1$ such that

$$\|U(a, b)\| \leq M_\omega e^{\omega(a-b)} \quad \text{for } (a, b) \in D.$$

Note that $\omega(U) = -\infty$ if I is compact. On $E = L^1(I, X)$ we now define

$$(T_0(t)f)(a) := \chi_I(a - t)U(a, a - t)f(a - t), \quad t \geq 0,\ a \in I.$$

It is not difficult to see that $(T_0(t))_{t \geq 0}$ is a C_0-semigroup on E and $\omega(T_0) = \omega(U)$, cf. Schnaubelt [128] or Räbiger et al. [115]. Let us denote its generator by G_0. The following representation of G_0 exhibits a relation between G_0 and the operator

$$Lf := -f' + A(\cdot)f, \quad \text{with domain}$$

$$D(L) := \{f \in E : f \in W^{1,1}(I, X),\ f(0) = 0, \qquad\qquad (19.10)$$

$$f(a) \in D(A(a)) \text{ for a.e. } a \in I,\ A(\cdot)f(\cdot) \in E\}.$$

For a proof we refer to Latushkin, Montgomery-Smith, and Randolph [84, Proposition 2.9].

Lemma 19.16. *Let* $(U(a,b))_{(a,b)\in D}$ *be an exponentially bounded evolution family on X. Assume that the function $U(\cdot,b)\varphi$ solves the problem*

$$u'(a) = A(a)u(a), \quad u(b) = \varphi \in Y_b \subseteq D(A(b)),$$

where $A(a)$ a linear operator with domain $D(A(a)) \subset X$, and Y_b is dense in X. Denote by $\mathcal{D}_{00}$ the linear span of functions of the form $\alpha(\cdot)U(\cdot,b)\varphi$ for $b \in I$, $\alpha \in C^1(I)$ with compact support contained in $(b,\infty)\cap I$, and $\varphi \in Y_b$. Then $\mathcal{D}_{00} \subset D(L)$, $\mathcal{D}_{00}$ is dense in E, and $T_0(t)\mathcal{D}_{00} \subset \mathcal{D}_{00}$. Moreover, G_0 is the closure of the operator L defined in (19.10) with domain $\mathcal{D}_{00}$.

Now let $V(\cdot) \in L^\infty(I, \mathcal{L}(X))$. Applying Theorem 11.5 and Proposition 11.6 we obtain that

$$G_V := G_0 - V \text{ with } D(G_V) = D(G_0) \tag{19.11}$$

generates a C_0-semigroup $(T_V(t))_{t\geq 0}$ on E given by

$$(T_V(t)f)(a) := \chi_I(a-t)U_V(a, a-t)f(a-t), \quad t \geq 0,\ a \in I,$$

where $(Vf)(a) := V(a)f(a)$, $a \in I$ and $(U_V(a,b))_{(a,b)\in D}$ is the exponentially bounded evolution family satisfying

$$
\begin{aligned}
U_V(a,b)\varphi &= U(a,b)\varphi - \int_b^a U(a,s)V(s)U_V(s,b)\varphi\, ds \\
&= U(a,b)\varphi - \int_b^a U_V(a,s)V(s)U(s,b)\varphi\, ds
\end{aligned}
\tag{19.12}
$$

for $(a,b) \in D$ and $\varphi \in X$. Since the resolvent of G_0 equals

$$
\begin{aligned}
R(\lambda, G_0)f(a) &= \int_0^\infty \chi_I(a-t)e^{-\lambda t}U(a, a-t)f(a-t)\, dt \\
&= \int_0^a e^{-\lambda(a-t)}U(a,t)f(t)\, dt
\end{aligned}
$$

for $\lambda > \omega(U)$, $a \in I$, and $f \in E$, we see that $D(G_0)$ consists of continuous functions vanishing at $a = 0$. In order to consider functions f with $f(0) \neq 0$, we introduce an extension G of G_0 defined by

$$
\begin{aligned}
Gf &:= G_0 f_0 + \omega e_\omega(\cdot)\varphi, \quad \text{with domain} \\
D(G) &:= \{f = f_0 + e_\omega(\cdot)\varphi : f_0 \in D(G_0), \varphi \in X\},
\end{aligned}
\tag{19.13}
$$

where

$$e_\omega(a)\varphi := e^{-\omega a}U(a,0)\varphi$$

for $\varphi \in X$ and $\omega > \max\{\omega(U), \omega(U_V)\} =: \omega_1$. We will further need the following auxiliary result.

Lemma 19.17. *The following holds for the operators G and G_V defined in (19.11) and (19.13).*

a) $e_\lambda \varphi \in D(G)$ *and* $G e_\lambda \varphi = \lambda e_\lambda \varphi$ *for* $\operatorname{Re} \lambda > \omega(U)$ *and* $\varphi \in X$;

b) $e_\lambda^V := e^{-\lambda \cdot} U_V(\cdot, 0) = e_\lambda - R(\lambda, G_V) V e_\lambda$ *for* $\operatorname{Re} \lambda > \omega_1$;

c) $\ker(\lambda - (G - V)) = \{e_\lambda^V \varphi : \varphi \in X\}$ *for* $\operatorname{Re} \lambda > \omega_1$.

Proof. Let $f := e_\lambda(\cdot) \varphi - e_\mu(\cdot) \varphi$ and $\xi(a) := e^{-\lambda a} - e^{-\mu a}$ for $\operatorname{Re} \lambda, \operatorname{Re} \mu > \omega(U)$ and $\varphi \in X$. We have

$$T_0(t) f(a) - f(a) = (\chi_I(a - t) \xi(a - t) - \xi(a)) U(a, 0) \varphi, \quad a \in I, \, t \ge 0.$$

Hence, $f \in D(G_0)$ and $G_0(e_\lambda(\cdot) \varphi - e_\mu(\cdot) \varphi) = \lambda e_\lambda(\cdot) \varphi - \mu e_\mu(\cdot) \varphi$. So, considering $e_\lambda = e_\lambda - e_\mu + e_\mu$ we obtain a).

Assertion b) follows from relations (19.12) and

$$R(\lambda, G_V) f(a) = \int_0^a e^{-\lambda(a - t)} U_V(a, t) f(t) \, dt.$$

Further, a) and b) imply that

$$
\begin{aligned}
(G - V) e_\lambda^V &= (G - V)(e_\lambda - R(\lambda, G_V) V e_\lambda) \\
&= \lambda e_\lambda - V e_\lambda - G_V R(\lambda, G_V) V e_\lambda \\
&= \lambda(e_\lambda - R(\lambda, G_V) V e_\lambda) = \lambda e_\lambda^V
\end{aligned}
$$

for $\operatorname{Re} \lambda > \omega_1$. Conversely, let $f \in \ker(\lambda - (G - V))$. Using a) and b) we obtain

$$f - e_\lambda^V f(0) = (f - e_\omega f(0)) + (e_\omega f(0) - e_\lambda f(0)) + (e_\lambda f(0) - e_\lambda^V f(0)) \in D(G_0).$$

Therefore, $0 = (\lambda - G_V)(f - e_\lambda^V f(0))$. This implies that $f = e_\lambda^V f(0)$, since $\lambda \in \rho(G_V)$. $\qquad \square$

We define now the concept of solutions of the following Cauchy problem with boundary perturbation:

$$
\begin{cases}
u'(t) = (G - V) u(t), \\
u(0) = f \in E, \\
u(t, 0) = B u(t) \in X, \quad t \ge 0,
\end{cases}
\tag{19.14}
$$

where $E = L^1(I, X)$, $B \in \mathcal{L}(E, X)$, and G and V are as before. A *classical solution* of problem (19.14) is a function $u \in C^1(\mathbb{R}_+, E)$ such that $u(t) \in D(G)$ and (19.14) holds for all $t \ge 0$. To prove the existence of classical solutions, we consider the product space $\mathcal{E} := X \times E$ with the maximum norm. On $\mathcal{E}$ we define the matrix operators

$$
\mathcal{B} := \begin{pmatrix} 0 & B \\ 0 & 0 \end{pmatrix} \quad \text{and} \quad \mathcal{G}_V := \begin{pmatrix} 0 & -\delta_0 \\ 0 & G - V \end{pmatrix},
$$

where $D(\mathcal{G}_V) := \{0\} \times D(G)$. Let us show that $\mathcal{G}_V$ is a Hille–Yosida operator on $\mathcal{E}$.

Lemma 19.18. *The operator $\mathcal{G}_V$ with domain $\{0\} \times D(G)$ is a Hille–Yosida operator on $\mathcal{E}$. Moreover,*

$$R(\lambda, \mathcal{G}_V) = \begin{pmatrix} 0 & 0 \\ e_\lambda^V & R(\lambda, G_V) \end{pmatrix}, \quad \operatorname{Re}\lambda > \omega_1.$$

Proof. For $\operatorname{Re}\lambda > \omega_1$ we set

$$R(\lambda) := \begin{pmatrix} 0 & 0 \\ e_\lambda^V & R(\lambda, G_V) \end{pmatrix}.$$

It follows directly that $R(\lambda) \in \mathcal{L}(\mathcal{E})$. Moreover, by Lemma 19.17, $R(\lambda)\mathcal{E} \subseteq D(\mathcal{G}_V)$, and $(\lambda - \mathcal{G}_V)R(\lambda) = I$. Hence, $\lambda - \mathcal{G}_V$ is surjective. On the other hand, from (19.11) and (19.13) it follows, upon applying Lemma 19.17.b), that

$$R(\omega)(\omega - \mathcal{G}_V)\begin{pmatrix} 0 \\ f \end{pmatrix} = \begin{pmatrix} 0 & 0 \\ e_\omega^V & R(\omega, G_V) \end{pmatrix}\begin{pmatrix} f(0) \\ (\omega - G_V)f_0 + Ve_\omega f(0) \end{pmatrix} = \begin{pmatrix} 0 \\ f \end{pmatrix}$$

for $f = f_0 + e_\omega f(0) \in D(G)$. Hence, $R(\omega) = R(\omega, \mathcal{G}_V)$ and $\mathcal{G}_V$ is closed. On the other hand, we see that $R(\lambda)(\lambda - \mathcal{G}_V)\begin{pmatrix} 0 \\ f_0 \end{pmatrix} = \begin{pmatrix} 0 \\ f_0 \end{pmatrix}$ for $f_0 \in D(G_0)$. If $(\lambda - \mathcal{G}_V)\begin{pmatrix} 0 \\ f \end{pmatrix} = 0$ for some $f \in D(G)$, then $f(0) = 0$, and so $f = 0$. Thus, $\lambda - \mathcal{G}_V$ is injective for $\operatorname{Re}\lambda > \omega_1$. Therefore, $R(\lambda) = R(\lambda, \mathcal{G}_V)$ for $\operatorname{Re}\lambda > \omega_1$.

Finally, we obtain

$$R(\lambda, \mathcal{G}_V)^k = \begin{pmatrix} 0 & 0 \\ R(\lambda, G_V)^{k-1}e_\lambda^V & R(\lambda, G_V)^k \end{pmatrix},$$

and since G_V is a generator, we infer that

$$\|R(\lambda, G_V)^{k-1}e_\lambda^V(\cdot)x\| \le \frac{M}{(\lambda - \omega)^{k-1}}\|e_\lambda^V(\cdot)x\| \le \frac{M^2}{(\lambda - \omega)^k}\|x\|$$

for $\lambda > \omega > \omega_1$, $x \in X$. Thus, $\mathcal{G}_V$ is a Hille–Yosida operator on $\mathcal{E}$. □

Remark 19.19. As a consequence, we obtain that the part $\mathcal{G}_{V,0}$ in $\mathcal{E}_0 := \{0\} \times E = \overline{D(\mathcal{G}_V)}$ generates a C_0-semigroup $\mathcal{T}_{V,0}(\cdot)$ in $\mathcal{E}_0$. In particular, we have

$$\mathcal{G}_{V,0}\begin{pmatrix} 0 \\ f \end{pmatrix} = \begin{pmatrix} 0 \\ G_V f \end{pmatrix}, \quad \text{with domain}$$

$$D(\mathcal{G}_{V,0}) = \left\{ \begin{pmatrix} 0 \\ f \end{pmatrix} \in \mathcal{E}_0 : f \in D(G_0) \right\}.$$

So, since $\mathcal{E}_0 \cong E$, we can identify $\mathcal{G}_{V,0}$ and $\mathcal{T}_{V,0}(\cdot)$ with G_V and $T_V(\cdot)$, respectively. Moreover, there exists the *extrapolated semigroup* $\mathcal{T}_{V,-1}(\cdot)$ on $\mathcal{E}_{-1}$ with generator $\mathcal{G}_{V,-1}$ with domain $D(\mathcal{G}_{V,-1}) = \mathcal{E}$, see Section 19.2.

We are now ready to state the main result of this section.

Theorem 19.20. *Suppose that* $(U(a,b))_{(a,b)\in D}$ *is an exponentially bounded evolution family on a Banach space* X, $V \in \mathrm{L}^\infty(I, \mathcal{L}(X))$, *and* $B \in \mathcal{L}(E, X)$. *Let* G *be the operator on* $E = \mathrm{L}^1(I, X)$ *defined in* (19.13). *Then the operator*

$$G_{VB}f = (G - V)f \quad with \quad D(G_{VB}) = \{f \in D(G) : f(0) = Bf\} \qquad (19.15)$$

generates a C_0-*semigroup* $(S_0(t))_{t\geq 0}$ *on* E *satisfying*

$$S_0(t)f = T_V(t)f + (\lambda - G_V) \int_0^t T_V(t-s)e_\lambda^V BS_0(s)f \, ds \qquad (19.16)$$

for $f \in E$, $t \geq 0$ *and* $\lambda > \omega_1$.

Proof. By Lemma 19.18 and Theorem 19.4, the part $\mathcal{G}_{VB}$ of the operator $\mathcal{G}_{V,-1} + \mathcal{B}$ in $\mathcal{E}_0$ generates a C_0-semigroup $(\mathcal{S}_0(t))_{t\geq 0}$ on $\mathcal{E}_0$ satisfying

$$\mathcal{S}_0(t)\begin{pmatrix} 0 \\ f \end{pmatrix} = \mathcal{T}_{V,0}(t)\begin{pmatrix} 0 \\ f \end{pmatrix} + \int_0^t \mathcal{T}_{V,-1}(t-s)\mathcal{B}\mathcal{S}_0(s)\begin{pmatrix} 0 \\ f \end{pmatrix} \, ds, \ f \in E.$$

Identifying $(\mathcal{S}_0(t))_{t\geq 0}$ with a C_0-semigroup $(S_0(t))_{t\geq 0}$ on E, we deduce that

$$\begin{pmatrix} 0 \\ S_0(t)f \end{pmatrix} = \begin{pmatrix} 0 \\ T_V(t)f \end{pmatrix} + \int_0^t \mathcal{T}_{V,-1}(t-s)\begin{pmatrix} BS_0(s)f \\ 0 \end{pmatrix} \, ds. \qquad (19.17)$$

Now, by Lemma 19.18, we have

$$\int_0^t \mathcal{T}_{V,-1}(t-s)\begin{pmatrix} BS_0(s)f \\ 0 \end{pmatrix} ds = \int_0^t \mathcal{T}_{V,-1}(t-s)(\lambda - \mathcal{G}_V)R(\lambda, \mathcal{G}_V)\begin{pmatrix} BS_0(s)f \\ 0 \end{pmatrix} ds$$

$$= (\lambda - \mathcal{G}_{V,-1}) \int_0^t \mathcal{T}_{V,-1}(t-s)\begin{pmatrix} 0 \\ e_\lambda^V BS_0(s)f \end{pmatrix} ds$$

$$= (\lambda - \mathcal{G}_{V,-1})\begin{pmatrix} 0 \\ \int_0^t T_V(t-s)e_\lambda^V BS_0(s)f \, ds \end{pmatrix}$$

$$=: (\lambda - \mathcal{G}_{V,-1})\begin{pmatrix} 0 \\ g \end{pmatrix}$$

for $\lambda > \omega_1$. On the other hand, from (19.17) we see that $(\lambda - \mathcal{G}_{V,-1})\begin{pmatrix} 0 \\ g \end{pmatrix} \in \mathcal{E}_0$ and thus, since $\mathcal{G}_{V,0}$ is the part of $\mathcal{G}_{V,-1}$ in $\mathcal{E}_0$, we also have $\begin{pmatrix} 0 \\ g \end{pmatrix} \in D(\mathcal{G}_{V,0})$, so identity (19.16) follows.

It remains to show (19.15). For $\begin{pmatrix} 0 \\ f \end{pmatrix} \in D(\mathcal{G}_{VB})$, we have

$$\mathcal{G}_{V,-1}\begin{pmatrix} 0 \\ f \end{pmatrix} + \begin{pmatrix} Bf \\ 0 \end{pmatrix} \in \mathcal{E}_0, \quad \text{and hence } \mathcal{G}_{V,-1}\begin{pmatrix} 0 \\ f \end{pmatrix} \in \mathcal{E}.$$

Since $\mathcal{G}_V$ is the part of $\mathcal{G}_{V,-1}$ in $\mathcal{E}$, we deduce that $f \in D(G)$ and

$$\mathcal{G}_{VB}\begin{pmatrix} 0 \\ f \end{pmatrix} = \begin{pmatrix} -f(0) + Bf \\ (G - V)f \end{pmatrix} \in \mathcal{E}_0.$$

This establishes (19.15) by identifying $\mathcal{G}_{VB}$ with G_{VB}. $\qquad \square$

19.5   Back to the Population Equation

Let us now return to our population model problem (19.1) and apply the abstract results developed in the previous sections. For convenience we repeat these equations:

$$
\begin{cases}
\partial_t u(t,a,x) + \partial_a u(t,a,x) = \kappa(a)\Delta u(t,a,x) - \mu(a,x)u(t,a,x), \\
\qquad\qquad\qquad\qquad\qquad\qquad t \geq 0,\ a \in I,\ x \in \Omega, \\
\partial_\nu u(t,a,x) = 0, \quad t \geq 0,\ a \in I,\ x \in \Gamma_1, \\
u(t,a,x) = 0, \quad t \geq 0,\ a \in I,\ x \in \Gamma_2, \\
u(t,0,x) = \displaystyle\int_I \beta(a,x)u(t,a,x)\,\mathrm{d}a, \quad t \geq 0,\ x \in \Omega, \\
u(0,a,x) = f(a,x), \quad a \in I,\ x \in \Omega.
\end{cases}
\tag{19.18}
$$

Assumptions 19.21. We introduce the following hypotheses.

a) $\Omega \subset \mathbb{R}^n$ is a bounded domain with C^2-boundary $\partial\Omega = \Gamma_1 \cup \Gamma_2$, where Γ_i are open and closed in $\partial\Omega$ with $\Gamma_1 \cap \Gamma_2 = \emptyset$.

b) $k \in \mathrm{C_b}(I)$ and $\kappa(a) \geq \delta > 0$ for all $a \in I$ and some constant δ.

c) $0 \leq \mu \in \mathrm{L}^\infty(I \times \Omega)$.

d) $0 \leq \beta \in \mathrm{C_{ub}}(I, \mathrm{L}^\infty(\Omega))$.

Let $X := \mathrm{L}^1(\Omega)$ and consider the evolution family

$$
U(a,b) = T_{\Delta_1}\left(\int_b^a \kappa(\tau)\mathrm{d}\tau\right), \quad (a,b) \in D := \{(a,b) \in I^2 : a \geq b\},
$$

given in Example 19.15. The associated positive C_0-semigroup on $E := \mathrm{L}^1(I \times \Omega) \cong \mathrm{L}^1(I, X)$ with generator G_0 is given by

$$
(T_0(t)f)(a) := \chi_I(a - t)U(a, a - t)f(a - t), \quad t \geq 0,\ a \in I.
\tag{19.19}
$$

We define the positive (bounded) multiplication operator

$$
(Vf)(a) := \mu(a,\cdot)f(a), \quad f \in E.
$$

Then, $G_V := G_0 - V$ with $D(G_V) = D(G_0)$ generates a positive C_0-semigroup given by

$$
(T_V(t)f)(a) := \chi_I(a - t)U_V(a, a - t)f(a - t), \quad t \geq 0,\ a \in I.
$$

Relation (19.12) yields

$$
U_V(a,b)\varphi = T_{\Delta_1}\left(\int_b^a \kappa(\tau)\mathrm{d}\tau\right)\varphi - \int_b^a T_{\Delta_1}\left(\int_s^a \kappa(\tau)\mathrm{d}\tau\right)\mu(s,\cdot)U_V(s,b)\varphi\,\mathrm{d}s
$$

$$
\tag{19.20}
$$

for $(a, b) \in D$ and $\varphi \in X$. Since $(T_V(t))_{t \geq 0}$ is positive, we see that $(U_V(a, b))_{(a,b) \in D}$ is positive as well and by relation (19.20) we have

$$0 \leq U_V(a, b) \leq T_{\Delta_1} \left(\int_b^a \kappa(\tau) d\tau \right), \quad (a, b) \in D. \tag{19.21}$$

Finally, we define the boundary operator

$$Bf := \int_I \beta(a, \cdot) f(a) da, \quad f \in E.$$

Assumption 19.21.d) implies that $0 \leq B \in \mathcal{L}(E, X)$. With these spaces and operators we now consider the abstract Cauchy problem in (19.14). We say that problem (19.18) admits a *generalized solution* u if $u \in C(\mathbb{R}_+, E)$ and solves problem (19.14).

The positivity of $(U_V(a, b))_{(a,b) \in D}$ and Theorem 19.20 yield existence and uniqueness of generalized solutions to problem (19.18).

Corollary 19.22. *Let Assumptions* 19.21 *be satisfied. Then the operator*

$$G_{VB}f := (G - V)f \text{ with } D(G_{VB}) := \left\{ f \in D(G) : f(0) = \int_I \beta(a, \cdot) f(a) da \right\}$$

generates a positive C_0-semigroup $(S_0(t))_{t \geq 0}$ on E satisfying the relation (19.16), *where G is the operator given by* (19.13).

In order to understand our notion of generalized solutions to problem (19.18) we have to determine the operator G. To give a partial answer we introduce the spaces

$$\mathcal{D} := \{ f \in E : f \in W^{1,1}(I, X), \ f(a) \in D(\Delta_1) \text{ for a.e. } a \in I, \ A_1(\cdot) f(\cdot) \in E \},$$
$$\mathcal{D}_0 := \{ f \in \mathcal{D} : f(0) = 0 \},$$
$$\mathcal{D}_1 := \{ f \in \mathcal{D} : f(0) \in D(\Delta_1) \},$$

and the operator $Lf := -f' + A_1(\cdot)f \in E$ with domain $\mathcal{D}$, where

$$A_1(a) := \kappa(a)\Delta_1 f(a), \ a \in I.$$

Proposition 19.23. *Let Assumptions* 19.21.a)–c) *be satisfied. Then G_0 is the closure in E of the operator L with domain $\mathcal{D}_0$, and $\mathcal{D}_1$ is dense in $D(G)$ endowed with the graph norm. In particular, G is the closure of L with domain $\mathcal{D}_1$.*

Proof. Let us fix $\lambda_0 > 0$. Lemma 9.31 implies that

$$\left\| \left(e^{-\lambda_0 t} T_{\Delta_1} \left(\int_{a-t}^a \kappa(\tau) d\tau \right) - I \right) (A_1(a - t) - \lambda_0)^{-1} \right\| \leq Mte^{\omega t} \tag{19.22}$$

for $(a, a - t) \in D$, $t \geq 0$ and some constants $M, \omega \in \mathbb{R}$. Moreover, we have

$$\lim_{t \to 0} \frac{1}{t} \left(e^{-\lambda_0 t} T_{\Delta_1} \left(\int_{a-t}^{a} \kappa(\tau) d\tau \right) - I \right) (A_1(a - t) - \lambda_0)^{-1} \varphi = \varphi \qquad (19.23)$$

for $(a, a - t) \in D$, $t > 0$ and $\varphi \in X$. For $a - t \geq 0$ and $f \in \mathcal{D}_0$ use relation (19.19) and deduce that

$$T_0(t)f(a) - f(a) = e^{\lambda_0 t} \left(e^{-\lambda_0 t} T_{\Delta_1} \left(\int_{a-t}^{a} \kappa(\tau) d\tau \right) - I \right)$$
$$\times (A_1(a - t) - \lambda_0)^{-1}(A_1(a - t) - \lambda_0)f(a - t)$$
$$+ e^{\lambda_0 t}(f(a - t) - f(a)) + (e^{\lambda_0 t} - 1)f(a).$$

Using (19.22), (19.23), and the Dominated Convergence Theorem (see Theorem A.23) we obtain that $\lim_{t \to 0} \frac{1}{t}(T_0(t)f - f) = Lf$ in E and so $\mathcal{D}_0 \subset D(G_0)$ and $G_0 f = Lf$ for any $f \in \mathcal{D}_0$. Since $\mathcal{D}_{00} \subset \mathcal{D}_0$, it follows from Lemma 19.16, that G_0 is the closure of L with domain $\mathcal{D}_0$.

On the other hand, let $f \in \mathcal{D}_1$. By writing $f = (f - e_\omega(\cdot)f(0)) + e_\omega(\cdot)f(0)$ and noting that $e_\omega(\cdot)f(0) \in \mathcal{D}$, since $f(0) \in D(\Delta_1)$, we have from (19.13) that $f \in D(G)$ and

$$Gf = L(f - e_\omega(\cdot)f(0)) + \omega e_\omega(\cdot)f(0) = Lf.$$

Thus, G is an extension of L with domain $\mathcal{D}_1$. Let now $f = f_0 + e_\omega(\cdot)f(0) \in D(G)$. Since G_0 is the closure of L with domain $\mathcal{D}_0$, we infer that there is $f_{0,k} \in \mathcal{D}_0$ such that $f_{0,k} \to f_0$ and $Lf_{0,k} \to G_0 f_0$ in E as $k \to \infty$. Moreover, there is $\varphi_k \in D(\Delta_1)$ such that $\lim_{k \to \infty} \varphi_k = f(0)$ in X. As above, since $Le_\omega(\cdot)\varphi_k = \omega e_\omega(\cdot)\varphi_k$, we see that $f_k := f_{0,k} + e_\omega(\cdot)\varphi_k \in \mathcal{D}_1$ converges to f in the graph norm of G. Thus, G is the closure of L with domain $\mathcal{D}_1$. $\qquad \square$

We end this chapter by studying the asymptotic behavior of generalized solutions to problem (19.18). For this purpose we first observe the following.

Lemma 19.24. *Let Assumptions 19.21.a)–c) be satisfied. Then* $\omega(U_V) = \omega(U) = -\infty$ *for* $a_m < \infty$*, while* $\omega(U_V) \leq \omega(U) < 0$ *for* $a_m = \infty$ *and* $\Gamma_1 = \emptyset$*.*

Proof. The first assertion is clear from the definition of $\omega(U)$ and the fact that $I = [0, a_m]$ is a compact interval. The second assertion follows from inequality (19.21) and the fact that Δ_1 with Dirichlet boundary conditions generates an exponentially stable semigroup on $L^1(\Omega)$. $\qquad \square$

In order to verify the conditions from Assumptions 19.6, we need the following regularity result.

Lemma 19.25. *Let Assumptions 19.21.a)–c) be satisfied and let* $a_m = \infty$*. Then, for any* $t_0 > 0$*, the following holds:*

$$\limsup_{\substack{t \to t_0 \\ a \in I}} \|U_V(a + t, a) - U_V(a + t_0, a)\| = 0.$$

Proof. Let $t_0 + 1 \geq t \geq t_0 > 0$, $a \geq 0$ and $\varphi \in X$. By (19.20), we have

$$U_V(a + t, a)\varphi - U_V(a + t_0, a)\varphi$$

$$= T_{\Delta_1}\left(\int_a^{a+t} \kappa(\tau)d\tau\right)\varphi - T_{\Delta_1}\left(\int_a^{a+t_0} \kappa(\tau)d\tau\right)\varphi$$

$$- \int_{a+t_0}^{a+t} T_{\Delta_1}\left(\int_s^{a+t} \kappa(\tau)d\tau\right)\mu(s,\cdot)U_V(s,a)\varphi\,ds$$

$$- \int_a^{a+t_0}\left(T_{\Delta_1}\left(\int_s^{a+t} \kappa(\tau)d\tau\right) - T_{\Delta_1}\left(\int_s^{a+t_0} \kappa(\tau)d\tau\right)\right)$$

$$\times \mu(s,\cdot)U_V(s,a)\varphi\,ds$$

$$=: I_1 + I_2 + I_3.$$

To estimate I_1, we recall that $T_{\Delta_1}(\cdot)$ can be extended to an analytic semigroup on X, which is equivalent to $T_{\Delta_1}(t)X \subset D(\Delta_1)$ and $\|\Delta_1 T_{\Delta_1}(t)\| \leq \frac{M}{t}e^{\omega t}$ for all $t > 0$ and some constants $M, \omega \in \mathbb{R}_+$. So, by Proposition 9.16, we obtain

$$\|T_{\Delta_1}(\tau_1)\varphi - T_{\Delta_1}(\tau_0)\varphi\| = \left\|\int_{\tau_0}^{\tau_1} \Delta_1 T_{\Delta_1}(s)\varphi\,ds\right\|$$

$$\leq M \int_{\tau_0}^{\tau_1} \frac{e^{\omega s}}{s}\,ds\|\varphi\|$$

$$\leq Me^{\omega(\tau_0+1)} \log\left(\frac{\tau_1}{\tau_0}\right)\|\varphi\|$$

for $0 < \tau_0 \leq \tau_1 \leq \tau_0 + 1$ and $\varphi \in X$. Consequently,

$$\|I_1\| \leq Me^{\omega(\|\kappa\|_\infty t_0+1)} \log\left(\frac{\int_a^{a+t} \kappa(\tau)d\tau}{\int_a^{a+t_0} \kappa(\tau)d\tau}\right)\|\varphi\|$$

$$\leq Me^{\omega(\|\kappa\|_\infty t_0+1)} \log\left(1 + \frac{\int_{a+t_0}^{a+t} \kappa(\tau)d\tau}{\int_a^{a+t_0} \kappa(\tau)d\tau}\right)\|\varphi\|$$

$$\leq Me^{\omega(\|\kappa\|_\infty t_0+1)} \log\left(1 + \frac{\|\kappa\|_\infty(t - t_0)}{\delta t_0}\right)\|\varphi\|,$$

where δ is the constant from Assumptions 19.21.b). Moreover, using the above estimate for $\|I_1\|$ we also obtain

$$\|I_3\| \leq M\|\mu\|_\infty t_0 e^{\omega(\|\kappa\|_\infty t_0+t_0+1)} \log\left(1 + \frac{\|\kappa\|_\infty(t - t_0)}{\delta t_0}\right)\|\varphi\|.$$

By the exponential boundedness of $U(\cdot,\cdot)$ and (19.21), we finally have

$$\|I_2\| \leq M_2\|\mu\|_\infty e^{\omega_2(t_0+1)}(t - t_0)\|\varphi\|$$

for some constants $M_2 \geq 1$ and $\omega_2 \in \mathbb{R}$. The case $t_0 \geq t > \varepsilon > 0$ can be treated in the same way. $\square$

From now on we assume that the Assumptions 19.21 are satisfied. We apply
the results of the previous sections to the operators

$$\mathcal{B} := \begin{pmatrix} 0 & B \\ 0 & 0 \end{pmatrix} \text{ and } \mathcal{G}_V := \begin{pmatrix} 0 & -\delta_0 \\ 0 & G - V \end{pmatrix}$$

on $\mathcal{E} = X \times E$, where $D(\mathcal{G}_V) := \{0\} \times D(G)$. By Lemma 19.18, $\mathcal{G}_V$ is a Hille–
Yosida operator and, by our assumptions, it is resolvent positive on $\mathcal{E}$. Moreover,
$0 \le \mathcal{B} \in \mathcal{L}(\mathcal{E}_0, \mathcal{E})$, where $\mathcal{E}_0 = \{0\} \times E \cong E$. Combining this with the spectral
decomposition developed in Chapter 14, we can describe the asymptotic behavior
of the generalized solution to problem (19.18) given by the positive C_0-semigroup
$(S_0(t))_{t \ge 0}$ on E generated by $\mathcal{G}_V$, see Corollary 19.22.

Theorem 19.26. *Let Assumptions 19.21 be satisfied. If $a_m < \infty$, or $a_m = \infty$ and
$\Gamma_1 = \emptyset$, then $\omega_{\mathrm{ess}}(S_0) < 0$. Therefore, the set $\{\lambda \in \sigma(G_{VB}) : \mathrm{Re}\,\lambda \ge 0\}$ is finite
(or empty) and consists of poles of $R(\cdot, G_{VB})$ of finite algebraic multiplicity. If
$\lambda_1, \ldots, \lambda_m$ are these poles with the corresponding spectral projections $P_1, \ldots, P_m$
and orders $k_1, \ldots, k_m$, then*

$$\left\| S_0(t) - \sum_{j=1}^{m} e^{\lambda_j t} \sum_{k=0}^{k_j - 1} \frac{t^k}{k!} (G_{VB} - \lambda_j)^k P_j \right\| \le M e^{-\varepsilon t}, \quad t \ge 0$$

for some constants $M \ge 1$ and $\varepsilon > 0$.

Proof. We only need to show that $\omega_{\mathrm{ess}}(S_0) < 0$, the rest then follows by Theorem
14.4. We shall proceed in two steps.

We start with the case when $a_m = \infty$. By Lemma 19.24 and Proposition
19.13, the assertion is proved as soon as we verify Assumptions 19.6.a) and b) for
the operators $\mathcal{B}$ and $\mathcal{G}_V$. We already know that $\mathcal{G}_V$ is a resolvent positive operator
on $\mathcal{E}$ and $0 \le \mathcal{B} \in \mathcal{L}(\mathcal{E}_0, \mathcal{E})$. For the mapping $t \mapsto \mathcal{B}\mathcal{T}_{V,0}(t) \in \mathcal{L}(\mathcal{E}_0, \mathcal{E})$ we have

$$\left\| (\mathcal{B}\mathcal{T}_{V,0}(t) - \mathcal{B}\mathcal{T}_{V,0}(t_0)) \begin{pmatrix} 0 \\ f \end{pmatrix} \right\|$$

$$= \| BT_V(t)f - BT_V(t_0)f \|_X$$

$$= \left\| \int_t^\infty \beta(a, \cdot) U_V(a, a - t) f(a - t) \mathrm{d}a - \int_{t_0}^\infty \beta(a, \cdot) U_V(a, a - t_0) f(a - t_0) \mathrm{d}a \right\|_X$$

$$\le \int_0^\infty \| \beta(a + t, \cdot) - \beta(a + t_0, \cdot) \|_\infty \| U_V(a + t, a) f(a) \| \mathrm{d}a \qquad (19.24)$$

$$+ \int_0^\infty \| \beta(a + t_0, \cdot) \|_\infty \| U_V(a + t, a) - U_V(a + t_0, a) \| \| f(a) \| \mathrm{d}a$$

$$\le \| f \| \sup_{a \ge 0} \Big(M e^{\omega t} \| \beta(a + t, \cdot) - \beta(a + t_0, \cdot) \|_\infty + \| \beta \|$$

$$\times \| U_V(a + t, a) - U_V(a + t_0, a) \| \Big)$$

for $t \geq t_0 > 0$ and $f \in E$, where $\|\beta\| = \|\beta\|_\infty$ is the supremum norm. By Assumptions 19.21 and Lemma 19.25, hypothesis 19.6.a) is verified. To verify 19.6.b), we use (19.21), which yields

$$0 \leq \mathcal{B}\mathcal{T}_{V,0}(\varepsilon)\begin{pmatrix} 0 \\ f \end{pmatrix} = \begin{pmatrix} \int_0^\infty \beta(a+\varepsilon,\cdot)U_V(a+\varepsilon,a)f(a)\mathrm{d}a \\ 0 \end{pmatrix}$$

$$\leq \begin{pmatrix} \int_0^\infty \beta(a+\varepsilon,\cdot)T_{\Delta_1}\left(\int_a^{a+\varepsilon}\kappa(\tau)\mathrm{d}\tau\right)f(a)\mathrm{d}a \\ 0 \end{pmatrix}$$

$$\leq \begin{pmatrix} \|\beta\|_\infty T_{\Delta_1}(\delta\varepsilon)L_\varepsilon f \\ 0 \end{pmatrix} =: \mathcal{K}_\varepsilon\begin{pmatrix} 0 \\ f \end{pmatrix},$$

where

$$L_\varepsilon f := \int_0^\infty T_{\Delta_1}\left(\int_a^{a+\varepsilon}(\kappa(\tau)-\delta)\mathrm{d}\tau\right)f(a)\mathrm{d}a, \quad f \in E.$$

By Example 19.15, $D(\Delta_1) \subset \mathrm{W}^{1,1}(\Omega)$, which by Rellich's theorem (see Theorem A.45) is compactly embedded in $\mathrm{L}^1(\Omega)$, hence $T_{\Delta_1}(\delta\varepsilon)$ is compact. Therefore, $L_\varepsilon \in \mathcal{L}(E,X)$ implies that $\mathcal{K}_\varepsilon \in \mathcal{L}(\mathcal{E}_0,\mathcal{E})$ is a compact operator. Finally, the continuity of the mapping $t \mapsto \mathcal{K}_\varepsilon \mathcal{T}_{V,0}(t) \in \mathcal{L}(\mathcal{E}_0,\mathcal{E})$ for $t > 0$ can be shown as in (19.24), which proves Assumption 19.6.b).

In the second step we consider $a_\mathrm{m} < \infty$. In this case we extend $A_1(\cdot)$, $V(\cdot)$, and β by setting $A_1(a) := A(a_\mathrm{m})$, $V(a) := -2a$, and $\beta(a,\cdot) := \beta(a_\mathrm{m},\cdot)$ for $a > a_\mathrm{m}$. This extension yields the evolution family $(U_V^\infty(a,b))_{a \geq b \geq 0}$ on X, where

$$U_V^\infty(a+t,a) := \mathrm{e}^{-t^2-2at}T_{\Delta_1}(\kappa(a_m)t)$$

for $a > a_\mathrm{m}$ and $t \geq 0$. Then, $(U_V^\infty(a,b))_{a \geq b \geq 0}$ satisfies the conclusion of Lemma 19.25, and by Lemma 19.24, $\omega(U_V^\infty) = -\infty$. Denote by

$$T_V^\infty(t)f(a) := \chi_{\mathbb{R}_+}(a-t)U_V^\infty(a,a-t)f(a-t), \quad t \geq 0, \ a \geq 0 \text{ a.e.},$$

the corresponding C_0-semigroup on $\mathrm{L}^1(\mathbb{R}_+,X)$ with generator G_V^∞. Setting

$$Pf := \chi_I f \in \mathrm{L}^1([0,a_\mathrm{m}],X) \text{ for } f \in \mathrm{L}^1(\mathbb{R}_+,X),$$

$$Jf(a) := \begin{cases} f(a) & \text{if } a \in [0,a_\mathrm{m}], \\ 0 & \text{if } a \in \mathbb{R}_+ \setminus [0,a_\mathrm{m}], \end{cases} \text{ for } f \in \mathrm{L}^1([0,a_\mathrm{m}],X),$$

we have $PT_V^\infty(t) = T_V(t)P$ and $PG_V^\infty = G_V P$. Further, we perturb the corresponding Hille–Yosida operator $\mathcal{G}_V^\infty$ on $X \times \mathrm{L}^1(\mathbb{R}_+,X)$ by the matrix operators

$$\mathcal{B}^\infty := \begin{pmatrix} 0 & B^\infty \\ 0 & 0 \end{pmatrix} \quad \text{and} \quad \widetilde{\mathcal{B}} := \begin{pmatrix} 0 & \widetilde{B} \\ 0 & 0 \end{pmatrix},$$

respectively, where

$$B^\infty f = \int_0^{a_\mathrm{m}} \beta(a,\cdot)\mathrm{d}a \quad \text{and} \quad \widetilde{B}f = \int_0^\infty \beta(a,\cdot)\mathrm{d}a, \quad f \in \mathrm{L}^1(\mathbb{R}_+,X).$$

This gives two C_0-semigroups $(S_0^\infty(t))_{t\geq 0}$ and $(\widetilde{S}_0(t))_{t\geq 0}$ on $L^1(\mathbb{R}_+, X)$, see Corollary 19.22. From the uniqueness of the solutions we have $S_0(t) = PS_0^\infty(t)J$, and thus

$$S_0(t) = \sum_{k=0}^{\infty} PU_k^\infty(t)J, \qquad (19.25)$$

where $U_k^\infty(t)$ are the coefficients of the Dyson–Phillips expansion of $S_0^\infty(t)$. On the other hand, we know from the first part of the proof that the remainder $\widetilde{R}_3(t)$ of $\widetilde{S}_0(t)$ is compact for $t \geq 0$ (see also the proof of Proposition 19.13). Using Lemma 19.11, we deduce that $0 \leq R_3^\infty(t) \leq \widetilde{R}_3(t)$, and hence

$$0 \leq R_3^\infty(t)S \leq \widetilde{R}_3(t)S, \quad t \geq 0,$$

for any $0 \leq S \in \mathcal{L}(L^1(\mathbb{R}_+, X))$. Since, by Lemma 19.10, any bounded linear operator on $L^1(\mathbb{R}_+ \times \Omega)$ can be written as a linear combination of positive operators, Lemma 19.9 implies that $(R_3^\infty(t)S)^2$ is compact for $t \geq 0$ and any $S \in \mathcal{L}(L^1(\mathbb{R}_+, X))$. Thus, the remainder $PR_3^\infty(t)J$ of the expansion in (19.25) is strictly power compact in $L^1([0, a_m], X)$. The assertion now follows from Theorem 19.5 and Lemma 19.24. $\qquad\qquad\square$

We end this chapter by a further generalization.

Remark 19.27. Using the theory of Miyadera perturbations one can see that all the results obtained in this section remain true if, instead of the boundedness of μ, one assumes the following

$$0 \leq \mu \in L_{\mathrm{loc,u}}^q(I, L^p(\Omega)) \quad \text{for} \quad p > \frac{d}{2} \quad \text{and} \quad q > \left(1 - \frac{d}{2p}\right)^{-1},$$

where $L_{\mathrm{loc,u}}^q(I)$ is the space of uniformly locally q-integrable functions on I endowed with the norm $\|\varphi\|_{L_{\mathrm{loc,u}}^q} := \sup_{s,s+1\in I} \|\chi_{[s.s+1]}\varphi\|_q$.

19.6   Notes and Remarks

The study of age-structured population models with diffusion was stimulated by the works of Gurtin and MacCamy [57, 58, 59, 60] and by Aronson [8]. For a survey, we refer to Anita [4] and the references therein.

The results of this chapter can be found in Rhandi [121]. The same study was conducted for more general diffusion with fertility rate β depending also on time, see Rhandi and Schnaubelt [122]. We refer also to Thieme [141, 140] for similar results. Problem (19.1) was solved in $L^2(I \times \Omega)$ by using the semigroup approach, see Chan and Guo [23], and Huyer [68].

A simpler model of the age-dependent population equation was already considered in an abstract form in Exercise 17.4.5. For a detailed study of this model we refer to Engel and Nagel [44, Sec. VI.4] and the references therein.

Appendix

Background Material from Linear Algebra and Functional Analysis

We collect here the necessary notation and results needed in different parts of the book. The book is written in a functional-analytic spirit, but for Part I only some knowledge of linear algebra and complex analysis is required, see Sections A.1–A.3.

Our main objects in Part II are operators on Banach spaces and we use many results and techniques from functional analysis and operator theory. There are many excellent sources and we refer to textbooks like Conway [26], Dunford and Schwartz [36], Lang [81], Reed and Simon [118], Rudin [125], Taylor and Lay [138], or Yosida [157]. However, for the convenience of the reader we introduce our notation and list important theorems in Sections A.4–A.8. Section 9.6 also uses notions from Section A.11.

For Part III already considerable knowledge of operator theory is assumed. We recall some definitions and facts in Sections A.9–A.11 and give references for the rest.

Some general knowledge of measure theory is assumed in Parts II and III. We refer to textbooks like Tao [137, 136], and Brezis [21].

A.1   Basic Linear Algebra

The set of $n \times n$ complex matrices is denoted by $M_n(\mathbb{C})$. We denote by $X := \mathbb{C}^n$ the usual Euclidean vector space and use the identification $M_n(\mathbb{C}) \cong \mathcal{L}(X)$. Here $\mathcal{L}(X)$ denotes the set of (continuous) linear transformations from X to X. Hence, we consider matrices as linear transformations or operators. We can therefore talk about the action of a matrix, or the *kernel* or *image* of a matrix. The latter will be denoted by $\ker A$ and $\operatorname{im} A$, respectively, for a matrix $A \in \mathcal{L}(X)$.

To be able to define convergence of vectors, there is a need for the notion of the length of a vector. A natural way to define the length of a vector in $\mathbb{R}^n$ is to use geometric intuition and define the length of $x = (x_1, x_2, \ldots, x_n)$ as

$$\|x\|_2 = \sqrt{x_1^2 + x_2^2 + \cdots + x_n^2},$$

the Euclidean length of a vector. However, it turns out that for different mathematical purposes different notions are more useful. This is crystallized in the abstract notion of the *norm* of a vector.

Definition A.1. A *norm* on X is a function $\| \cdot \|\colon X \to \mathbb{R}$ with the following properties.

a) For every $x \in X$ we have $\|x\| \geq 0$, and $\|x\| = 0 \iff x = 0$;

b) $\|\lambda x\| = |\lambda|\|x\|$ for all $x \in X$ and $\lambda \in \mathbb{C}$;

c) $\|x + y\| \leq \|x\| + \|y\|$ for all $x, y \in X$.

Here we list some norms on the space $X = \mathbb{C}^n$.

Example A.2. For $x = (x_1, x_2, \ldots, x_n) \in \mathbb{C}^n$ the following formulas define norms on $\mathbb{C}^n$:

a) $\|x\|_2 := \sqrt{|x_1|^2 + |x_2|^2 + \cdots + |x_n|^2}$ (the *Euclidean* or 2-*norm*);

b) $\|x\|_1 := |x_1| + |x_2| + \cdots + |x_n|$ (the *taxicab* or 1-*norm*);

c) $\|x\|_\infty := \max\{|x_1|, |x_2|, \ldots, |x_n|\}$ (the *maximum* or ∞-*norm*);

d) $\|x\|_p := (|x_1|^p + |x_2|^p + \cdots + |x_n|^p)^{1/p}$ for $p \in \mathbb{R}$, $p \geq 1$ (the *p-norm*).

The p-norm is obviously a generalization of the Euclidean and taxicab norms. It is also not difficult to show that $\lim_{p\to\infty} \|x\|_p = \|x\|_\infty$.

The Euclidean norm is induced by the scalar product:

$$\|x\|_2 = \sqrt{(x \mid x)}.$$

Here $(\cdot \mid \cdot)$ denotes the usual scalar product on $\mathbb{C}^n$, i.e., for $x = (x_1, x_2, \ldots, x_n)$ and $y = (y_1, y_2, \ldots, y_n)$ we have

$$(x \mid y) := x_1\overline{y_1} + x_2\overline{y_2} + \cdots + x_n\overline{y_n}.$$

Hölder's inequality regarding the scalar product and the p-norm is very useful. For the case $p = q = 2$ it is actually the well-known *Cauchy–Schwarz inequality*.

Lemma A.3 (Hölder's Inequality). *Let either $\frac{1}{p} + \frac{1}{q} = 1$ or $p = 1$, $q = \infty$ hold. Then*

$$|(x \mid y)| \leq \|x\|_p\|y\|_q$$

for all $x, y \in \mathbb{C}^n$.

Any norm defines a metric $d_{\|\cdot\|}(x, y) := \|x - y\|$ on a vector space and hence the notions of convergence, continuity, etc., can be defined accordingly.

There are also many matrix norms on $\mathcal{L}(X)$. We recall some of them.

Example A.4. For an $n \times n$ matrix $A = (\alpha_{ij}) \in \mathcal{L}(X)$ the following formulas define norms on $\mathcal{L}(X)$:

    a) $\|A\|_1 := \max_{1 \leq j \leq n} \sum_{i=1}^{n} |\alpha_{ij}|$ (the *column norm*);

    b) $\|A\|_\infty := \max_{1 \leq i \leq n} \sum_{j=1}^{n} |\alpha_{ij}|$ (the *row norm*);

    c) $\|A\|_{\max} := \max_{1 \leq i,j \leq n} |\alpha_{ij}|$ (the *max norm*).

A.2 Reducing Subspaces and Projections

We recall here some important geometric notions. Taking $A \in \mathcal{L}(X)$, a subspace $Y \subset X$ is called *invariant* under A, if

$$AY := \{Ay \ : \ y \in Y\} \subset Y.$$

A subspace $Y \subset X$ is called *reducing* if there is a subspace $Z \subset X$ such that $X = Y \oplus Z$ and both Y and Z are invariant under A. Here the symbol "$\oplus$" denotes the direct sum of two subspaces, meaning that for every $x \in X$ there exists exactly one $y \in Y$ and one $z \in Z$ such that $x = y + z$. In this case we also say that Z *complements* Y. It is important to note that, if the subspace Y reduces A, there will, in general, be infinitely many subspaces complementing Y which are not invariant under A. So some care is needed when choosing the subspace Z. It readily follows from the definition that if a subspace Y reduces A, then the corresponding complementing subspace reduces A as well.

Let us illustrate the notion of a reducing subspace on two examples.

Examples A.5.

    a) Consider $X := \mathbb{C}^2$ and the matrix $A := \begin{pmatrix} 1 & 0 \\ 0 & 2 \end{pmatrix}$. It is easily seen that the subspaces

$$Y := \left\{ \begin{pmatrix} t \\ 0 \end{pmatrix} \ : \ t \in \mathbb{C} \right\},$$

$$Z := \left\{ \begin{pmatrix} 0 \\ t \end{pmatrix} \ : \ t \in \mathbb{C} \right\}$$

are invariant and that $X = Y \oplus Z$. Hence, these subspaces reduce the matrix A.

    b) Consider $X := \mathbb{C}^2$ and the matrix $A := \begin{pmatrix} 1 & 1 \\ 0 & 1 \end{pmatrix}$. Then the subspace

$$Y := \left\{ \begin{pmatrix} t \\ 0 \end{pmatrix} \ : \ t \in \mathbb{C} \right\}$$

is invariant. However, there is no other one-dimensional invariant subspace for A. Hence, Y is not a reducing subspace.

The existence of reducing subspaces is really convenient because it allows us to investigate the matrix on smaller subspaces. This is due to the fact that the matrix acts independently on each of its reducing subspaces. Reducing subspaces are of course closely related to projections.

Definition A.6. An operator $P \in \mathcal{L}(X)$ is called a *projection* if $P^2 = P$.

Proposition A.7. *An operator $P \in \mathcal{L}(X)$ is a projection if and only if $X = \ker P \oplus \operatorname{im} P$ and $P|_{\operatorname{im} P} = I$.*

Proof. First note that $P|_{\operatorname{im} P} = I$ implies $P(Px) = Px$, i.e., $P^2 = P$.

Conversely, let $x \in X$ be given. Take $y \in \operatorname{im} P$ such that $y := Px$ and let $z := x - y$. Then $x = y + z$, $y \in \operatorname{im} P$ and $Pz = Px - Py = Px - P^2x = 0$. Hence $z \in \ker P$. We can also deduce that if $y \in \operatorname{im} P$, then $Py = y$.

Finally, we only have to show that the obtained decomposition is unique. But if $x = y' + z'$ with $y' \in \operatorname{im} P$ and $z' \in \ker P$, it follows that $Px = Py' + Pz' = y' = y$. Hence $y' = y$ and $z' = z$.   □

Proposition A.8. *If $P \in \mathcal{L}(X)$ is a projection, then $Q := I - P$ is also a projection. Further, $\ker P = \operatorname{im} Q$ and $\operatorname{im} P = \ker Q$.*

Proof. Clearly, using that P is a projection, we see that $Q^2 = (I - P)(I - P) = I - 2P + P^2 = I - P = Q$. Hence, Q is also a projection. We also see that

$$PQ = P(I - P) = P - P^2 = 0,$$
$$QP = (I - P)P = P - P^2 = 0.$$

Hence $\operatorname{im} Q \subset \ker P$ and $\operatorname{im} P \subset \ker Q$. To show that here actually equality takes place, take $z \in \ker P$. Then $Qz = z - Pz = z$, which implies $z \in \operatorname{im} Q$. Similarly, if $y \in \ker Q$, then $Py = y - Qy = y$, implying $y \in \operatorname{im} P$.   □

It is important to know that for every direct sum decomposition there is a corresponding projection.

Proposition A.9. *Assume that $Y, Z \subset X$ are subspaces such that $X = Y \oplus Z$. Then there is a unique projection $P \in \mathcal{L}(X)$ such that $\operatorname{im} P = Y$ and $\ker P = Z$.*

Proof. Let us start from the decomposition $X = Y \oplus Z$. We define, in the spirit of Proposition A.7, $Px := y$, where $x = y + z$ is the unique decomposition with the property $y \in Y$ and $z \in Z$. Then for $y \in Y$, we obtain $Py = y$ and for $z \in Z$ we see that $Pz = 0$. Hence, $P^2x = Py = y = Px$, meaning that P is indeed a projection.   □

Finally, we close this summary with a connection between reducing subspaces and operators.

Proposition A.10. *Let $A \in \mathcal{L}(X)$ be given. The following are equivalent.*

(i) *A subspace $Y \subset X$ reduces A with complementing reducing subspace $Z \subset X$.*
(ii) *The projection $P \in \mathcal{L}(X)$ with $Y = \operatorname{im} P$, $Z = \ker P$ commutes with A, i.e.,*
$$PA = AP.$$

Proof. (i) $\Longrightarrow$ (ii): If $X = Y \oplus Z$ with A-invariant subspaces Y and Z, then the projection P with range Y and kernel Z commutes with A. In fact, if $x \in X$ and $x = y + z$ is the unique decomposition of x into components $y \in Y$ and $z \in Z$, then

$$APx = Ay = PAy = PAy + PAz = PAx.$$

Here we used that for $y \in Y$, $Ay \in Y$ and hence $PAy = Ay$. Further, we also used that for $z \in Z$, $Az \in Z$ and hence $PAz = 0$.

(ii) $\Longrightarrow$ (i): If $PA = AP$ and $\operatorname{im} P = PX =: Y$, then

$$AY = APX = PAX \subset PX = Y$$

and

$$A(I - P)X = (I - P)AX \subset (I - P)X = \ker P. \qquad \square$$

Let us recall also a basic property of linear mappings on finite-dimensional spaces.

Proposition A.11. *For an operator $A \in \mathcal{L}(X)$, where $\dim X < \infty$, the following are equivalent:*

$$A \text{ is injective } \Longleftrightarrow A \text{ is surjective } \Longleftrightarrow A \text{ is bijective.}$$

A.3 Interpolation Polynomials

We recall some basic facts on the derivatives of polynomials with complex coefficients.

If two complex functions f, g are at least m-times differentiable at a point $z_0 \in \mathbb{C}$, then fg is also m-times differentiable, and

$$(fg)^{(m)}(z_0) = \sum_{i=0}^{m} \binom{m}{i} f^{(m-i)}(z_0) g^{(i)}(z_0), \qquad (A.1)$$

where $f^{(j)}$ denotes the jth derivative of the function f and $\binom{m}{i} = \dfrac{m!}{i!(m-i)!}$.

Lemma A.12. *Let $z_1 \in \mathbb{C}$ be fixed. A polynomial p has the form $p(z) = (z - z_1)^m q(z)$ for some $m \in \mathbb{N}$ and $q \in \mathbb{C}[x]$ if and only if*

$$p^{(i)}(z_1) = 0 \quad \text{for } i = 0, 1, \ldots, m - 1.$$

Proof. We proceed by induction. For $m = 1$, the statement is well known from elementary algebra. Assume now that the desired identity holds up to some $m = k$. By this induction assumption, $p(z) = (z - z_1)^k q_1(z)$ for some $q_1 \in \mathbb{C}[x]$. Formula (A.1) yields the identity

$$p^{(k)}(z) = \sum_{i=0}^{k} \binom{k}{i} \frac{k!}{i!} (z - z_1)^i q_1^{(i)}(z) = (z - z_1) p_1(z) + k! q_1(z), \qquad (A.2)$$

where p_1 is a suitable chosen polynomial, more precisely,

$$p_1(z) = \sum_{i=1}^{k} \binom{k}{i} \frac{k!}{i!} (z - z_1)^{i-1} q_1^{(i)}(z).$$

Relation (A.2) shows that $p^{(k)}(z_1) = 0$ if and only if $q_1(z_1) = 0$. Hence, using again the well-known fact from elementary algebra, we infer that this last equation is equivalent to the fact that

$$q_1(z) = (z - z_1) q(z)$$

for some polynomial $q \in \mathbb{C}[x]$, proving the assertion.   □

We collect some information on *interpolation polynomials*. Given m distinct complex numbers $z_1, \ldots, z_m$ and another collection of complex numbers $w_1, \ldots, w_m$, one can easily find a polynomial p of degree $m - 1$ with $p(z_i) = w_i$, $i = 1, \ldots, m$,

$$p(z) := \sum_{k=1}^{m} w_k \prod_{\substack{1 \le j \le m \\ j \ne k}} \frac{z - z_j}{z_k - z_j},$$

which is known as the *Langrange polynomial*. The problem becomes more difficult if we would also like to prescribe the values of the derivatives of p at points $z_1, \ldots, z_m$. We obviously need higher-order polynomials to fulfill this task. If for any z_i the values of p and of its first $\nu_i - 1$ derivatives are given, $i = 1, \ldots, m$, then the minimal degree of the appropriate interpolation polynomial is $\nu_1 + \cdots + \nu_m - 1$ and can be achieved using *Hermite interpolation*. For more information we refer to the monographs by Lancaster [80, Section 5.2] and Meyer [94, Example 7.9.4].

A.4   Function Spaces

We list here some classical sequence and function spaces. In the following, J is a real interval, $\mathbb{K}$ denotes $\mathbb{R}$ or $\mathbb{C}$, and Ω, depending on the context, is a domain in $\mathbb{R}^n$, a locally compact Hausdorff space, or a set endowed with a σ-algebra and a measure. This should be clear from the context. The symbol X always stands for a Banach space. The canonical sequence spaces are:

$$\ell^\infty(\mathbb{N}, X) := \left\{ (x_k) \subset X : \sup \|x_k\| < \infty \right\}, \quad \|(x_k)\|_\infty := \sup_{k \in \mathbb{N}} \|x_k\|,$$

$$c(\mathbb{N}, X) := \left\{ (x_k) \subset X : \lim_{k \to \infty} x_k \text{ exists} \right\} \subset \ell^\infty(\mathbb{N}, X) =: \ell^\infty(X),$$

$$c_0(\mathbb{N}, X) := \left\{ (x_k) \subset X : \lim_{k \to \infty} x_k = 0 \right\} \subset c(\mathbb{N}, X) =: c(X),$$

$$\ell^\infty := \ell^\infty(\mathbb{N}, \mathbb{C}),$$

$$c := c(\mathbb{N}, \mathbb{C}),$$

$$c_0 := c_0(\mathbb{N}, \mathbb{C}),$$

$$\ell^p := \ell^p(\mathbb{N}, \mathbb{C}) := \left\{ (x_k) \subset \mathbb{C} : \sum_{k \in \mathbb{N}} |x_k|^p < \infty \right\},$$

$$\|(x_k)\|_p := \left(\sum_{k \in \mathbb{N}} |x_k|^p \right)^{1/p}, \quad p \in [1, \infty).$$

The space ℓ^2 is a Hilbert space with the scalar product

$$((x_k), (y_k)) := \left(\sum_{k \in \mathbb{N}} x_k y_k \right)^{1/2}.$$

Recall the *Cauchy–Schwarz inequality* (compare with Lemma A.3):

$$\|(x_k y_k)\|_1 \le \|(x_k)\|_2 \|(y_k)\|_2 \tag{A.3}$$

for all $(x_k), (y_k) \in \ell^2$.

Further, here are some spaces of continuous functions:

$$C(\Omega) := \{f : \Omega \to \mathbb{K} : f \text{ is continuous}\}$$
$$\text{if } \Omega = K \text{ is compact: } \|f\|_\infty := \sup_{s \in K} |f(s)|,$$

$$C_0(\Omega) := \{f \in C(\Omega) : f \text{ vanishes at infinity}\},$$

$$C_b(\Omega) := \{f \in C(\Omega) : f \text{ is bounded}\},$$

$$C_c(\Omega) := \{f \in C(\Omega) : f \text{ has compact support}\},$$

$$\mathrm{BUC}(\Omega) := \{f \in C(\Omega) : f \text{ is bounded and uniformly continuous}\},$$

$$\mathrm{AC}(J) := \{f : J \to \mathbb{K} : f \text{ is absolutely continuous}\},$$

$$C^k(J) := \{f \in C(J) : f \text{ is } k\text{-times continuously differentiable}\},$$

$$C^\infty(J) := \{f \in C(J) : f \text{ is infinitely many times differentiable}\}.$$

The usual spaces of integrable functions are:

$$L^p(\Omega, \mu) := \{f : \Omega \to \mathbb{K} \ : \ f \text{ is } p\text{-integrable on } \Omega\},$$

$$\|f\|_p := \left(\int_\Omega |f|^p(s) \, d\mu(s)\right)^{\frac{1}{p}}, p \in [1, \infty),$$

$$L^\infty(\Omega, \mu) := \{f : \Omega \to \mathbb{K} \ : \ f \text{ is measurable and } \mu\text{-essentially bounded}\},$$

$$\|f\|_\infty := \text{essup}|f|,$$

$$L^1_{\text{loc}}(\Omega, \mu) := \{f : \Omega \to \mathbb{K} \ : \ f \text{ measurable, } f|_K \in L^1(K, \mu) \text{ for compact } K \subset \Omega\},$$

where μ is a positive measure defined on the σ-algebra of all Borel sets of Ω.

The following inequality regarding L^p-norms is very useful (compare with Lemma A.3).

Lemma A.13 (Hölder's Inequality). *Let either $\frac{1}{p} + \frac{1}{q} = 1$ or $p = 1$, $q = \infty$ hold. Then for every $f \in L^p(\Omega, \mu)$ and $g \in L^q(\Omega, \mu)$, $fg \in L^1(\Omega, \mu)$ and*

$$\|fg\|_1 \le \|f\|_p \|g\|_q.$$

We state here a similar inequality for convolutions.

Lemma A.14 (Young's Inequality). *Let $1 \le p, q, r \le \infty$ such that $\frac{1}{r} = \frac{1}{p} + \frac{1}{q} - 1$. Then for every $f \in L^p(\Omega, \mu)$ and $g \in L^q(\Omega, \mu)$, $f * g \in L^r(\Omega, \mu)$ and*

$$\|f * g\|_r \le \|f\|_p \|g\|_q.$$

A.5 The Strong Operator Topology

We will not give the definition of the strong operator topology, but just point out what *convergence* and *boundedness* mean in this setting.

Let X, Y be Banach spaces and let $(T_k) \subset \mathcal{L}(X, Y)$ be a sequence of bounded linear operators between X and Y. We say that the sequence (T_k) *converges strongly* to $T \in \mathcal{L}(X, Y)$ if

$$T_k x \longrightarrow Tx \quad \text{holds in } Y \text{ as } k \longrightarrow \infty, \text{ for all } x \in X.$$

Thus strong convergence of a sequence of operators is pointwise convergence on the domain.

A subset $\mathcal{K} \subset \mathcal{L}(X, Y)$ is called *strongly bounded* (or bounded pointwise) if for all $x \in X$ we have

$$\sup\{\|Tx\| \ : \ T \in \mathcal{K}\} < \infty.$$

Next, we list some classical results relating these two notions.

Theorem A.15 (Uniform Boundedness Principle). *Let* X, Y *be Banach spaces and suppose* $\mathcal{K} \subset \mathcal{L}(X, Y)$ *is strongly bounded, i.e., for all* $x \in X$ *we have*

$$\sup\{\|Tx\| : T \in \mathcal{K}\} < \infty.$$

Then $\mathcal{K}$ *is* uniformly bounded, *that is*

$$\sup\{\|T\| : T \in \mathcal{K}\} < \infty.$$

This theorem has the following important consequence.

Theorem A.16. *Let* X, Y *be Banach spaces, and let* $(T_k) \subset \mathcal{L}(X, Y)$ *be a sequence such that* $(T_k x)$ *converges in* Y *for all* $x \in X$. *Then*

$$Tx := \lim_{k \to \infty} T_k x$$

defines a bounded linear operator on X.

Theorem A.17. *Let* X, Y *be Banach spaces,* $T \in \mathcal{L}(X, Y)$ *and* $(T_k) \subset \mathcal{L}(X, Y)$ *be a norm bounded sequence. Then the following assertions are equivalent.*

(i) *For every* $x \in X$ *we have* $T_k x \to Tx$ *in* X.
(ii) *There is a dense subspace* $D \subset X$ *such that for all* $x \in D$ *we have* $T_k x \to Tx$ *in* X.
(iii) *For every compact set* $K \subset X$ *we have* $T_k x \to Tx$ *in* X, *uniformly for* $x \in K$.

By adapting the classical proof of the product rule of differentiation and using the theorem above one can prove the next result.

Theorem A.18 (Product rule). *Let* $u : [a, b] \to X$ *be differentiable, and let* $F : [a, b] \to \mathcal{L}(X, Y)$ *be strongly continuous such that for every* $t \in [a, b]$ *the mapping*

$$Fu : s \longmapsto F(s)u(t) \in Y$$

is differentiable. Then $s \mapsto F(s)u(s) \in Y$ *is differentiable and*

$$(Fu)'(t) = F'(t) \cdot u(t) + F(t) \cdot u'(t).$$

A.6 Some Classical Theorems

Here we collect some classical results we refer to in the text. Their proofs can be found in all standard textbooks. We start by a fundamental result from topology due to P. Urysohn [145]. Note that both assertions stated in this lemma hold for any compact Hausdorff space Ω.

Lemma A.19 (Urysohn). *For a topological space Ω the following assertions are equivalent.*

(i) *For every pair of disjoint closed sets $A, B \subset \Omega$ there are open neighbourhoods $U \supset A$ and $V \supset B$ that are also disjoint.*

(ii) *For every pair of nonempty disjoint closed sets $A, B \subset \Omega$ there is a continuous map $f : \Omega \to [0, 1]$ such that $f(A) = 0$ and $f(B) = 1$.*

We continue with some results from functional analysis. The first one is in its present form due to M.H. Stone [133].

Theorem A.20 (Stone–Weierstrass). *Let Ω be a compact Hausdorff space, and let $\mathcal{A}$ be a unital sub-algebra of $\mathrm{C}(\Omega)$ which is closed under the conjugation operation and separates points (i.e., for every distinct $x_1, x_2 \in \Omega$, there exists at least one $f \in \mathcal{A}$ such that $f(x_1) \neq f(x_2)$). Then $\mathcal{A}$ is dense in $\mathrm{C}(\Omega)$.*

Next we recall that an operator with dense domain and closed graph is bounded if and only if it is everywhere defined. This was shown by S. Banach [10, page 41].

Theorem A.21 (Closed Graph Theorem). *Let X be a Banach space and let $A : X \to Y$ be a linear operator with dense domain $D(A)$ in X and such that its graph is a closed subspace of $X \times Y$. Then A is bounded if and only if $D(A) = X$.*

The following result is a generalization of geometric series and is named after C. Neumann [105].

Theorem A.22 (Neumann Series). *Let X be a Banach space and let $A \in \mathcal{L}(X)$ with $\|A\| < 1$. Then the series $\sum_{k=0}^{\infty} A^k$ converges in the operator norm and*

$$\sum_{k=0}^{\infty} A^k = (I - A)^{-1}. \tag{A.4}$$

We also state a couple of results from measure theory. The first one is a famous result due to H. Lebesgue [85].

Theorem A.23 (Lebesgue Dominated Convergence Theorem). *Let (Ω, μ) be a measurable space and $(f_k) \subset \mathrm{L}^1(\Omega, \mu)$ a sequence such that*

a) $\lim_{k \to \infty} f_k(x) = f(x)$ *a.e. on Ω for some measurable function f,*

b) *there exists $g \in \mathrm{L}^1(\Omega, \mu)$ such that $|f_k(x)| \leq g(x)$ a.e. on Ω for all $k \in \mathbb{N}$.*

Then $f \in \mathrm{L}^1(\Omega, \mu)$ and

$$\lim_{k \to \infty} \int_{\Omega} f_k \mathrm{d}\mu = \int_{\Omega} f \mathrm{d}\mu.$$

Next we recall a result allowing to change the order of the integration provided that the function is nice enough, which originates in the works of G. Fubini [49] and L. Tonelli [143].

Theorem A.24 (Fubini). *Let (Ω_1, μ_1) and (Ω_2, μ_2) be σ-finite measurable spaces and $f\colon \Omega_1 \times \Omega_1 \to \mathbb{R}$ a measurable function. If any of the three integrals*

$$\int_{\Omega_1} \left(\int_{\Omega_2} |f| \, \mathrm{d}\mu_2 \right) \mathrm{d}\mu_1, \quad \int_{\Omega_2} \left(\int_{\Omega_1} |f| \, \mathrm{d}\mu_1 \right) \mathrm{d}\mu_2, \quad \iint_{\Omega_1 \times \Omega_2} |f| \, \mathrm{d}(\mu_1 \times \mu_2),$$

is finite, then all of them are finite and

$$\int_{\Omega_1} \left(\int_{\Omega_2} f \, \mathrm{d}\mu_2 \right) \mathrm{d}\mu_1 = \int_{\Omega_2} \left(\int_{\Omega_1} f \, \mathrm{d}\mu_1 \right) \mathrm{d}\mu_2 = \iint_{\Omega_1 \times \Omega_2} f \, \mathrm{d}(\mu_1 \times \mu_2).$$

Let μ and ν be two σ-finite measures on Ω. We call the measure ν *absolutely continuous* with respect to μ if $\mu(H) = 0$ implies $\nu(H) = 0$ for all measurable sets H. Every absolutely continuous measure can be represented by an integral, as the following result by O. Nikodym [106] shows.

Theorem A.25 (Radon–Nikodým). *If a σ-finite measure ν is absolutely continuous with respect to a σ-finite measure μ, then there exists $f \in \mathrm{L}^1(\Omega, \mu)$ such that*

$$\nu(H) = \int_H f \mathrm{d}\mu$$

for all measurable sets $H \subset \Omega$.

A.7   Riemann Integral

Denote by $C([a, b]; X)$ the space of continuous X-valued functions on $[a, b]$, which becomes a Banach space when endowed with the supremum norm. For a continuous function $u \in C([a, b]; X)$ we define its *Riemann integral* via approximation by means of Riemann sums. Let us briefly sketch how to do this. For $P_k := \{a = t_1 < t_2 < \cdots < t_k = b\} \subset [a, b]$ we set

$$\delta(P_k) := \max\{t_{j+1} - t_j : j = 1, \ldots, k - 1\},$$

and call P_k a *partition* of $[a, b]$ and $\delta(P_k)$ the *mesh* of P_k. We define the *Riemann sum* of u corresponding to the partition P_k by

$$S(P_k, u) := \sum_{j=1}^{k-1} u(t_j)(t_{j+1} - t_j),$$

where k is the number of elements in P_k. From the uniform continuity of u on the compact interval $[a, b]$ it follows that there exists $x_0 \in X$ such that $S(P_k, u)$ converges to x_0 if $\delta(P_k) \to 0$. More precisely, for all $\varepsilon > 0$ there is $\delta > 0$ such that

$$\|S(P_k, u) - x_0\| < \varepsilon$$

whenever $\delta(P_k) < \delta$. We call this $x_0 \in X$ the *Riemann integral* of u and denote it by

$$\int_a^b u(s) \, \mathrm{d}s.$$

The Riemann integral enjoys all the usual properties known for scalar-valued functions. Some of them are collected in the next proposition.

Proposition A.26. *For a continuous function* $u : [a, b] \to X$ *the following hold.*

a) *For every sequence of partitions* P_k *with* $\delta(P_k) \to 0$, *the Riemann sums* $S(P_k, u)$ *converge to the Riemann integral of* u.

b) *The Riemann integral is a bounded linear operator on the space* $\mathrm{C}([a, b]; X)$ *with values in* X.

c) *If* $T \in \mathcal{L}(X, Y)$, *then*

$$T \int_a^b u(s) \, \mathrm{d}s = \int_a^b Tu(s) \, \mathrm{d}s.$$

d) *The function*

$$v(t) := \int_a^t u(s) \, \mathrm{d}s$$

is differentiable with derivative u.

e) *If* $u : [a, b] \to X$ *is continuously differentiable, then*

$$u(b) - u(a) = \int_a^b u'(s) \, \mathrm{d}s$$

holds.

For the proof of these assertions one can take the standard route valid for scalar-valued functions.

A.8   Dual Spaces and Adjoint Operators

Let X be a Banach space. A linear mapping $\varphi : X \to \mathbb{C}$ is called a *linear functional* or a *linear form*. We shall use the notation $\varphi(f) = \langle f, \varphi \rangle$. A linear functional $\varphi : X \to \mathbb{C}$ is *bounded* if there is a constant $M \geq 0$ such that

$$\|\varphi(f)\| \leq M\|f\| \quad \text{for all } f \in X.$$

The set

$$X^* := \{\varphi : \varphi \text{ is a bounded linear functional on } X\}$$

is a linear space and becomes a Banach space with the *functional norm*

$$\|\varphi\| := \sup_{\substack{f \in X \\ \|f\| \le 1}} |\varphi(f)| = \sup_{\substack{f \in X \\ \|f\| \le 1}} |\langle f, \varphi \rangle|.$$

If $\varphi \in X^*$, then

$$|\langle f, \varphi \rangle| \le \|\varphi\| \cdot \|f\|$$

holds for all $f \in X$. The space X^* is called the *dual space* of X. That X^* is large enough is actually the content of the Hahn–Banach theorem (see Hahn [63] and Banach [9]).

Theorem A.27 (Hahn–Banach). *Let X be a Banach space, and let X^* be its dual space. Then the following assertions are true.*

a) *For $f \in X$, $f \ne 0$, there is a $\varphi \in X^*$ with $\varphi(f) = \|f\|$ and $\|\varphi\| = 1$. Or, which is the same, for every $0 \ne f \in X$ there is a $\varphi \in X^*$ with $\varphi(f) = \|f\|^2 = \|\varphi\|^2$.*
b) *For $f, g \in X$ one has $f = g$ if and only if $\langle f, \varphi \rangle = \langle g, \varphi \rangle$ for all $\varphi \in X^*$.*
c) *A subspace Y is dense in X if and only if the zero functional is the only bounded linear functional that vanishes on Y.*

Note however that in many examples the dual space can be determined, and this fundamental theorem is not needed.

We use another important result that holds in Hilbert spaces and is due to F. Riesz [123] and Fréchet [47].

Theorem A.28 (Riesz–Fréchet). *Let H be a Hilbert space with scalar product $(\cdot|\cdot)$ and H^* be its dual. For any $f \in H^*$ there exists a unique $x \in H$ such that $f(y) = (y|x)$ for all $y \in H$. Moreover, $\|f\|_{H^*} = \|x\|_H$.*

Every positive linear functional on the space of continuous functions on a compact space can be represented by a measure. The original version of this result was first obtained by F. Riesz [124].

Theorem A.29 (Riesz Representation Theorem). *Let K be a compact Hausdorff topological space and $X = C(K)$. For every positive linear functional $f^* \in X^*$ there exists a unique regular Borel measure μ such that*

$$\langle f, f^* \rangle = \int_K f \mathrm{d}\mu \quad \text{for all } f \in X.$$

Unbounded linear forms have always dense kernels as the following proposition shows.

Proposition A.30. *The kernel of any unbounded linear form on a Banach space X is dense in X.*

Let $\varphi_k, \varphi \in X^*$. We call φ_k *weak*-convergent* to φ if for all $f \in X$

$$\langle f, \varphi_k - \varphi \rangle \longrightarrow 0 \quad \text{as } k \longrightarrow \infty.$$

The functional φ is called the *weak*-limit* of the sequence, which, if exists, is unique. We call φ a *weak*-accumulation* point of the sequence (φ_k) if for all $f \in X$ and $\varepsilon > 0$ there is a subsequence (φ_{k_ℓ}) with

$$|\langle f, \varphi_{k_\ell} - \varphi \rangle| \leq \varepsilon \quad \text{for all } \ell \in \mathbb{N}.$$

Obviously, if (φ_k) has a subsequence weak*-converging to φ, then φ is an accumulation point of the sequence. The converse implication is in general not true. The next rather weak formulation of a central result from functional analysis (see Alaoglu [1]) suffices for our purposes.

Theorem A.31 (Banach–Alaoglu). *Let X be a Banach space and consider its dual space X^*. Let*

$$\mathcal{B}^* := \left\{ \varphi \in X^* : \|\varphi\| \leq 1 \right\}$$

be the unit ball in X^*. Then every sequence $(\varphi_k) \subset \mathcal{B}^*$ has a weak*-accumulation point in $\mathcal{B}^*$. If X is reflexive or separable, then every sequence $(\varphi_k) \subset \mathcal{B}^*$ has a weak*-convergent subsequence with limit in $\mathcal{B}^*$.

We define the *adjoint operator* A^* of a densely defined linear operator A on a Banach space X by

$$\begin{aligned} D(A^*) = \{ x \in X^* \ : \ &\text{there is } y^* \in X^* \text{ such that} \\ &\langle Ax, x^* \rangle = \langle x, y^* \rangle \text{ for all } x \in D(A) \}, \\ A^* x^* = y^*, \quad x^* \in D(A^*). \end{aligned}$$

We note that A^* is well defined since $D(A)$ is dense in X.

From the following we see that the spectra of the operator and its adjoint coincide, as long the operator is nice enough.

Proposition A.32. *For a closed, densely defined linear operator A and $\lambda \in \mathbb{C}$ one has*

$$\lambda \in \rho(A) \iff \lambda \in \rho(A^*),$$

and in this case $R(\lambda, A)^* = R(\lambda, A^*)$.

A.9 Spectrum, Essential Spectrum, and Compact Operators

Some general notions and basic results concerning the spectrum of a linear operator on a Banach space were already collected in Section 9.7.

We recall here just a version of the *spectral mapping theorem* (see Engel and Nagel [43, Theorem IV.3.7]).

Theorem A.33 (Spectral Mapping Theorem for the Point Spectra). *Let $(T(t))_{t\geq 0}$ be a C_0-semigroup with generator A on Banach space X. Then*

$$\sigma_{\mathrm{p}}(T(t)) \setminus \{0\} = e^{t\sigma_{\mathrm{p}}(A)}, \quad t \geq 0.$$

A bounded operator $S \in \mathcal{L}(X)$ is called a *Fredholm operator* if there is $T \in \mathcal{L}(X)$ such that both operators $I - TS$ and $I - ST$ are compact. We denote by

$$\sigma_{\mathrm{ess}}(S) := \mathbb{C} \setminus \rho_{\mathrm{F}}(S)$$

the *essential spectrum* of S, where

$$\rho_{\mathrm{F}}(S) := \{\lambda \in \mathbb{C} : (\lambda - S) \text{ is a Fredholm operator}\}.$$

Let $\mathcal{K}(X)$ denote the set of all compact operators on X. The *Calkin algebra* $\mathcal{C}(X) := \mathcal{L}(X)/\mathcal{K}(X)$ equipped with the quotient norm

$$\|S\|_{\mathrm{ess}} := \|S + \mathcal{K}(X)\| = \mathrm{dist}(S, \mathcal{K}(X)) = \inf\{\|S - K\| : K \in \mathcal{K}(X)\}$$

is a Banach algebra with unit. The essential spectrum of $S \in \mathcal{L}(X)$ can also be defined as the spectrum of $S + \mathcal{K}(X)$ in this Banach algebra. This implies that, for $S \in \mathcal{L}(X)$, $\sigma_{\mathrm{ess}}(S)$ is non-empty and compact.

For $S \in \mathcal{L}(X)$ we define the *essential spectral radius* by

$$\mathrm{r}_{\mathrm{ess}}(S) := \mathrm{r}(S + \mathcal{K}(X)) = \max\{|\lambda| : \lambda \in \sigma_{\mathrm{ess}}(S)\}.$$

Since $(S + \mathcal{K}(X))^k = S^k + \mathcal{K}(X)$ for $k \in \mathbb{N}$, we have $\mathrm{r}_{\mathrm{ess}}(S) = \lim_{k\to\infty} \|S^k\|_{\mathrm{ess}}^{1/k}$, and consequently

$$\mathrm{r}_{\mathrm{ess}}(S + K) = \mathrm{r}_{\mathrm{ess}}(S) \quad \text{for every } K \in \mathcal{K}(X).$$

Denote by

$$\mathrm{Pol}(S) := \{\lambda \in \mathbb{C} : \lambda \text{ is a pole of finite algebraic multiplicity of } R(\cdot, S)\}.$$

One can prove that $\mathrm{Pol}(S) \subset \rho_{\mathrm{F}}(S)$ and that an element of the unbounded connected component of $\rho_{\mathrm{F}}(S)$ is either in $\rho(S)$ or is a pole of finite algebraic multiplicity. For details concerning the essential spectrum we refer to Kato [73, IV.5.6] and to Gohberg, Goldberg, and Kaashoek [53, Chap. XVII].

We now give a characterization of the essential spectral radius.

Proposition A.34. *For $S \in \mathcal{L}(X)$ the essential spectral radius is*

$$\mathrm{r}_{\mathrm{ess}}(S) = \inf\{r > 0 : \lambda \in \sigma(S), |\lambda| > r \text{ and } \lambda \in \mathrm{Pol}(S)\}.$$

Proof. If we set

$$a := \inf\{r > 0 : \lambda \in \sigma(S), |\lambda| > r \text{ and } \lambda \in \text{Pol}(S)\},$$

then for all $\varepsilon > 0$ there is an $r_\varepsilon > 0$ such that

$$\{\lambda \in \sigma(S) : |\lambda| > r_\varepsilon\} \subset \text{Pol}(S)$$

and $r_\varepsilon - \varepsilon \leq a$. On the other hand, there is a $\lambda_0 \in \sigma_{\text{ess}}(S)$ with $r_{\text{ess}}(S) = |\lambda_0|$. Supposing that $r_{\text{ess}}(S) > r_\varepsilon$, we have $\lambda_0 \in \text{Pol}(S)$. This implies that $\lambda_0 \in \rho_F(S)$, which is a contradiction. Hence $r_{\text{ess}}(S) \leq r_\varepsilon \leq a + \varepsilon$ and $r_{\text{ess}}(S) \leq a$.

To show the other inequality, note that

$$\{\lambda \in \sigma(S) : |\lambda| > r_{\text{ess}}(S)\} \subset \rho_F(S).$$

Therefore,

$$\{\lambda \in \sigma(S) : |\lambda| > r_{\text{ess}}(S)\} \subset \text{Pol}(S).$$

Consequently $a \leq r_{\text{ess}}(S)$ and the proposition in proved. $\square$

There are many notions generalizing compactness of an operator. An operator $B \in \mathcal{L}(X)$ is called *strictly power compact* if there is a $k \in \mathbb{N}$ such that $(BT)^k$ is compact for all $T \in \mathcal{L}(X)$.

The following theorem gives the relationship between the essential spectra of the perturbed and the unperturbed semigroups (see Voigt [149, Corollary 1.4 and Theorem 2.2]).

Theorem A.35. *Let A be the generator of a C_0-semigroup $(T(t))_{t \geq 0}$ on a Banach space X and $B \in \mathcal{L}(X)$. Let $(S(t))_{t \geq 0}$ the C_0-semigroup generated by $A + B$. Assume that there exists $m \in \mathbb{N}$ and a sequence $(t_k) \subset \mathbb{R}_+$, $t_k \to \infty$, such that the remainder $R_m(t_k) := \sum_{j=m}^\infty U_j(t_k)$ of the Dyson–Phillips expansion in (11.6) at t_k is strictly power compact for all $k \in \mathbb{N}$. Then*

$$r_{\text{ess}}(S(t_k)) = r_{\text{ess}}\left(\sum_{j=0}^{m-1} U_j(t_k)\right), \quad \text{and hence } \omega_{\text{ess}}(S) \leq \omega_0(T).$$

An operator $T \in \mathcal{L}(X)$ is said to be *weakly compact* if for every norm bounded sequence (f_k) in X the sequence (Tf_k) has a weakly convergent subsequence in X. Every weakly compact operator is strictly power compact. Moreover, the following holds (see Dunford and Schwartz [36, Corollary VI.8.13]).

Proposition A.36. *If X is an L^1-space, then the product of any two weakly compact operators on X is compact. In particular, every weakly compact operator on X is strictly power compact.*

For the following result we refer to Aliprantis and Burkinshaw [2, Theorem 5.25 and Theorem 5.31].

Proposition A.37. *Let (Ω, μ) be a σ-finite, positive measure space and S, T be two bounded linear operators on $\mathrm{L}^1(\Omega, \mu)$. The following assertions hold.*

a) *The set of all weakly compact operators is a norm-closed subset of $\mathcal{L}(\mathrm{L}^1(\Omega, \mu))$.*

b) *If T is weakly compact and $0 \leq S \leq T$, then S is also weakly compact.*

De Pagter [109] proved the following fundamental result on the spectral radius of positive irreducible compact operators.

Theorem A.38. *Let E be a Banach lattice. If $0 \leq T \in \mathcal{L}(E)$ is an irreducible compact operator, then $r(T) > 0$.*

We also state here a version of the Jacobs–de Leeuw–Glicksberg splitting theorem, referring for the proof and related results to the monographs Eisner, Farkas, Haase, and Nagel [40, Chapter 16] or Eisner [39, Section I.6]. Recall that a semigroup $(T(t))_{t \geq 0}$ on a Banach space X is called *relatively (weakly) compact* if for each $x \in X$ the set $\{T(t)x : t \geq 0\}$ is relatively (weakly) compact in X.

Theorem A.39 (Jacobs–de Leeuw–Glicksberg). *Let $(T(t))_{t \geq 0}$ be a C_0-semigroup with generator A on a Banach space X. Assume that $(T(t))_{t \geq 0}$ is relatively weakly compact. Then there is a projection $P \in \mathcal{L}(X)$ commuting with $T(t)$ such that $X = \operatorname{im} P \oplus \ker P$ and*

a) $\operatorname{im} P = \overline{\operatorname{span}}\{x \in D(A) : \exists \alpha \in \mathbb{R} \text{ such that } Ax = i\alpha x\}$,

b) $\ker P = \{x \in X : 0 \text{ belongs to the weak closure of } \{T(t)x : t \geq 0\}\}$.

Moreover, P belongs to the weak operator closure of $\{T(t) : t \geq 0\}$.

A more precise description of $\ker P$ is given by the following result in the case where $\sigma(A) \cap i\mathbb{R}$ is countable.

Proposition A.40. *Let $(T(t))_{t \geq 0}$ be a uniformly bounded C_0-semigroup with generator A on a Banach space X. If $\sigma(A) \cap i\mathbb{R}$ is countable. Then the following assertions are equivalent.*

(i) $(T(t))_{t \geq 0}$ *is relatively weakly compact.*

(ii) $(T(t))_{t \geq 0}$ *is relatively compact.*

In this case the projection P from Theorem A.39 satisfies

$$\ker P = \{x \in X : \lim_{t \to \infty} \|T(t)x\| = 0\}$$

holds.

Assuming also positivity, the following technical lemma reveals some more information on $\operatorname{im} P$, see Schaefer [126, Proposition III.11.5].

Lemma A.41. *Let E be a Banach lattice and $P \in \mathcal{L}(E)$ a strictly positive projection. Then $\operatorname{im} P$ is a sublattice of E.*

A.10   Bochner Integral, Laplace and Fourier Transforms

A general reference for the following facts is the monograph by Arendt et al. [6].

Let X be a (complex) Banach space and $I \subset \mathbb{R}$ an interval. A function $F : I \to X$ is called a *step function* if there are $m \in \mathbb{N}$, and Lebesgue measurable sets $I_i \subset I$, and $x_i \in X$ for all $i = 1, \ldots, n$, such that

$$F(t) = \sum_{i=1}^{m} x_i \chi_{I_i}. \tag{A.5}$$

A function $F : I \to X$ is called *measurable* if there is a sequence $F_k : I \to X$ of step functions such that

$$F(t) = \lim_{k \to \infty} F_k(t)$$

for almost every $t \in I$. Here "almost every" is to be understood in the sense of the Lebesgue measure on I.

We say that $F \in \mathrm{L}^p(I, X)$ if $F : I \to X$ is measurable and $\|F(\cdot)\| \in \mathrm{L}^p(I, \mathbb{R})$ for $p \in [1, \infty)$ and define its norm by

$$\|F\|_p := \left(\int_I \|F(t)\|^p \, \mathrm{d}t \right)^{1/p}.$$

Using more or less straightforward arguments one can show that $\mathrm{L}^p(I, X)$ enjoys properties similar to the scalar-valued Lebesgue spaces. In particular, if H is a Hilbert space, then $\mathrm{L}^2(I, H)$ becomes a Hilbert space with the scalar product

$$(F|G) := \int_I (F(t)|G(t)) \, \mathrm{d}t.$$

The integral of a step function F given by (A.5) is defined as

$$\int_I F := \sum_{i=0}^{m} x_i \lambda(I_i),$$

where λ is the Lebesgue measure. A function $F : I \to X$ is said to be *Bochner integrable* if there is a sequence of step functions $F_k : I \to X$ such that

$$\int_I \|F(t) - F_k(t)\| \, \mathrm{d}t \longrightarrow 0 \quad \text{as } k \longrightarrow \infty.$$

The integral of a Bochner integrable function can then be defined as

$$\int_I F := \lim_{k \to \infty} \int_I F_k.$$

It can be shown that this integral is well defined and that if F is Bochner integrable, then $F \in \mathrm{L}^1(I, X)$ and

$$\left\| \int_I F \right\| \leq \int \|F(t)\| \, \mathrm{d}t.$$

The *Fourier transform* of a function $F \in \mathrm{L}^1(\mathbb{R}, X)$ is

$$\mathcal{F}F(s) := \int_{-\infty}^{\infty} \mathrm{e}^{-ist} F(t) \, \mathrm{d}t \quad \text{for } s \in \mathbb{R},$$

and the *inverse Fourier transform* of a function $G \in \mathrm{L}^1(\mathbb{R}, X)$ is

$$\mathcal{F}^{-1}G(t) := \int_{-\infty}^{\infty} \mathrm{e}^{ist} G(s) \, \mathrm{d}s \quad \text{for } t \in \mathbb{R}.$$

It can be shown (for example, by taking scalar products and reducing to the scalar case) that if H is a Hilbert space, then for $F \in \mathrm{L}^1(\mathbb{R}, H) \cap \mathrm{L}^2(\mathbb{R}, H)$, we have

$$\int_{\mathbb{R}} \|\mathcal{F}(F)(t)\|^2 \, \mathrm{d}t = 2\pi \int_{\mathbb{R}} \|F(t)\|^2 \, \mathrm{d}t,$$

yielding the vector-valued version of *Plancherel's theorem*.

Theorem A.42. *The Fourier transform extends uniquely to a bounded linear operator on $\mathrm{L}^2(\mathbb{R}, H)$, and the operator $\frac{1}{\sqrt{2\pi}}\mathcal{F}$ is unitary.*

As in the scalar case, we can define *convolution* (see also Lemma A.14).

Theorem A.43 (Hausdorff–Young). *Suppose $K : \mathbb{R} \to \mathcal{L}(X)$ satisfies $K(\cdot)x \in \mathrm{L}^1(\mathbb{R}, X)$ for every $x \in X$, and take $\varphi \in \mathrm{L}^p(\mathbb{R}, X)$ for some $1 \leq p < \infty$. Defining*

$$(K * \varphi)(t) := \int_0^t K(t - s)\varphi(s)\mathrm{d}s,$$

we see that $K * \varphi \in \mathrm{L}^p(\mathbb{R}, X)$ and

$$\|K * \varphi\|_{\mathrm{L}^p} \leq \|K\|_{\mathrm{L}^1} \cdot \|\varphi\|_{\mathrm{L}^p}. \tag{A.6}$$

If $F : \mathbb{R}_+ \to X$ is measurable and exponentially bounded (meaning that there are $M \geq 0$ and $\omega \in \mathbb{R}$ such that $|F(t)| \leq M\mathrm{e}^{\omega t}$), then we can define its *Laplace transform* analogously to the scalar case as

$$\mathcal{L}(F)(\lambda) := \int_0^{\infty} \mathrm{e}^{-\lambda t} F(t) \, \mathrm{d}t$$

for $\operatorname{Re} \lambda > \omega$.

If $(T(t))_{t \geq 0}$ is a C_0-semigroup of type (M, ω) with generator A, then

$$R(\lambda, A)f = \mathcal{L}(T(\cdot)f)(\lambda)$$

for $\operatorname{Re} \lambda > \omega$, see Proposition 9.33.

A.11   Distributions and Sobolev Spaces

In this last part of the Appendix we briefly collect some basic facts on distributions and Sobolev spaces. They are needed for applications of semigroup theory to elliptic and parabolic problems. As a reference, we suggest Brezis [21, Chapters 7 and 8].

Let $\alpha = (\alpha_1, \ldots, \alpha_n) \in \mathbb{N}^n$ be a multi-index with $|\alpha| := \sum_{j=1}^n \alpha_j$. We denote by D_j the first derivative with respect to x_j and we write $D^\alpha := D_1^{\alpha_1} \cdots D_n^{\alpha_n}$. For $f \in C(\Omega)$ the set

$$\operatorname{supp} f := \overline{\{x \in \Omega : f(x) \neq 0\}}$$

is called *the support* of f. For the *test functions* we take $C_c^\infty(\Omega)$, i.e., the space of all complex-valued C^∞-functions on Ω having compact support, where $\Omega \subseteq \mathbb{R}^n$ is an arbitrary domain (open set). We say that a sequence $(\varphi_m) \subset C_c^\infty(\Omega)$ *converges in* $C_c^\infty(\Omega)$ to $\varphi \in C_c^\infty(\Omega)$ if there is a compact set $K \subset \Omega$ such that $\operatorname{supp} \varphi_m \subset K$ for all $m \in \mathbb{N}$ and

$$\lim_{m \to \infty} \sup_{x \in K} |D^\alpha \varphi_m(x) - D^\alpha \varphi(x)| = 0 \quad \text{for all } \alpha \in \mathbb{N}^n.$$

We denote by $C_c^\infty(\Omega)^*$ the space of all *distributions*, i.e., all linear forms $T : C_c^\infty(\Omega) \ni \varphi \mapsto T(\varphi) \in \mathbb{C}$ such that $\lim_{m \to \infty} T(\varphi_m) = T(\varphi)$ whenever φ_m converges in $C_c^\infty(\Omega)$ to $\varphi \in C_c^\infty(\Omega)$.

Any $f \in L^1_{\text{loc}}(\Omega)$ can be identified with a distribution via

$$\langle \varphi, f \rangle := \int_\Omega \varphi(x) f(x) \, \mathrm{d}x, \quad \varphi \in C_c^\infty(\Omega).$$

Let us now define the distributional derivative of a locally integrable function. Take $\alpha \in \mathbb{N}^n$ and $f, g \in L^1_{\text{loc}}(\Omega)$ such that

$$\int_\Omega \varphi g \, \mathrm{d}x = (-1)^{|\alpha|} \int_\Omega f D^\alpha \varphi \, \mathrm{d}x$$

for all $\varphi \in C_c^\infty(\Omega)$. Then we call the function g the D^α *derivative of f in the sense of distributions* and write $g := D^\alpha f$.

For $1 \leq p \leq \infty$ and $m \in \mathbb{N}$ we define the *Sobolev space*

$$W^{m,p}(\Omega) := \{f \in L^p(\Omega) : D^\alpha f \in L^p(\Omega), \, 0 \leq |\alpha| \leq m\}.$$

Equipped with the norm

$$\|f\|_{W^{m,p}} := \begin{cases} \left(\sum_{0 \leq |\alpha| \leq m} \|D^\alpha f\|_{L^p}^p \right)^{1/p} & \text{if } 1 \leq p < \infty, \\ \max_{0 \leq |\alpha| \leq m} \|D^\alpha f\|_{L^\infty} & \text{if } p = \infty, \end{cases}$$

the space $W^{m,p}(\Omega)$ is a Banach space. Here we simply write $L^p(\Omega)$ when the underlying measure is the Lebesgue measure. Further, we set

$$W_0^{m,p}(\Omega) := \text{ the closure of } C_c^\infty(\Omega) \text{ in } W^{m,p}(\Omega).$$

Usually we write $H^m(\Omega) := W^{m,2}(\Omega)$ and $H_0^m(\Omega) := W_0^{m,2}(\Omega)$.

Definition A.44. A bounded domain $\Omega \subset \mathbb{R}^n$ is called *of class C^k* ($k \in \mathbb{N}$ or $k = \infty$) if the following conditions are satisfied.

a) There is a finite open cover

$$\partial\Omega \subset \bigcup_{j=1}^N U_j$$

of the boundary $\partial\Omega$ such that the intersection $\Omega \cap U_j$ can be described as the graph of a function $g_j \in C^k(\overline{Q})$, where Q is a cube $\{|x_i| < a,\ i = 1, \dots, n-1\}$. More precisely, if $x = (x_1, \dots, x_n) \in \Omega \cap U_j$, then $x_n = g_j(x_1, \dots, x_{n-1})$ with a suitable change of coordinates.

b) There is $\eta > 0$ such that

$$\{x : g_j(x') - \eta < x_n < g_j(x'),\ x' \in Q\} \subset \mathbb{R}^n \setminus \overline{\Omega}$$

and

$$\{x : g_j(x') < x_n < g_j(x') + \eta,\ x' \in Q\} \subset \Omega.$$

We recall the Rellich–Sobolev embedding theorem.

Theorem A.45. *Suppose that $\Omega \subset \mathbb{R}^n$ is a bounded domain of class C^1. Then we have the following compact injections:*

$$W^{1,p}(\Omega) \subset \begin{cases} L^q(\Omega) \text{ for all } q \in [1, p^*), \text{ where } \frac{1}{p^*} = \frac{1}{p} - \frac{1}{n} & \text{if } p < n, \\ L^q(\Omega) \text{ for all } q \in [1, \infty) & \text{if } p = n, \\ C(\overline{\Omega}) & \text{if } p > n. \end{cases}$$

In particular, $W^{1,p}(\Omega) \subset L^p(\Omega)$ with compact injective embedding for all $p \in [1, \infty)$.

For smooth domains the Sobolev space $W_0^{1,p}(\Omega)$ coincides with the set of all functions in $W^{1,p}(\Omega)$ vanishing on the boundary of Ω.

Theorem A.46. *Let $\Omega \subset \mathbb{R}^n$ be a bounded domain of class C^1. If $f \in W^{1,p}(\Omega) \cap C(\overline{\Omega})$, $1 \le p < \infty$, then the following assertions are equivalent.*

(i) *$f = 0$ on $\partial\Omega$.*
(ii) *$f \in W_0^{1,p}(\Omega)$.*

The following interpolation inequality is in fact valid for any domain $\Omega \subset \mathbb{R}^n$. Here we present only the case when Ω is an interval in $\mathbb{R}$.

Proposition A.47. *Let $I \subset \mathbb{R}$ be an open interval and $f \in W_0^{2,p}(I)$. Then for any $\varepsilon > 0$ there is a constant $C(\varepsilon) > 0$ such that*

$$\|f'\|_p \leq \varepsilon \|f''\|_p + C(\varepsilon)\|f\|_p.$$

We mention also the following order theoretic property of $W^{1,p}(\Omega)$ for $p \in [1, \infty]$.

Proposition A.48. *If $f \in W^{1,p}(\Omega)$ for $p \in [1, +\infty]$, then the functions f^+, f^-, and $|f|$ belong to $W^{1,p}(\Omega)$ and*

$$Df^+ = \begin{cases} Df & \text{if } f > 0, \\ 0 & \text{if } f \leq 0, \end{cases}$$

$$Df^- = \begin{cases} Df & \text{if } f < 0, \\ 0 & \text{if } f \geq 0, \end{cases}$$

$$D|f|, = \begin{cases} Df & \text{if } f > 0, \\ 0 & \text{if } f = 0, \\ -Df & \text{if } f < 0. \end{cases}$$

To conclude, we recall the *Schwartz space* of rapidly decreasing functions

$$\mathcal{S}(\mathbb{R}^n) := \left\{ f \in C^\infty(\mathbb{R}^n) : \sup_{x \in \mathbb{R}^n} |x^\alpha D^\beta f(x)| < \infty \text{ for all multi-indices } \alpha, \beta \in \mathbb{N}^n \right\}.$$
$$\text{(A.7)}$$

By using the Schwartz space $\mathcal{S}(\mathbb{R}^n)$ as *test functions* instead of the space $C_c^\infty(\Omega)$ and proceeding as above, one obtains the subspace $\mathcal{S}(\mathbb{R}^n)^* \subset C_c^\infty(\Omega)^*$ of *tempered distributions*. These distributions are useful when applying the (scalar-valued) Fourier transform, see (9.7). The Fourier transform maps $\mathcal{S}(\mathbb{R}^n)$ into itself and is an algebra homomorphism. Moreover, the Fourier transform $\mathcal{F}$ is also an isomorphism from $\mathcal{S}(\mathbb{R}^n)^*$ onto $\mathcal{S}(\mathbb{R}^n)^*$. We refer to the books by Hörmander [67] and Rauch [117] for more information.

Bibliography

[1] Alaoglu, L.: Weak topologies of normed linear spaces. Ann. of Math. **41**, 252–267 (1940)

[2] Aliprantis, C., Burkinshaw, O.: Positive Operators. Springer-Verlag (2006)

[3] Amann, H.: Ordinary Differential Equations. An Introduction to Nonlinear Analysis, *de Gruyter Stud. Math.*, vol. 13. de Gruyter (1990)

[4] Aniţa, S.: Analysis and Control of Age-Dependent Population Dynamics, *Mathematical modelling – theory and applications*, vol. 11. Kluwer Academic Publishers (2000)

[5] Arendt, W.: Resolvent positive operators. Proc. London Math. Soc. **54**, 321–349 (1987)

[6] Arendt, W., Batty, C., Hieber, M., Neubrander, F.: Vector-valued Laplace Transforms and Cauchy Problems. Birkhäuser Verlag (2011)

[7] Arendt, W., Rhandi, A.: Perturbation of positive semigroups. Arch. Math. **56**, 107–119 (1991)

[8] Aronson, D.G.: The role of diffusion in mathematical population biology: Skellam revisited. In: Mathematics in biology and medicine (Bari, 1983), *Lecture Notes in Biomath.*, vol. 57, pp. 2–6. Springer, Berlin (1985)

[9] Banach, S.: Sur les fonctionelles linéaires. Studia Math. **1**, 211–216 (1929)

[10] Banach, S.: Théorie des Opérations Linéaires. Warszawa (1932). Reprint by Chelsea Publ. Comp.

[11] Banasiak, J., Arlotti, L.: Perturbations of Positive Semigroups with Applications. Springer-Verlag (2006)

[12] Banasiak, J., Namayanja, P.: Asymptotic behaviour of flows on reducible networks. J. Networks Heterogeneous Media **9**, 197–216 (2014)

[13] Bátkai, A., Piazzera, S.: Semigroups for Delay Equations, *Research Notes in Mathematics*, vol. 10. A K Peters, Ltd., Wellesley, MA (2005)

[14] Bayazit, F., Dorn, B., Kramar Fijavž, M.: Asymptotic periodicity of flows in time-depending networks. J. Networks Heterogeneous Media **8**, 843–855 (2013)

[15] Bayazit, F., Dorn, B., Rhandi, A.: Flows in networks with delay in the vertices. Math. Nachr. **285**, 1603–1615 (2012)

[16] Belleni-Morante, A.: Applied Semigroups and Evolution Equations. Oxford University Press (1979)

[17] Belleni-Morante, A., McBride, A.: Applied Nonlinear Semigroups. John Wiley & Sons (1998)

[18] Berman, A., Plemmons, R.: Nonnegative Matrices in the Mathematical Sciences, *Classics in Applied Mathematics*, vol. 9. SIAM (1994)

[19] Bondy, A., Murty, U.: Graph Theory, *Graduate Texts in Math.*, vol. 244. Springer-Verlag (2008)

[20] Boulite, S., Bouslous, H., El Azzouzi, M., Maniar, L.: Approximate positive controllability of positive boundary control systems. Positivity **18**(2), 375–393 (2014)

[21] Brezis, H.: Functional analysis, Sobolev Spaces and Partial Differential Equations. Universitext. Springer, New York (2011)

[22] Cazenave, T., Haraux, A.: An Introduction to Semilinear Evolution Equations, *Oxford Lecture Series in Mathematics and its Applications*, vol. 13. The Clarendon Press, Oxford University Press, New York (1998). Translated from the 1990 French original by Yvan Martel and revised by the authors

[23] Chan, W., Guo, B.: On the semigroups for age-dependent population dynamics with spatial diffusion. J. Math. Anal. Appl. **184**, 190–199 (1994)

[24] Ciarlet, P.G.: Discrete maximum principle for finite-difference operators. Aequationes Math. **4**, 338–352 (1970)

[25] Clark, S., Latushkin, Y., Montgomery-Smith, S., Randolph, T.: Stability radius and internal versus external stability in Banach spaces: An evolution semigroup approach. SIAM J. Control Optim. **38**(6), 1757–1793 (2000)

[26] Conway, J.: A Course in Functional Analysis, *Graduate Texts in Math.*, vol. 96. Springer-Verlag (1985)

[27] Cramer, D., Latushkin, Y.: Gearhart-Prüss theorem in stability for wave equations: a survey. In: Evolution Equations, *Lecture Notes in Pure and Appl. Math.*, vol. 234, pp. 105–119. Dekker, New York (2003)

[28] Datko, R.: Extending a theorem of A.M. Liapunov to Hilbert space. J. Math. Anal. Appl. **32**, 610–616 (1970)

[29] Derndinger, R.: Über das Spektrum positiver Generatoren. Math. Z. **172**, 281–293 (1980)

[30] Dickson, J.G.: The Wild Turkey: Biology and Management. Stackpole Books, Harrisburg, PA (1992)

[31] Diekmann, O., Heesterbeek, J.A.P.: Mathematical Epidemiology of Infectious Diseases. Wiley Series in Mathematical and Computational Biology. John Wiley & Sons Ltd., Chichester (2000). Model building, analysis and interpretation

[32] Ding, J., Zhou, A.: Nonnegative Matrices, Positive Operators, and Applications. Hackensack, NJ: World Scientific (2009)

[33] Dorn, B.: Semigroups for flows in infinite networks. Semigroup Forum **76**, 341–356 (2008)

[34] Dorn, B., Keicher, V., Sikolya, E.: Asymptotic periodicity of recurrent flows in infinite networks. Math. Z. **263**, 69–87 (2009)

[35] Dorn, B., Kramar Fijavž, M., Nagel, R., Radl, A.: The semigroup approach to transport processes in networks. Phys. D **239**(15), 1416–1421 (2010)

[36] Dunford, N., Schwartz, J.: Linear Operators I. General Theory. Interscience Publisher (1958)

[37] Dyson, F.: The radiation theories of Tomonaga, Schwinger, and Feynman. Phys. Rev. **75**, 486–502 (1949)

[38] Edeko, N., Kühner, V.: One-parameter Koopman semigroups on L^p and $C(K)$ spaces (2016). arXiv:1511.02342

[39] Eisner, T.: Stability of Operators and Operator Semigroups, *Operator Theory: Advances and Applications*, vol. 209. Birkhäuser Verlag, Basel (2010)

[40] Eisner, T., Farkas, B., Haase, M., Nagel, R.: Operator Theoretic Aspects of Ergodic Theory, *Graduate Texts in Math.*, vol. 272. Springer-Verlag (2015)

[41] Engel, K., Klöss, B., Kramar Fijavž, M., Nagel, R., Sikolya, E.: Maximal controllability for boundary control problems. Appl. Math. Opt. **62**, 205–227 (2010)

[42] Engel, K., Kramar Fijavž, M., Nagel, R., Sikolya, E.: Vertex control of flows in networks. J. Networks Heterogeneous Media **3**, 709–722 (2008)

[43] Engel, K.J., Nagel, R.: One-Parameter Semigroups for Linear Evolution Equations, *Graduate Texts in Math.*, vol. 194. Springer-Verlag (2000)

[44] Engel, K.J., Nagel, R.: A Short Course on Operator Semigroups. Universitext. Springer-Verlag (2006)

[45] Farina, L., Rinaldi, S.: Positive Linear Systems. Pure and Applied Mathematics (New York). Wiley-Interscience, New York (2000)

[46] Feller, W.: An Introduction to Probability Theory and its Applications. Vol. I. Third edition. John Wiley & Sons Inc., New York (1968)

[47] Fréchet, M.: Sur les ensembles de fonctions et les opérations linéaires. C. R. Acad. Sci., Paris **144**, 1414–1416 (1907)

[48] Frobenius, G.: Über Matrizen aus nicht negativen Elementen. Sitzungsber. Preuß. Akad. Wiss. Phys.-Math. Kl. pp. 456–477 (1909)

[49] Fubini, G.: Sugli integrali multipli. Rom. Acc. L. Rend. (5) **16**(1), 608–614 (1907)

[50] Fuchs, L.: Partially Ordered Algebraic Systems. Addison-Wesley (1963)

[51] Gearhart, L.: Spectral theory for contraction semigroups on Hilbert spaces. Trans. Amer. Math. Soc. **236**, 385–394 (1978)

[52] Godsil, C.D., Royle, G.: Algebraic Graph Theory, *Graduate Texts in Math.*, vol. 207. Springer-Verlag (2008)

[53] Gohberg, I., Goldberg, S., Kaashoek, M.: Classes of Linear Operators I, *Oper. Theory Adv. Appl.*, vol. 49. Birkhäuser Verlag (1990)

[54] Goldstein, J.: Semigroups of Operators and Applications. Oxford University Press (1985)

[55] Greiner, G.: Spectral properties and asymptotic behavior of the linear transport equation. Math. Z. **185**, 167–177 (1984)

[56] Greiner, G., Nagel, R.: On the stability of strongly continuous semigroups of positive operators on $L^2(\mu)$. Ann. Scuola Norm. Sup. Pisa Cl. Sci. **10**, 257–262 (1983)

[57] Gurtin, M.: A system of equations for age-dependent population diffusion. J. Theor. Bioi. **40**, 389–392 (1973)

[58] Gurtin, M., MacCamy, R.: On the diffusion of biological population. Math. Biosci. **38**, 35–49 (1977)

[59] Gurtin, M., MacCamy, R.: Diffusion models for age structured populations. Math. Biosci. **54**, 49–59 (1981)

[60] Gurtin, M., MacCamy, R.: Product solutions and asymptotic behaviour in age-dependent population diffusion. Math. Biosci. **62**, 157–167 (1982)

[61] Gustafson, K.: A perturbation lemma. Bull. Amer. Math. Soc. **72**, 334–338 (1966)

[62] Haase, M.: The Functional Calculus for Sectorial Operators, *Operator Theory: Advances and Applications*, vol. 169. Birkhäuser Verlag, Basel (2006)

[63] Hahn, H.: Über lineare Gleichungssysteme in linearen Räumen. J. Reine Angew. Math. **157**, 214–229 (1927)

[64] Hieber, M.: A characterization of the growth bound of a semigroup via Fourier multipliers. In: Evolution equations and their applications in physical and life sciences (Bad Herrenalb, 1998), *Lecture Notes in Pure and Appl. Math.*, vol. 215, pp. 121–124. Dekker, New York (2001)

[65] Hille, E.: Functional Analysis and Semigroups, *Amer. Math. Soc. Coll. Publ.*, vol. 31. Amer. Math. Soc. (1948)

[66] Hille, E., Phillips, R.: Functional Analysis and Semigroups, *Amer. Math. Soc. Coll. Publ.*, vol. 31. Amer. Math. Soc. (1957)

[67] Hörmander, L.: The analysis of linear partial differential operators. I. Distribution theory and Fourier analysis. 2nd ed. Berlin etc.: Springer-Verlag (1990)

[68] Huyer, W.: Semigroup formulation and approximation of a linear age-dependent population problem with spatial diffusion. Semigroup Forum **49**, 99–114 (1994)

[69] Jacob, B., Zwart, H.J.: Linear port-Hamiltonian Systems on Infinite-dimensional Spaces, *Operator Theory: Advances and Applications*, vol. 223. Birkhäuser/Springer Basel AG, Basel (2012). Linear Operators and Linear Systems

[70] Kakutani, S.: Concrete representation of abstract (M)-spaces. (A characterization of the space of continuous functions.) Ann. Math. (2) **42**, 994–1024 (1941)

[71] Kalauch, A.: On maximum principles for M-operators. Linear Algebra Appl. **371**, 209–224 (2003)

[72] Kato, T.: Schrödinger operators with singular potentials. Israel Journal of Mathematics **13**(1), 135–148 (1972)

[73] Kato, T.: Perturbation Theory for Linear Operators, *Grundlehren Math. Wiss.*, vol. 132, 2nd edn. Springer-Verlag (1980)

[74] Keicher, V., Nagel, R.: Positive semigroups behave asymptotically as rotation groups. Positivity **12**(1), 93–103 (2008)

[75] Klenzendorf, S.: Population dynamics of virginia's hunted black bear population. Ph.D. thesis, Virginia Polytechnic Institute (2002)

[76] Kramar Fijavž, M., Sikolya, E.: Spectral properties and asymptotic periodicity of flows in networks. Math. Z. **249**, 139–162 (2005)

[77] Kreĭn, M.G., Rutman, M.A.: Linear operators leaving invariant a cone in a Banach space. Uspehi Matem. Nauk (N. S.) **3**(1(23)), 3–95 (1948)

[78] Kunszenti-Kovács, D.: Network perturbations and asymptotic periodicity of recurrent flows in infinite networks. SIAM J. Discrete Math. **23**, 1561–1574 (2009)

[79] Kunszenti-Kovács, D.: Perturbations of finite networks and asymptotic periodicity of flow semigroups. Semigroup Forum **79**, 229–243 (2009)

[80] Lancaster, P.: Theory of Matrices. Academic Press New York (1969)

[81] Lang, S.: Real and Functional Analysis, *Graduate Texts in Math.*, vol. 142. Springer-Verlag (1993)

[82] Langville, A., Meyer, C.: Google's PageRank and Beyond: The Science of Search Engine Rankings. Princeton University Press (2006)

[83] Lasota, A., Mackey, M.: Chaos, Fractals and Noise. Stochastic Aspects of Dynamics, *Appl. Math. Sci.*, vol. 97. Springer-Verlag (1994)

[84] Latushkin, Y., Montgomery-Smith, S., Randolph, T.: Evolutionary semigroups and dichotomy of linear skew-product flows on locally compact spaces with Banach fibers. J. Differential Equations **125**, 73–116 (1996)

[85] Lebesgue, H.: Intégrale, longueur, aire. Annali di Mat. (3) **7**, 231–359 (1902)

[86] Lefkovitch, L.P.: The study of population growth in organisms grouped by stages. Biometrics **21**, 1–18 (1965)

[87] Leslie, P.: On the use of matrices in certain population mathematics. Biometrika **33**, 183–212 (1945)

[88] Liapunov, A.: Stability of Motion. Ph.D. thesis, Kharkov (1892). English translation, Academic Press, 1966.

[89] Liu, Z., Zheng, S.: Semigroups Associated with Dissipative Systems, *Chapman & Hall/CRC Research Notes in Mathematics*, vol. 398. Chapman & Hall/CRC, Boca Raton, FL (1999)

[90] MacCluer, C.: The Many Proofs and Applications of Perron's Theorem. SIAM Rev. **42**, 487–498 (2000)

[91] Mátrai, T., Sikolya, E.: Asymptotic behavior of flows in networks. Forum Math. **19**, 429–461 (2007)

[92] Megginson, R.E.: An Introduction to Banach Space Theory, *Graduate Texts in Mathematics*, vol. 183. Springer-Verlag, New York (1998)

[93] Mehrmann, V.: Kontrolltheorie. Lecture notes, 2004

[94] Meyer, C.: Matrix Analysis and Applied Linear Algebra. SIAM (2000)

[95] Meyer-Nieberg, P.: Banach Lattices. Springer-Verlag (1991)

[96] Minc, H.: Nonnegative Matrices. John Wiley & Sons, Inc. (1988)

[97] Mokhtar-Kharroubi, M.: Mathematical Topics in Neutron Transport Theory: New Aspects. World Scientific (1997)

[98] Mokhtar-Kharroubi, M.: Spectral theory for neutron transport. In: J. Banasiak, M. Mokhtar-Kharroubi (eds.) Evolutionary Equations with Applications to Natural Sciences, *Lect. Notes in Math.*, vol. 2126. Springer-Verlag (2015)

[99] Moler, C., Loan, C.v.: Nineteen dubious ways to compute the exponential of a matrix, twenty-five years later. SIAM Rev. **45**, 3–49 (2003)

[100] Mugnolo, D.: Semigroup Methods for Evolution Equations on Networks. Understanding Complex Systems. Springer-Verlag (2014)

[101] Nagel, R. (ed.): One-parameter Semigroups of Positive Operators, *Lect. Notes in Math.*, vol. 1184. Springer-Verlag (1986)

[102] Nagel, R., Sinestrari, E.: Inhomogeneous Volterra integrodifferential equations for Hille–Yosida operators. In: K. Bierstedt, A. Pietsch, W. Ruess, D. Vogt (eds.) Functional Analysis, *Lect. Notes in Pure and Appl. Math.*, vol. 150, pp. 51–70. Marcel Dekker, Proceedings Essen 1991 (1993)

[103] Neerven, J.v.: The Asymptotic Behaviour of Semigroups of Linear Operators, *Oper. Theory Adv. Appl.*, vol. 88. Birkhäuser Verlag (1996)

[104] Neuberger, J.W.: Lie generators for one parameter semigroups of transformations. J. Reine Angew. Math. **258**, 133–136 (1973)

[105] Neumann, C.: Zur Theorie des logarithmischen und des Newton'schen Potentials. Math. Ann. **11**, 558–567 (1877)

[106] Nikodym, O.: Sur une généralisation des intégrales de M. J. Radon. Fundam. Math. **15**, 131–179 (1930)

[107] Norris, J.: Markov chains. Cambridge: Cambridge University Press (1997)

[108] Ouhabaz, E.M.: Analysis of Heat Equations on Domains, *London Mathematical Society Monographs Series*, vol. 31. Princeton University Press, Princeton, NJ (2005)

[109] Pagter, B. de: Irreducible compact operators. Math. Z. **192**, 149–153 (1986)

[110] Pazy, A.: Semigroups of Linear Operators and Applications to Partial Differential Equations, *Appl. Math. Sci.*, vol. 44. Springer-Verlag (1983)

[111] Perron, O.: Zur Theorie der Matrizen. Math. Ann. **64**, 248–263 (1907)

[112] Phillips, R.: Perturbation theory for semi-groups of linear operators. Trans. Amer. Math. Soc. **74**, 199–221 (1953)

[113] Phillips, R.S.: Semigroups of positive contraction operators. Czechoslovak Math. J. **12**, 294–313 (1962)

[114] Prüss, J.: On the spectrum of C_0-semigroups. Trans. Amer. Math. Soc. **284**, 847–857 (1984)

[115] Räbiger, F., Rhandi, A., Schnaubelt, R., Voigt, J.: Non-autonomous Miyadera perturbations. Differential Integral Equations **13**, 341–368 (2000)

[116] Radl, A.: Transport processes in networks with scattering ramification nodes. J. Appl. Funct. Anal. **3**, 461–483 (2008)

[117] Rauch, J.: Partial differential equations. New York etc.: Springer-Verlag (1991)

[118] Reed, M., Simon, B.: Methods of Modern Mathematical Physics I. Functional Analysis. Academic Press, New York (1972)

[119] Reed, M., Simon, B.: Methods of Modern Mathematical Physics II. Fourier Analysis and Self-Adjointness. Academic Press, New York (1975)

[120] Rhandi, A.: Dyson–Phillips expansion and unbounded perturbation of linear C_0-semigroups. J. Comput. Appl. Math. **44**, 339–349 (1992)

[121] Rhandi, A.: Positivity and stability for a population equation with diffusion on L^1. Positivity **2**, 101–113 (1998)

[122] Rhandi, A., Schnaubelt, R.: Asymptotic behaviour of a non-autonomous population equation with diffusion in L^1. Discrete Contin. Dynam. Systems **3**, 663–683 (1999)

[123] Riesz, F.: Sur une espèce de géométrie analytique des systèmes de fonctions sommables. C. R. Acad. Sci., Paris **144**, 1409–1411 (1907)

[124] Riesz, F.: Sur les opérations fonctionnelles linéaires. C. R. Acad. Sci., Paris **149**, 974–977 (1910)

[125] Rudin, W.: Functional Analysis. McGraw-Hill (1973)

[126] Schaefer, H.: Banach Lattices and Positive Operators, *Grundlehren Math. Wiss.*, vol. 215. Springer-Verlag (1974)

[127] Schanbacher, T.: Aspects of positivity in control theory. SIAM J. Control Optim. **27**(3), 457–475 (1989)

[128] Schnaubelt, R.: Sufficient conditions for exponential stability and dichotomy of evolution families. Forum Math. **11**, 543–566 (1999)

[129] Seneta, E.: Non-negative matrices and Markov chains. Revised reprint of the 2nd ed. New York, NY: Springer (2006)

[130] Sikolya, E.: Flows in networks with dynamic ramification nodes. J. Evol. Equ. **5**, 441–463 (2005)

[131] Stein, E.: Harmonic Analysis: Real-variable Methods, Orthogonality, and Oscillatory Integrals. Princeton University Press (1993)

[132] Stein, E., Weiss, G.: Introduction to Fourier Analysis on Euclidean Spaces. Princeton University Press (1971)

[133] Stone, M.H.: Applications of the theory of Boolean rings to general topology. Trans. Amer. Math. Soc. **41**(3), 375–481 (1937)

[134] Stoyan, G.: On maximum principles for monotone matrices. Linear Algebra Appl. **78**, 147–161 (1986)

[135] Tanabe, H.: Functional Analytic Methods for Partial Differential Equations. Marcel Dekker (1997)

[136] Tao, T.: An Epsilon of Room, I: Real Analysis, *Graduate Studies in Mathematics*, vol. 117. American Mathematical Society, Providence, RI (2010)

[137] Tao, T.: An Introduction to Measure Theory, *Graduate Studies in Mathematics*, vol. 126. American Mathematical Society, Providence, RI (2011)

[138] Taylor, A., Lay, D.: Introduction to Functional Analysis, 2nd edn. John Wiley & Sons (1980)

[139] Teschl, G.: Ordinary differential equations and dynamical systems, *Graduate Studies in Mathematics*, vol. 140. American Mathematical Society, Providence, RI (2012)

[140] Thieme, H.: Positive perturbations of operator semigroups: growth bounds, essential compactness, and asynchronous exponential growth. Discrete Contin. Dynam. Systems **4**, 735–764 (1998)

[141] Thieme, H.R.: Quasi-compact semigroups via bounded perturbation. In: Advances in mathematical population dynamics – molecules, cells and man. Papers from the 4th international conference, Rice Univ., Houston, TX, USA, May 23–27, 1995, pp. 691–711. Singapore: World Scientific Publishing (1997)

[142] Tikhonov, A.: Ein Fixpunktsatz. Math. Ann. **111**, 767–776 (1935)

[143] Tonelli, L.: Sull'integrazione per parti. Rom. Acc. L. Rend. (5) **18**(2), 246–253 (1909)

[144] Ulmet, M.: Properties of semigroups generated by first order differential operators. Results Math. **22**, 821–832 (1992)

[145] Urysohn, P.: Über die Mächtigkeit der zusammenhängenden Mengen. Math. Ann. **94**, 262–295 (1925).

[146] Varga, R.S.: On a discrete maximum principle. SIAM J. Numer. Anal. **3**, 355–359 (1966)

[147] Vidav, I.: Spectra of perturbed semigroups with applications to transport theory. J. Math. Anal. Appl. **30**, 264–279 (1970)

[148] Voigt, J.: On the perturbation theory for strongly continuous semigroups. Math. Ann. **229**, 163–171 (1977)

[149] Voigt, J.: A perturbation theorem for the essential spectral radius of strongly continuous semigroups. Monatsh. Math. **90**, 153–161 (1980)

[150] Voigt, J.: Positivity in time-dependent linear transport theory. Acta Appl. Math. **2**, 311–331 (1984)

[151] Voigt, J.: Spectral properties of the neutron transport equation. J. Math. Anal. Appl. **106**, 140–153 (1985)

[152] Voigt, J.: On resolvent positive operators and positive strongly continuous semigroups on AL-spaces. Semigroup Forum **38**, 263–266 (1989)

[153] Weis, L.: The stability of positive semigroups on L_p-spaces. Proc. Amer. Math. Soc. **123**, 3089–3094 (1995)

[154] West, D.: Introduction to Graph Theory. Prentice-Hall (2001)

[155] Wielandt, H.: Unzerlegbare, nicht-negative Matrizen. Math. Z. **52**, 642–648 (1950)

[156] Yosida, K.: On the differentiability and the representation of one-parameter semigroups of linear operators. J. Math. Soc. Japan **1**, 15–21 (1948)

[157] Yosida, K.: Functional Analysis, *Grundlehren Math. Wiss.*, vol. 123. Springer-Verlag (1965)

[158] Zabczyk, J.: Mathematical Control Theory. Systems & Control: Foundations & Applications. Birkhäuser Verlag (1992)

Index

Printed in the United States
By Bookmasters